Remote Sensing of Vegetation Function and Traits

Remote Sensing of Vegetation Function and Traits

Guest Editors

Tawanda W. Gara
Cletah Shoko
Timothy Dube

MDPI

Basel • Beijing • Wuhan • Barcelona • Belgrade • Novi Sad • Cluj • Manchester

Guest Editors
Tawanda W. Gara
Environmental Science and Management
California State Polytechnic University Humboldt
Arcata, CA
USA

Cletah Shoko
School of Geography, Archaeology and Environmental Studies
University of Witwatersrand
Johannesburg
South Africa

Timothy Dube
Department of Earth Sciences
University of The Western Cape
Belville
South Africa

Editorial Office
MDPI AG
Grosspeteranlage 5
4052 Basel, Switzerland

This is a reprint of the Special Issue, published open access by the journal *Remote Sensing* (ISSN 2072-4292), freely accessible at: https://www.mdpi.com/journal/remotesensing/special_issues/Traits.

For citation purposes, cite each article independently as indicated on the article page online and as indicated below:

Lastname, A.A.; Lastname, B.B. Article Title. *Journal Name* **Year**, *Volume Number*, Page Range.

ISBN 978-3-7258-6464-5 (Hbk)
ISBN 978-3-7258-6465-2 (PDF)
https://doi.org/10.3390/books978-3-7258-6465-2

Contents

About the Editors

Tawanda Gara

Tawanda Gara (Dr) is an Assistant Professor of Geospatial Science in the Department of Environmental Science and Management, California State Polytechnic University Humboldt, United States. He holds a PhD in Quantitative Remote Sensing and MSc in Geo-Information Science and Earth Observation for Natural Resources Management from the University of Twente, Netherlands. Tawanda also holds a Master of Philosophy and B.A (Hons) in Geography and Environmental Science from the University of Zimbabwe. Tawanda's research interests lie in GIScience; earth observation; and spatial statistics and their application in natural resources management, specifically forestry, agriculture, and the environment. To date, he has published over 40 peer-reviewed articles, and he has reviewed manuscripts for several international journals, including *Remote Sensing* (MDPI), the *International Journal of Applied Earth Observations and Geoinformation*, and *Agricultural and Forest Meteorology*. Tawanda carried out his postdoctoral research at the University of Maine, School of Forest Resources, working on a NASA-funded project on monitoring the health and productivity of forests using remote sensing. Tawanda has participated in training graduate students on remote sensing of forests' essential biodiversity variables (EBVs)—a training program administered by the European Space Agency-European Facility for Airborne Research (ESA-EUFAR) and VITO across Europe. He has been involved in several projects on environmental impact assessments (EIAs), remote sensing of crop quality and yield, and monitoring of vegetation health. He was a key investigator on a Kenya Wildlife Services (KWS) and International Fund for Animal Welfare (IFAW) project on linking African elephants' movement patterns to landscape fragmentation and forage abundance in the Amboseli ecosystem, Kenya.

Cletah Shoko

Cletah Shoko (Dr) is a Senior Lecturer in Geospatial Science at the University of the Witwatersrand, School of Geography, Archaeology, and Environmental Studies. She holds a PhD in Environmental Science and an MSc in Bioresources Systems from the University of KwaZulu-Natal, South Africa. Her research focuses on leveraging geospatial modeling to understand and address environmental changes, with an emphasis on sustainability. Specializing in the modeling, assessment, and monitoring of water and vegetation resources, she contributes to the understanding of the dynamics of environmental change. Her work is pivotal in developing geospatial models that provide essential insights for the sustainable management of biodiversity and water resources. Driven by a commitment to sustainability and informed by cutting-edge geospatial science, Dr. Shoko has published over 40 articles in leading geospatial journals. Her research plays a crucial role in advancing our knowledge of environmental changes and contributing to sustainable development practices.

Timothy Dube

Timothy Dube (Prof) is the Director of the Institute for Water Studies at the University of the Western Cape and a recognized expert in Geospatial Science for Water Resources and Environmental Management. He also serves as an Adjunct Professor at the United Nations University Institute for Integrated Management of Material Fluxes, Germany. He has an extensive research career spanning earth observation, geospatial sciences, and global environmental change. He holds a Ph.D. in Environmental Science, a Master of Science in GIScience and Earth Observation for Water Resources and Environmental Management, and a B.A (Hons) in Geography and Environmental

Science. Timothy has made significant contributions in advancing EO applications in climate change monitoring, hydrology, and environmental sustainability, leveraging cutting-edge satellite and in situ EO technologies. His expertise further extends to hydro-ecological modeling, land degradation assessment, big data, and remote sensing applications in agriculture and biodiversity conservation. Throughout his career, Timothy has collaborated on multiple international research initiatives, including the Global Monitoring for Environment and Security (GMES and Africa) projects, the Non-Perennial Rivers Research Project, and the South African National Research Foundation–Dutch-NWO-funded Water–Energy–Food Nexus Project. Beyond research, Timothy is deeply engaged in capacity building and EO training, working with institutions of higher learning and research organizations across Africa. His work plays a crucial role in bridging scientific expertise and policy development to support evidence-based decision-making in Africa and beyond. As a leading authority in earth observation and water science, he has made substantial editorial contributions, serving as an Editorial Board Member and reviewer for high-impact scientific journals. To date, he has authored over 300 peer-reviewed journal articles, book chapters, and books.

Preface

Understanding vegetation's structure and function is essential for understanding ecosystems, biodiversity, and environmental change. While field-based methods are accurate, they are spatially disabled. Remote sensing provides scalable, repeatable, and non-invasive approaches to capture vegetation traits and ecosystem functions. With advances in hyperspectral imaging, LiDAR, satellites, and UAVs, vegetation function and traits can now be observed across multiple spatial and temporal scales. This Special Issue showcases global interdisciplinary research connecting remote sensing, plant science, and environmental management.

Tawanda W. Gara, Cletah Shoko, and Timothy Dube
Guest Editors

remote sensing

MDPI

Article

A Novel Approach to Match Individual Trees between Aerial Photographs and Airborne LiDAR Data

Yi Xu [1,*], Tiejun Wang [1], Andrew K. Skidmore [1,2] and Tawanda W. Gara [3]

1 Faculty of Geo-Information Science and Earth Observation (ITC), University of Twente, P.O. Box 217, 7500 AE Enschede, The Netherlands; t.wang@utwente.nl (T.W.); a.k.skidmore@utwente.nl (A.K.S.)
2 School of Natural Sciences, Macquarie University, 12 Wally's Walk, Sydney, NSW 2109, Australia
3 Department of Environmental Science and Management, California State Polytechnic University Humboldt, Arcata, CA 95521, USA; tawanda.gara@humboldt.edu
* Correspondence: y.xu-3@utwente.nl

Abstract: Integrating multimodal remote sensing data can optimize the mapping accuracy of individual trees. Yet, one issue that is not trivial but generally overlooked in previous studies is the spatial mismatch of individual trees between remote sensing datasets, especially in different imaging modalities. These offset errors between the same tree on different data that have been geometrically corrected can lead to substantial inaccuracies in applications. In this study, we propose a novel approach to match individual trees between aerial photographs and airborne LiDAR data. To achieve this, we first leveraged the maximum overlap of the tree crowns in a local area to determine the correct and the optimal offset vector, and then used the offset vector to rectify the mismatch on individual tree positions. Finally, we compared our proposed approach with a commonly used automatic image registration method. We used pairing rate (the percentage of correctly paired trees) and matching accuracy (the degree of overlap between the correctly paired trees) to measure the effectiveness of results. We evaluated the performance of our approach across six typical landscapes, including broadleaved forest, coniferous forest, mixed forest, roadside trees, garden trees, and parkland trees. Compared to the conventional method, the average pairing rate of individual trees for all six landscapes increased from 91.13% to 100.00% ($p = 0.045$, t-test), and the average matching accuracy increased from 0.692 ± 0.175 (standard deviation) to 0.861 ± 0.152 ($p = 0.017$, t-test). Our study demonstrates that the proposed tree-oriented matching approach significantly improves the registration accuracy of individual trees between aerial photographs and airborne LiDAR data.

Keywords: remote sensing; multimodal images; object-based image analysis; spatial mismatch; registration noise; intersection over union

1. Introduction

Information on individual trees is essential for forest management, biodiversity conservation, and assessment of ecosystem services [1–3]. Meanwhile, as a meaningful unit of analysis at a fine scale [4], individual tree level information is crucial for accurate estimation of essential biodiversity variables (EBVs) [5–7], such as biomass [8–10], structural and chemical properties [11,12], as well as functional diversity [13,14]. However, traditional tree inventory tends to be carried out at the plot or stand level [15–18] since obtaining individual tree level information over large areas is usually impractical or impossible via ground survey methods due to extremely high cost and labor-intensive fieldwork [19–21].

To complement in situ measurements, remote sensing has been commonly used to study individual trees at a local or landscape level over the past two decades [22–24]. A critical step before studying individual trees with remote sensing data is delineating individual tree crowns [22,25,26]. High spatial resolution optical remote sensing data (e.g., satellite images, aerial photographs, and drone images) have a proven potential to effectively segment individual trees [3,23,27–39]. However, segmenting individual trees using optical images often leads to over- or under-segmentation of trees as objects in different scenarios, caused by spectral heterogeneity, noise pixels (intensity variation), shadows, as well as observation and illumination angles for imaging [30,31,33,40].

Light detection and ranging (LiDAR) can render three-dimensional (3D) structural information of trees [41–44]. Recently, with the rapid advancement of LiDAR sensors and the availability of LiDAR data, LiDAR technology has gradually become an indispensable active remote sensing technique for delineating individual trees [45–48]. Small-footprint airborne [49–52] and drone (UAV)-based LiDAR data [53–55] are the primary sources of LiDAR data that have been widely used to segment individual trees. Compared to optical sensors, LiDAR is not impacted by illumination artifacts such as the shading of shorter trees by their taller neighbors [25,56,57]. For that reason, the problem of under-segmentation has been effectively resolved with LiDAR by some existing individual tree segmentation algorithms [58–60]. However, these algorithms do detect multiple peaks in tree crowns, resulting in varying degrees of over-segmentation [61,62]. To address this problem, recent studies have attempted to improve the accuracy of individual tree delineation by integrating optical and LiDAR data [63,64]. These studies have shown that the accuracy of delineating individual trees can be considerably improved by fusing both spectral (optical images) and structural features (LiDAR data). Consequently, the fusion of multimodal remote sensing data for delineating individual trees has gained increasing attraction in recent years.

Local mismatches (misalignments) between images are often overlooked when integrating multimodal remote sensing data (e.g., fusion of optical and LiDAR data) to identify individual trees [65,66]. However, local mismatch still occurs and is referred to as registration noise/error [67,68] due to dissimilarities in acquisition circumstances and imaging principles of the sensor with different modes [69,70]. Moreover, images in different modes are more likely to cause complicated nonlinear deformation and displacement of ground objects between images [71–73]. The positional discrepancy between multimodal datasets has a significant effect on subsequent applications through error propagation [74–77]. Area- and feature-based matching are two methods commonly utilized to align remotely sensed images [78].

The area-based matching processes the intensity of corresponding image areas and searches for the best matching similarity in the window template, but the classic methods are not robust enough for multimodal image with apparent radiometric difference [79,80]. Although some normalized cross-correlation (NCC) (one of area-based methods) variants like automatic registration of remote sensing images (ARRSI) and orientation phase consistency histograms (HOPC) were proposed to allow NCC framework to realize image registration with nonlinear radiometric difference [81,82], they were limited to invariance of intensity mapping matching process and specific geometric and radiometric constraints, respectively [81,83]. In terms of feature-based matching, the most common method is to explore tie points between the reference image and candidate image by manual selection or automatic techniques [84]. Yet, the manual selection is labor intensive, time consuming, and prone to subjective errors, which is only suitable for small areas or medium- and low-resolution remote sensing imagery with spatially consistent offset. In terms of automatic approach, the scale-invariant feature transform (SIFT) [85] and variations thereof [86–89] are the most commonly used automatic algorithms [90,91]. These algorithms aim to detect the homonymy points based on local features in images, which is hence difficult or even not available to be applied in multimodal images due to the influence of nonlinear radiometric differences [73,92].

In recent years, researchers have attempted to use deep learning techniques to register multimodal images [93]. Two main strategies have been explored to leverage deep learning for image registration [72]. The first strategy is deep iterative methods that employ neural networks to compute advanced similarity metrics between image pairs and guide iterative optimization for accurate registration [94]. Examples of deep iterative methods include stacked autoencoders [95–97], convolutional neural networks (CNNs) [98–100], and reinforcement learning [101]. The second method is deep transformation estimation that directly predicts geometric transformation parameters or deformative fields. This method can be supervised or unsupervised, depending on the training strategy [102–105]. Supervised methods need ground-truth data and use various architectures and loss functions for training. Unsupervised methods rely on traditional similarity metrics and regularization for loss functions. They also use spatial transformer networks for end-to-end transformation learning and adapt to multimodal images. Deep learning-based methods have achieved remarkable progress in remote sensing image registration, but they also face challenges to practicality and generalizability [106]. They require substantial labeled data and computational resources for training and inference. However, traditional approaches, which are known for their efficiency and simplicity, struggle to match the high accuracy showcased by deep learning models. Therefore, considering the trade-offs between the advantages and challenges of deep learning, it is critical to further refine and expedite image registration processes, especially for multimodal images.

In this study, we propose a novel tree-oriented approach to match individual trees between aerial photographs and airborne LiDAR data. To achieve this, we first leverage the maximum overlap of the tree crowns in a local area to determine the correct and the optimal offset vector, and then use the offset vector to rectify the mismatch on individual tree positions. We used pairing rate (the percentage of correctly paired trees) and matching accuracy (the degree of overlap between the correctly paired trees) to assess the effectiveness between our proposed approach in matching individual trees and compare it with a commonly used automatic image registration method.

2. Methods

2.1. Proposed Tree-Oriented Matching Approach

When processing images, pixels are first grouped as an object, based on either spectral similarity or external variables (such as ownership and geological unit) as a new analysis unit, which is a more effective and meaningful alternative way to studying targets than using pixels directly [107–109], i.e., object-based image analysis (OBIA) [107,110]. The results of delineating an individual trees boundary fits the concept of OBIA, and some structural traits may be more easily retrieved at an individual tree level (e.g., tree height, crown size, and the diameter at breast height) [111,112]. Therefore, we proposed an individual tree-oriented matching approach to improve the matching accuracy of individual trees from aerial photographs and airborne LiDAR-derived CHMs. The flow diagram of the approach is shown in Figure 1 and followed by a detailed description of each step (Algorithm 1).

2.2. Automatic Image Registration Workflow in ENVI

To compare with the traditional method, we also used the *Automatic Image Registration Workflow* module in mainstream commercial software ENVI (version 5.6.3 (c) 2022 L3Harris Geospatial Solutions, Inc., Manila, Philippines) to geometrically align two images [113]. We chose the mutual information matching method (developed for images with different modalities) to automatically generate tie points and kept other parameters as default. After tie points were generated, we removed those with clear errors and then registered the warp image with the remaining points. Finally, we delineated bounding boxes of individual trees from the registered images to compare the effectiveness of both methods.

Algorithm 1: The pseudocode of proposed tree-oriented matching approach.

Matching individual trees is
Initialize:
minimum and maximum crown area ratio: *min_car, max_car;* maximum offset: *max_offset*
Input: canopy height models (CHMs) # reference dataset;
    aerial photographs (ARPs) # candidate dataset
 # Step 1: Mark individual trees with bounding boxes
(The bounding boxes were delineated manually in this study)
CHM_trees <- mark_trees(CHMs)
ARP_trees <- mark_trees(ARPs)
 # Step 2 Iterate over a tree in CHMs as a reference tree (Figure 1a)
for reference_tree in CHM_trees:
    search_center <- reference_tree.center
    # Step 3: Choose candidate trees according to *search_center* and *max_offset* (Figure 1b)
    (To filter those trees whose center exceeds the maximum offset of the reference tree)
    candidate_trees <- ARP_trees.within_radius(search_center, max_offset)
    reference_trees <- CHM_trees.within_radius(search_center, max_offset)
    # Step 4: Filter candidate trees based on crown area ratio (Figure 1c)
    (To filter those trees whose crown area is much larger or smaller than that of the reference tree)
    candidate_trees <- candidate_trees.filter(min_car, max_car, reference_tree.crown_area)
    # Step 5: Calculate offset vectors for each candidate tree and rectify tree locations (Figure 1d)
    (To rectify the candidate trees according to the offset vectors)
    for candidate_tree in candidate_trees:
        offset_vector <- reference_tree.center() - candidate_tree.center()
        candidate_rectified_trees.append(candidate_trees.rectify_location(offset_vector))
    # Step 6: Calculate NIoU and select the correct offset vector (Figure 1e)
    (To choose the correct offset vector for the currently selected reference tree that maximizes the sum of NIoU between the selected reference trees and the candidate trees)
    correct_offset_vectors.append(max(sum(NIoU(reference_trees, candidate_rectified_trees))))
 # Step 7: Rectify tree locations based on the correct offset vectors (Figure 1f)
(To rectify the trees in aerial photographs with the correct offset vectors)
for correct_offset_vector in correct_offset_vectors:
    ARP_rectified_trees.append(ARP_trees.rectify_location(correct_offset_vector))
 # Step 8: Determine the final offset vector (Figure 1g)
(To determine the final offset vector that maximizes the sum of NIoU between individual trees in CHMs and the rectified trees in aerial photographs)
final_offset_vector <- max(sum(NIoU(CHM_trees, ARP_rectified_trees)))

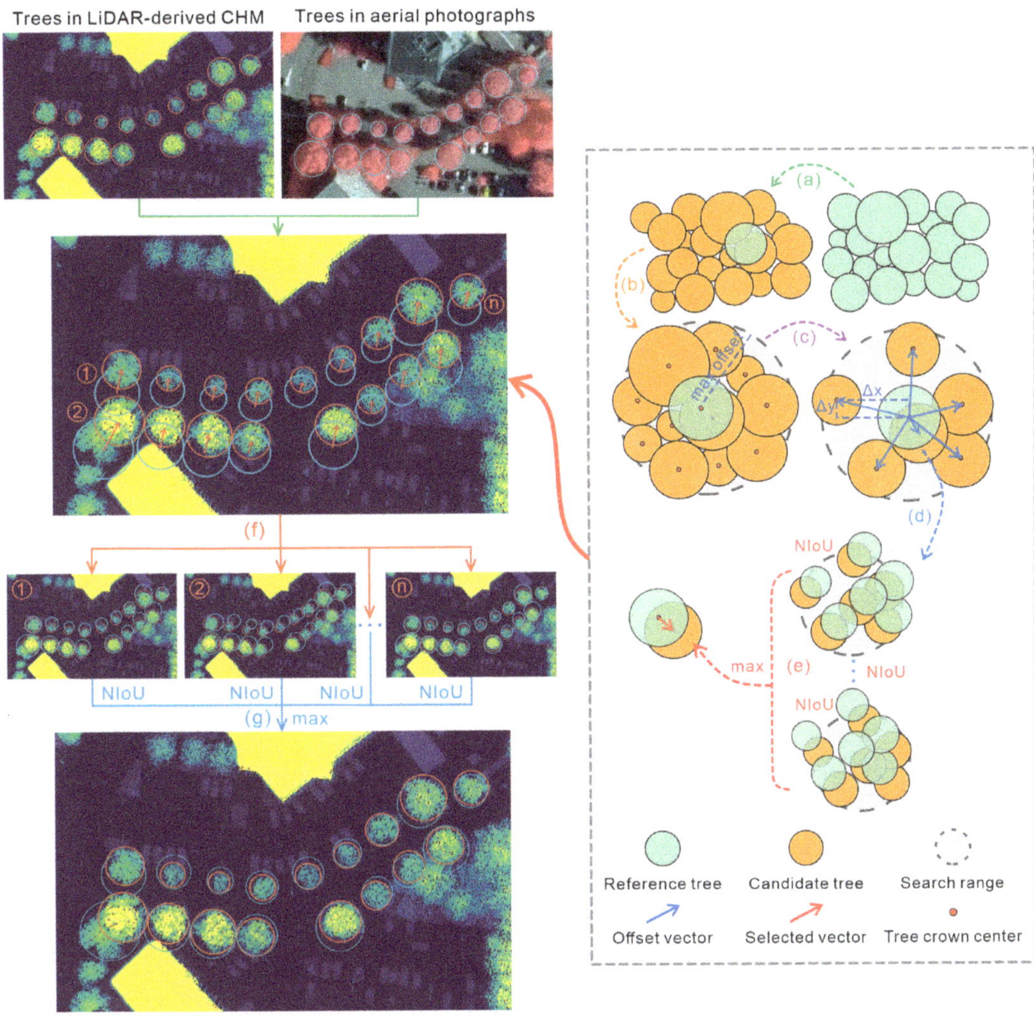

Figure 1. The flow diagram of the proposed tree-oriented matching approach.

3. Experiments

3.1. Experimental Site

The experimental site is Arnhem, the Netherlands (Figure 2). Arnhem is a city and municipality situated in the central part of the Netherlands, containing many parks and forests. The north corner of the municipality is part of the Veluwe—the largest forest area in the Netherlands. Following the definition of the United Nations' Food and Agriculture Organization (FAO) [114] and CORINE land cover classification scheme [115], we divided the forest landscapes in the study area into broadleaved forest (>75% cover of broadleaved trees), coniferous forest (>75% coniferous trees formation), and mixed forest (composed of more than 25% coniferous and broadleaved trees). Furthermore, due to some differences in the layout, density, species richness, and context of groups of individual trees in the city, we also chose three more landscape types—roadside trees, garden trees, and parkland trees. Specifically, roadside trees, regularly lined the road, are often composed of the same tree species with similar height and crown size in a local area (4–16 trees/100 m of linear feature [116]); garden trees, planted in the garden in front or behind a house or buildings, are often of various species and vary in height and crown size (but contain > 25 trees ha^{-1} [117]); and parkland trees, planted based on the layout of parkland, are species abundant, usually having large widely spaced trees separated by grass areas (>18 trees ha^{-1}). Samples of individual trees were chosen from these six landscapes to represent the diversity and complexity of tree distribution, with the level of tree cover used to assess the effectiveness and robustness of our proposed approach.

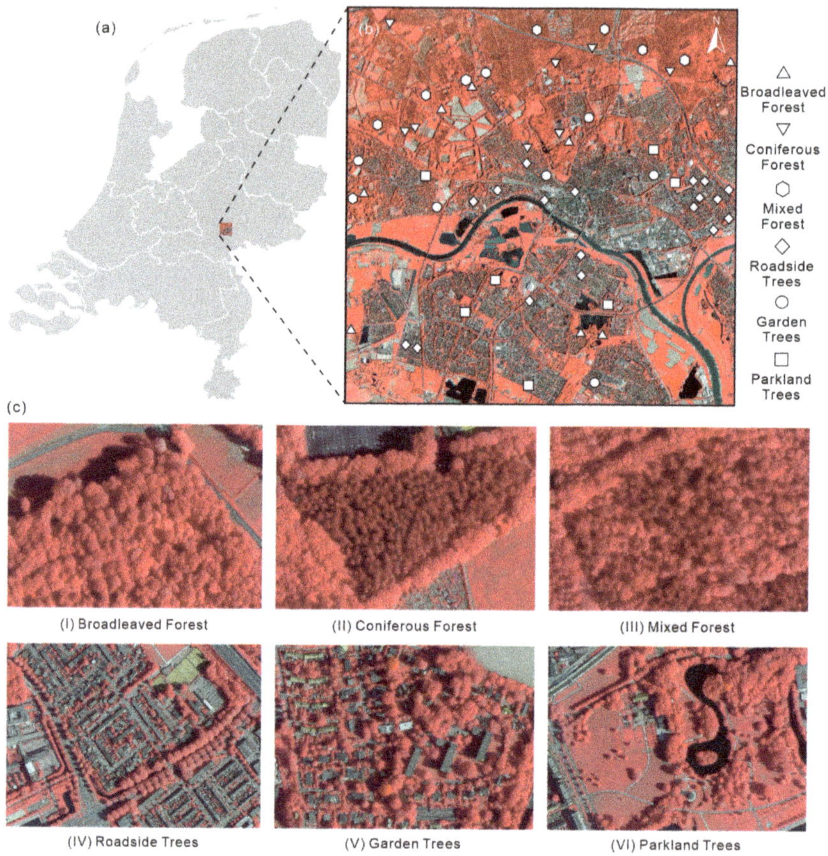

Figure 2. Location of the experimental site and the distribution of sample trees within the six typical landscape areas. (**a**) The location of the city of Arnhem in the Netherlands, (**b**) the distribution of sample trees selected in the typical landscape areas, and (**c**) sample landscapes from the aerial photographs of the study area.

3.2. *Aerial Photographs*

The Dutch government conducts country-wide aerial photography campaigns twice a year with one winter (leaf-off season) and one summer (leaf-on season) since 2012. The aerial photographs acquired in summer with four bands (red, green, blue, and infrared) have a spatial resolution of 25 cm, while the aerial photographs of winter with a spatial resolution of 8 cm have only three bands (red, green, and blue). Both datasets are openly accessible through the web map service (WMS) (opendata.beeldmateriaal.nl). The aerial photographs used in this study were captured during the summer in 2020. There are two reasons for choosing the aerial photographs acquired in 2020. On the one hand, higher image clarity and stability help us delineate tree canopies (the comparison is shown in

Figure S1). On the other hand, it can better simulate real-world applications since data from different sensors are often acquired on different dates and mounted on various platforms.

3.3. Airborne LiDAR Data

The airborne LiDAR point cloud data were from the 3rd national airborne LiDAR flight campaign in the Netherlands (i.e., Actueel Hoogtebestand Nederland 3, AHN3), which can be openly accessible via the online repository PDOK (app.pdok.nl/ahn3-downloadpage). The AHN3 data of our study area were acquired during the leaf-on season in 2018 (ahn.nl/kwaliteitsbeschrijving). The average point density of AHN3 is between 6 and 10 points m^{-2}, and the area of individual trees can reach approximately 40 points m^{-2} owing to the penetration capability of LiDAR.

3.4. Generation of Canopy Height Models from Airborne LiDAR Data

Canopy height models (CHMs) derived from LiDAR point cloud data have been commonly used to generate the boundaries of individual tree crowns [49,118,119], which is obtained through cutting a digital surface model (DSM) with a digital terrain model (DTM) [120–122]. To maintain the same spatial resolution as aerial photographs, we processed AHN3 LiDAR point cloud data to generate CHMs with a spatial resolution of 25 cm. To generate CHMs that could show clearer boundary of tree crowns, we conducted the following processing chain for the LiDAR point cloud. The first step was to normalize the LiDAR data, aiming to eliminate the influence of topographic relief on the elevation value of point, and those points classified as "unclassified" were extracted from the original AHN3 data. To perform normalization, the elevation of the nearest ground point to each point is subtracted from its elevation value. Then, the points with elevation less than 25% of the mean elevation in clipped point cloud dataset were eliminated, and the outliers were removed [123,124]. Finally, inverse distance weighting (IDW) function was chosen to generate the CHMs, a widely employed deterministic model in spatial interpolation that relies on the first law of geography [125–127]. The above processing was all implemented with the software LiDAR360 (version 4.1).

3.5. Delineating Individual Tree Crowns from Aerial Photographs and CHMs

In this study, we used manually delineated individual trees as samples to verify and test our proposed approach, which can effectively avoid the omission and commission error of automatic detection algorithms for tree detection. To improve the reliability of the crown boundaries in aerial photographs and CHMs, airborne images with higher spatial resolution (8 cm) and LiDAR point cloud data with 3D spatial structure information were employed as reference images. To guarantee the representativeness of the samples, we uniformly selected individual trees in space and number according to different landscape types. Specifically, we first delineated the boundaries of individual trees in the aerial photographs. Then, we further delineated individual trees' boundaries in CHMs according to the boundaries sketched in aerial photographs. The number of individual trees sketched in each landscape and the area of each landscape are shown in Table 1.

There are some small differences in the number of delineated individual trees between aerial photographs and LiDAR-derived CHMs, with more tree crowns sketched in CHMs than in aerial photographs. We chose trees that could be delineated in both the aerial photograph and CHM to validate the proposed approach, and meanwhile assigned a same ID number to the same trees by editing the vector in QGIS for both datasets, establishing which tree pairs matched correctly. In other words, a pair of trees are considered correctly matched when they have the same ID number. The percentage of correctly matched trees is used to evaluate the matching accuracy of individual trees for our proposed approach.

Table 1. The number of individual trees manually delineated from aerial photographs and LiDAR-derived CHMs, as well as its corresponding number of sample plots and total area in each landscape.

Landscape	Number of Trees in Aerial Photographs	Number of Trees in CHMs	Number of Sample Plots	Total Area of Sample Plots (ha)
Broadleaved Forest	788	829	8	11.93
Coniferous Forest	739	831	8	7.99
Mixed Forest	676	725	8	8.86
Roadside Trees	792	808	18	17.31
Garden Trees	680	776	7	16.07
Parkland Trees	768	783	7	18.84

3.6. Visualization of Individual Tree Mismatches between Aerial Photographs and CHMs

As shown in Figure 3, there are some examples of the spatial mismatch of individual trees in the manually delineated datasets, which are caused by local displacement between aerial photograph and CHMs. These mismatches are spatially inconsistent in both distance and orientation, and consequently cannot be directly addressed through traditional global geographic calibration for images [69,128–130].

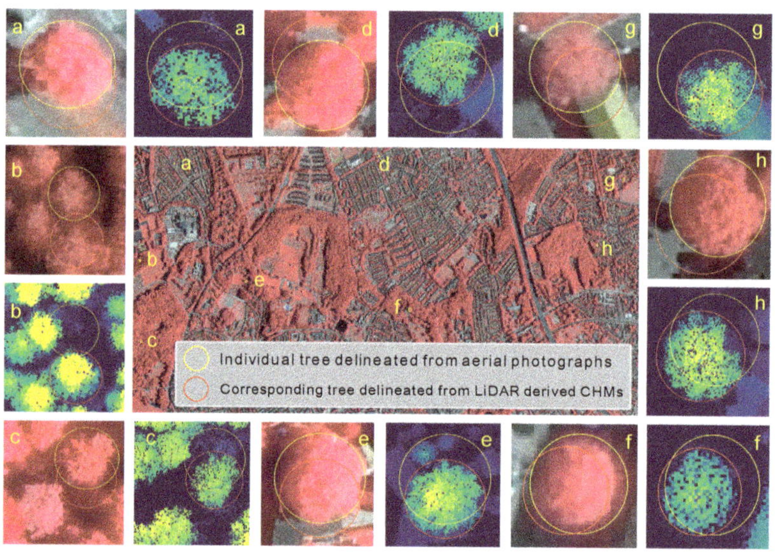

Figure 3. Examples of the spatial mismatches between individual trees resulting from local displacement between aerial photographs and the canopy height models (CHMs) generated from the airborne LiDAR point cloud data in the study area. Yellow and red circles with the same letter have the same geographical location.

3.7. Statistical Description on Mismatch of Individual Trees

We used crown area ratio and offset to explore the mismatch of individual trees between aerial photographs and CHMs (airborne LiDAR point cloud data were considered as the reference data in this study) (Figure 4). We defined the crown area ratio as the ratio of the crown area of an individual tree in the aerial photographs to the crown area of the corresponding tree in CHMs. The offset is defined as the distance between centers of two matching trees in aerial photographs and CHMs. We also performed a statistical description on the original offset of individual trees in each landscape (Figure S2).

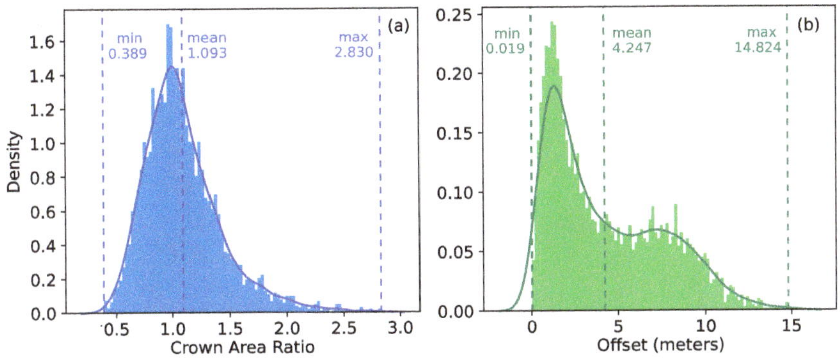

Figure 4. The statistical description on the mismatch of individual trees between aerial photographs and CHMs before rectifying the offset. (**a**) Density histograms of the crown area ratio of individual trees in aerial photographs to the corresponding trees in CHMs, and (**b**) density histograms of the offset between the center of two matching trees separately from aerial photographs and CHMs. The bar, vertical dotted line, and curve represent the density, mean, and density distribution curve of corresponding metrics, respectively.

3.8. Normalized Intersection over Union

The intersection over union (IoU, Figure 5a) is commonly used to evaluate the performance of an object detector for a specific dataset [131–133]. In the case of object detection and segmentation, IoU quantifies the degree of overlap between the target bounding box and the predicted bounding box [134–136]. The larger the overlapping area, the greater the IoU. The bounding boxes are usually the same size in most applications about object matching. While in our study, we found that the bounding box size between the tree in the aerial photograph and the corresponding tree in the CHMs was not always the same in most pairs of individual trees (Figure 4a), leading to discrepancies when evaluating the matching accuracy with IoU. In response, we proposed a normalized intersection over union (NIoU) to eliminate the impact of area difference on IoU by multiplying the ratio of the area of the larger tree to that of the smaller one (Figure 5b). The value of NIoU is between 0 and 1, the closer the value is to 1, the higher the degree of overlap between the two trees. We computed the IoU and NIoU for a roadside tree sample individually to show the difference in measuring the overlap between trees in both datasets (Figure 6). In this study, NIoU served two functions, i.e., (1) screening out the correct or optimal offset vector in determining both the correct offset of a particular tree and the final offset of trees in a region; (2) evaluating the matching accuracy after rectifying the offset of individual trees in two images.

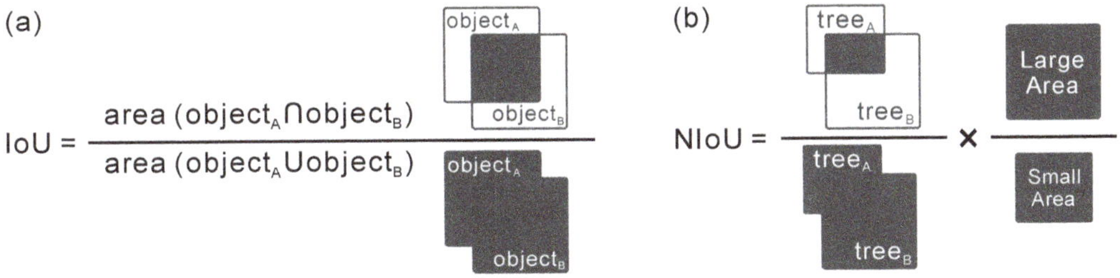

Figure 5. The formula and schematic diagram of intersection over union (IoU (**a**)) and normalized intersection over union (NIoU (**b**)).

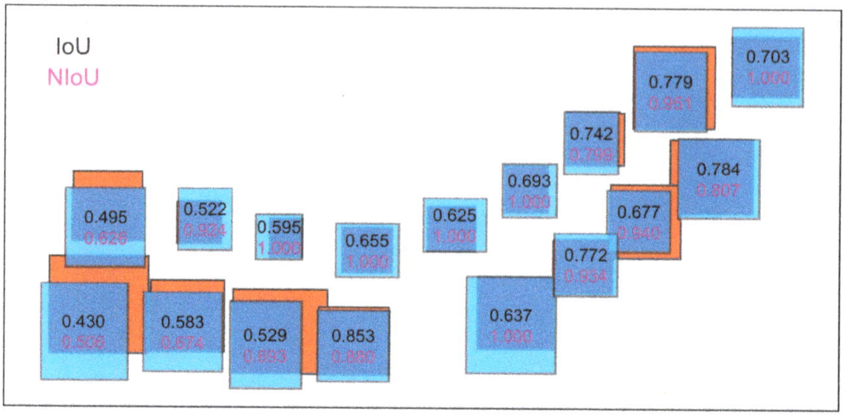

Figure 6. Comparison of IoU and NIoU in measuring overlap between trees. The numbers over the bounding boxes represent the corresponding values.

3.9. Accuracy Assessment

In our preliminary experiment, we found that some trees were assigned to non-matching trees when calculating the NIoU, especially in coniferous forest (Figure 3b). This is due to the fact that the algorithm considers the tree with maximum NIoU as the matching tree when calculating neighbor candidate trees with the reference tree. Therefore, we first used the pairing rate (the percentage of correctly paired trees) to screen out those trees that de facto were not pairwise matching with the reference tree (Equation (1)).

$$\text{pairing rate} = \frac{N_{same\_ID}}{N} \times 100\% \quad (1)$$

where N_{same_ID} represents the number of tree pairs with the same ID number when calculating the NIoU of individual trees between two data, while N is the total number of tree pairs. And then, we define matching accuracy (Equation (2)) as the mean NIoU calculated from the remaining matching tree pairs to measure the degree of overlap between the correctly paired trees.

$$\text{matching accuracy} = \frac{\sum NIoU}{N_{same\_ID}} \quad (2)$$

where $\Sigma NIoU$ indicates the sum of NIoU calculated from each matching tree pair. Moreover, the density histograms with curves of NIoU calculated from matching tree pairs were exploited to compare the difference more intuitively between the results before and after rectifying the offset.

3.10. Parameter Tuning

The key parameter for our approach is the number of reference trees used for calculating, selecting, and determining the final offset vector. Hence, we tested the pairing rate and matching accuracy as the number of reference trees increases. We tested the number of reference trees from one to ten. Depending on the number of reference trees we needed to test, we randomly selected the specific number of trees from CHMs as reference tree to calculate their correct offset vectors, and then chose the final one to rectify the trees in aerial photographs (the core steps are the same as our proposed approach in Section 2.1). We repeated the above experimental steps with 20 different random seeds in all samples of six landscapes (Figure 1b) and calculated the average value of the pairing rate, matching accuracy, and their standard deviation.

4. Results

4.1. Parameter Tuning

As shown in Figure 7a, when the number of reference trees reaches three, the pairing rate is close to saturation and its standard deviation (SD) no longer significantly changes. While the matching accuracy and its SD becomes steady after the number greater than eight (Figure 7b). Therefore, in accordance with the "maximum principle", we set the number of reference trees to be eight to apply our proposed method if there are too many reference trees for selecting the correct offset vectors.

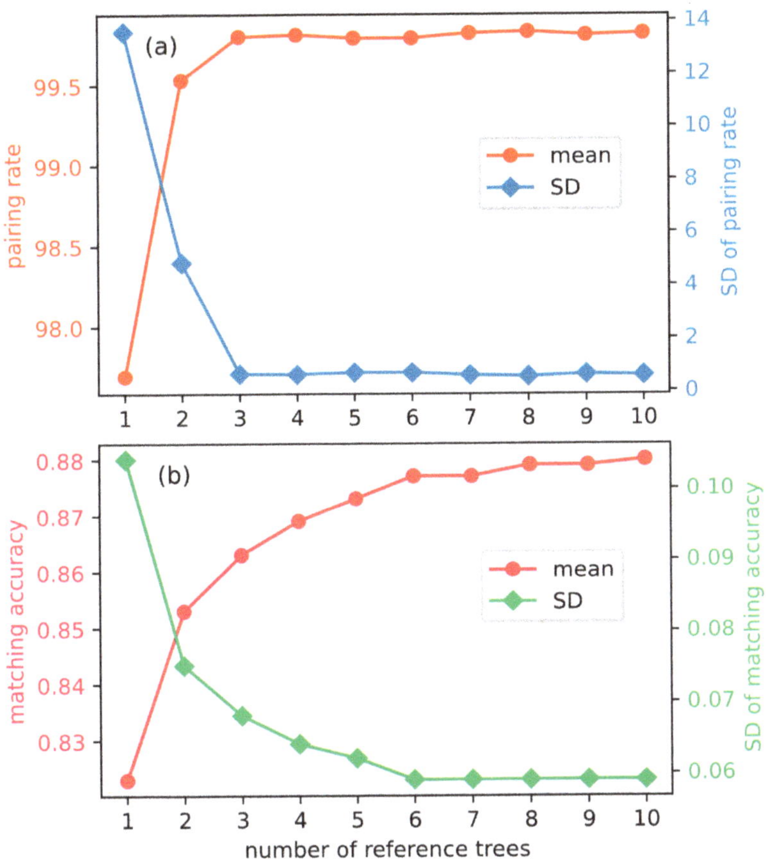

Figure 7. The evaluation indices and their standard deviation (SD) of individual trees using our proposed matching approach as the number of reference trees increases. (**a**) pairing rate; (**b**) matching accuracy.

4.2. Pairing Rate after Using Our Proposed Approach

The pairing rate of the individual trees in all six landscapes before and after rectifying the offset is presented in Figure 8, which shows that all trees were correctly paired after offset rectification using our proposed approach. The mean pairing rate of all six landscapes increased from 71.13% (SD = ±29.83) to 100.00% (SD = 0). Specifically, coniferous forest had the largest improvement with an increase of 76.54%, followed by broadleaved forest (55.03%), mixed forest (37.96%), and garden trees (3.70%), while the individual tree pairs in roadside trees and parkland trees were totally paired even before offset rectification.

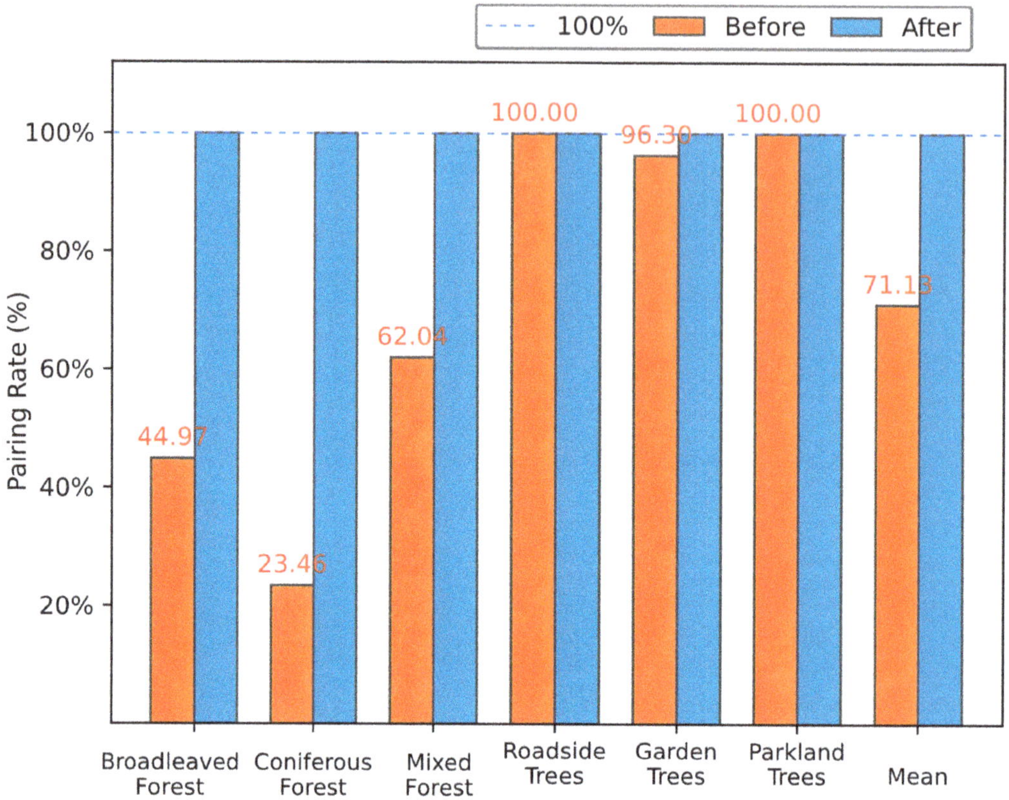

Figure 8. The pairing rate of the individual trees in all six landscapes before and after offset rectification.

4.3. Matching Accuracy after Using Our Proposed Approach

Figure 9 details the density of NIoU calculated from matching tree pairs in all six landscapes before and after rectifying the offset. In all six landscapes, the matching accuracy (mean NIoU) improved significantly ($p < 0.001$, ANOVA), and coniferous forest demonstrated the greatest improvement with an increase of 0.546, followed by mixed forest (0.469), broadleaved forest (0.391), roadside trees (0.218), garden trees (0.178), and parkland trees (0.091). The optimal matching accuracy of individual trees before and after offset rectification was respectively in parkland trees (0.796) and roadside trees (0.919), which was also the only landscape with matching accuracy greater than 0.9. Before applying our approach, the mean matching accuracy of landscapes in city (Figure 9d–f) was 0.718, which was significantly higher than that in forest (0.362). While after rectifying the offset, the six landscapes with matching accuracy in descending order were sequentially roadside trees (0.919), parkland trees (0.887), broadleaved forest (0.853), garden trees (0.835), mixed forest (0.823), and coniferous forest (0.815). Overall, the mean matching accuracy of all six landscapes significantly rose from 0.642 ± 0.264 (SD) to 0.861 ± 0.152 (SD).

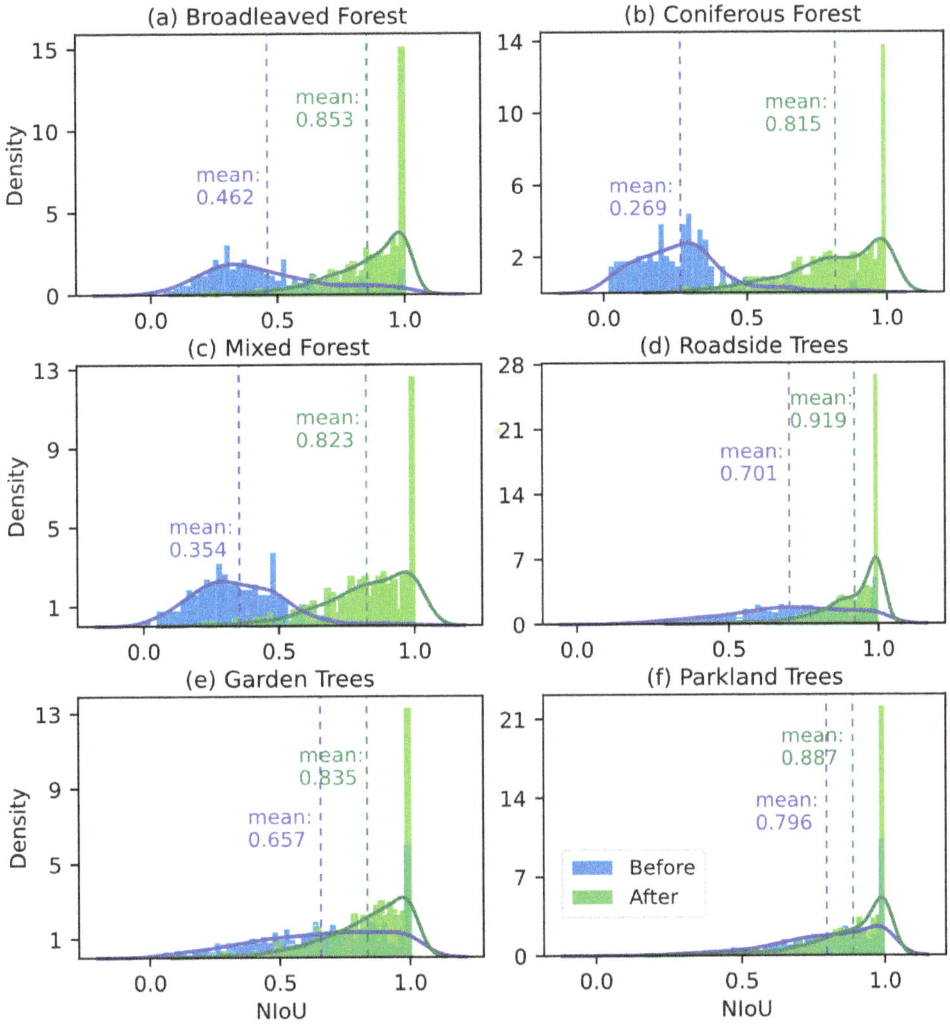

Figure 9. The overlaid density histograms with density curves of NIoU before and after offset rectification. The bar and curve represent the density and its distribution curve of NIoU, respectively, while the vertical dash line is the mean NIoU (i.e., matching accuracy). The blue elements denote the results before offset rectification, while the results after rectifying the offset are shown in green.

4.4. Visualization of Pairing Rate and Matching Accuracy before and after Rectification

We selected a representative sample from each landscape to visually illustrate the pairing and matching effect of individual trees before and after offset rectification (Figure 10). As observed in Figure 10, the pairing rate of the individual trees in all six landscapes was improved to 100% following application of the new matching algorithm described above, while the pairing rate was 14.65%, 52.60%, and 88.59% in broadleaved, coniferous, and mixed forest, respectively, before rectifying the offset (Figure 10a–c). Meanwhile, mean matching accuracy increased to 0.839. Regarding the roadside, garden, and parkland trees, although the pairing rate has not increased significantly, the matching accuracy of roadside and garden trees has been substantially raised (Figure 10d,e). Furthermore, the matching accuracy in parkland trees was slightly improved after offset rectification (Figure 10f).

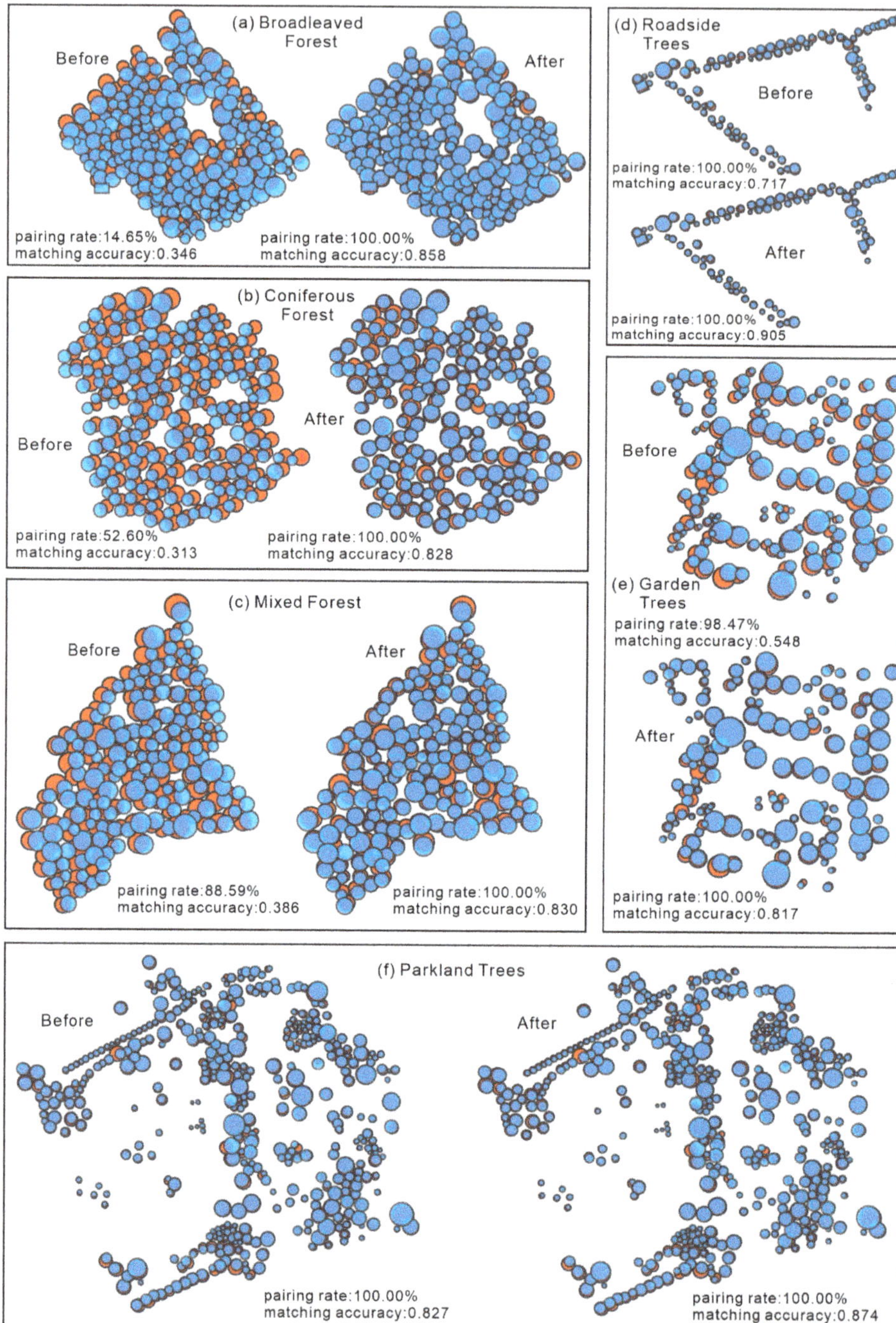

Figure 10. Instance of visualization of the pairing and matching effects of individual trees in all six landscapes before and after offset rectification.

4.5. Comparison of Results Using Image Registration with Our Proposed Approach

We compared the pairing rate and matching accuracy of individual trees after registering images with ENVI automatic registration module and matching with our proposed approach, respectively. As shown in Table 2, through our proposed approach, the mean pairing rate is up to 100.00%, while about 10% of the trees remain mismatched when using the conventional image registration method. In terms of matching accuracy, our proposed approach achieved better results and had smaller standard deviations. It is noted that the matching accuracy in urban areas (roadside trees, garden trees, and parkland trees) has a certain degree of decline, although the pairing rate increases. Moreover, comparing the evaluation indices of the traditional method and our proposed approach, the p-values of the paired t-test for pairing rate and matching accuracy were 0.045 and 0.017, respectively.

Table 2. The pairing rate, matching accuracy, and its standard deviation (SD) of each landscape using image registration module (ENVI) and our proposed approach.

Landscape	Image Registration (ENVI)		Our Proposed Approach	
	Pairing Rate	Matching Accuracy (SD)	Pairing Rate	Matching Accuracy (SD)
Broadleaved Forest	88.98%	0.710 (0.196)	100.00%	0.853 (0.175)
Coniferous Forest	99.75%	0.788 (0.186)	100.00%	0.815 (0.178)
Mixed Forest	100.0%	0.797 (0.181)	100.00%	0.823 (0.171)
Roadside Trees	79.13%	0.656 (0.153)	100.00%	0.919 (0.100)
Garden Trees	92.94%	0.682 (0.206)	100.00%	0.835 (0.199)
Parkland Trees	87.06%	0.676 (0.157)	100.00%	0.887 (0.147)
Mean	91.31%	0.692 (0.175)	100.00%	0.861 (0.152)

5. Discussion

5.1. Influence of Registration Noise on Pairing Rate

Results of this study show that the pairing rate of individual trees in the deciduous, coniferous, and mixed forest classes was clearly lower than that for roadside, garden, and parkland trees (Figure 8). We can readily infer that registration noise exerts a greater influence on forest classes, which might be caused by the quantity and quality of ground control points available for geo-correcting aerial photographs. Compared to roadside, garden, and parkland trees, it is more challenging to locate highly precise control points for fine registration in forested areas (Figure 2), and this leads to a larger offset between individual trees of two data. In general, the lower the density of trees, the higher the matching accuracy and vice versa [137]. It is not a unique problem to this study as there is generally a lack of tie points in areas that are difficult to survey (e.g., mountains, grasslands, and forests) [138]. There are only a few trees in the garden class that were not calculated from matching tree pairs, probably due to the larger variance in canopy and tree height from a higher abundance of tree species (Figure 2c(V)).

5.2. Effectiveness of the Proposed Approach

The improvement of pairing rate was more prominent in coniferous forest, probably since coniferous trees typically had large spaces between adjacent trees than broadleaved trees [25,139,140]. As shown in Figure 3b, coniferous forests were prone to be reference trees to compute the NIoU with a non-matching tree once the offset exceeds the diameter of the tree (especially in forest with high-density trees). In comparison, broadleaved forest and mixed forest had a relatively better pairing rate. Moreover, the improvement of matching accuracy in forest was more evident than that in landscapes of city. The matching accuracy of individual trees in forest before rectifying the offset was much lower than for roadside, garden, and parkland trees. After offset rectification, the matching accuracy reached an asymptotic and highest level for all six landscapes, demonstrating the effectiveness of our proposed matching approach in improving matching accuracy regardless of landscape type (see Figure 9).

5.3. Matching Accuracy in Different Landscapes

As described above for the pairing rate, the most dramatic improvement in the matching accuracy was for coniferous forest. Except for the greater offsets in forests, smaller crowns of coniferous trees had smaller NIoU values, as expected from the NIoU formula (Figure 5). Therefore, in forests, the matching accuracy of broadleaved forest ranks first, followed by mixed forest, and finally coniferous forest. For roadside, garden, and parkland trees, the maximum increment and the maximum value of matching accuracy after rectifying the offsets were for the roadside trees. Several possible reasons may have contributed. Most roadside trees consist of broadleaved trees which were usually the same species, similar in tree age (i.e., tree height and crown size) [141–143]. The regular layout of roadside trees made the offset simple and spatially consistent [144,145], reducing the local registration noise for roadside trees.

The best matching accuracy was observed in different landscapes before and after offset rectification. Before rectifying the offsets between individual trees from two datasets, the best matching accuracy was observed in parkland trees, one possible reason is that the design and context of parklands provide richer features for geographic registration of aerial photographs, as well as parklands had large widely spaced trees separated by grass areas [146,147]. Moreover, although garden trees were characterized by highly accurate ground control points due to their (mostly) solitary and isolated nature and growing environments surrounded by roads and buildings, tree matching results were not as satisfactory as expected, probably stemming from the species-rich areas with trees varying in height and crown size [148–151], complicating the degree and orientation of offsets between different trees.

5.4. Analyzing the Results Using Conventional Image Registration Approach

Compared with that before image registration, the matching accuracy has been improved for forest landscape (broadleaved forest, coniferous forest, and mixed forest), but there has been a certain degree of decline in urban landscapes (roadside trees, garden trees, and parkland trees) (Table 2). Meanwhile, the pairing rate of the corresponding landscape had a similar trend. The above phenomenon was mainly caused by unevenly distributed, low precision, and insufficient control points (CPs) that were found by ENVI automatic program. In this study, we chose landscapes in small patches to verify our proposed method, while it was difficult to find suitable and sufficient CPs to align images with high accuracy [152–155]. Taking the landscape roadside trees as an instance, its area is always narrow and long, thus the unevenly distributed points can easily change the shape of the image to be registered, resulting in serious distortion of the edge of the image. Therefore, though many methods have been proposed to solve multimodal images registration problems, they are usually suitable for large-scale images due to the uneven distribution and limited quantity of CPs [156,157], which is opposed to solving local registration noise.

In our study, aerial photographs and CHMs (different modalities) make it more difficult to find high-precision CPs. On the one hand, the values in different images represent different meanings, e.g., individual trees with high optical reflectance do not always correspond to high altitude in LiDAR height products and vice versa. On the other hand, different viewpoints also result in different shapes for the same objects, e.g., area (aerial photographs) to line (CHMs), line (aerial photographs) to point (CHMs). In addition, the characteristics of ground features change with time. Moreover, pixel values in CHMs are not always geographically true values and some values are generated by interpolation of adjacent values due to missing or insufficient points of the corresponding location in LiDAR point cloud data [158–160]. Therefore, looking for evenly distributed and high-quality CPs between multimodal images requires a large amount of manual intervention, while the method is usually not applicable for large-scale area and fine-grained object studies. In conclusion, in the face of different area sizes, our proposed matching approach is more robust.

5.5. Choosing a Suitable Threshold for Specific Applications

In the object detection domain, a predicted bounding box is widely considered to be correctly detected if its IoU with the ground truth bounding box is greater than 0.5 [1,133,134]. In this study, we used NIoU to determine whether the candidate tree was correctly matched with the reference one, and we found the widely accepted threshold of NIoU (i.e., 0.5) was not always optimal. Meanwhile, other factors like stand density and crown size of individual trees ought to be considered. For instance, in roadside trees where there was sufficient space between adjacent trees, we found an NIoU of around 0.4 or less. In a nutshell, the choice of an appropriate NIoU threshold for a specific application was required, which [161] also recommended. Last but not the least, in theory, the smaller the area of individual trees to be matched, the higher the matching accuracy. However, the corresponding amount of computation will also increase substantially, which should not be overlooked for large areas.

5.6. Possible Challenges and Improvements in Practical Application

First, one apparent challenge comes from the quality of individual tree products. Unlike the manual delineated trees used in this study, the over- and under- segmentation are unavoidable phenomena when using algorithm to automatically extract boundary information of individual trees, which can easily lead to some trees not having the corresponding trees in another dataset. While the rapid advancement of deep learning promises opportunities to reduce the errors caused by the phenomena. Another possible challenge may arise in plantation forests, the offset vector might be wrongly chosen when the spatial layout of trees is similar. At this point, we can expand the range of candidate trees to alleviate achieving the wrong offset vector. Moreover, we can make an improvement of our proposed approach by sightly modifying. If we need to achieve a more accurate or even completely corresponding offset vector for each tree, we can narrow down the trees that used to confirm which offset vector is optimal. However, it would cause a large amount of computation and is not suitable for large-scale use, like national or continental scales.

6. Conclusions

In this study, we proposed a novel tree-oriented matching algorithm that improved the pairing rate and matching accuracy of individual trees derived from aerial photographs and airborne LiDAR point cloud data. Our results demonstrated that the proposed approach effectively increases the pairing rate and matching accuracy of individual trees that were manually delineated from aerial photographs and LiDAR-derived CHMs. Compared to the traditional registration method, the average pairing rate of individual trees for all six landscapes increased from 91.13% to 100.00% ($p = 0.045$, t-test), which suggested that our proposed approach could perfectly solve the problem of matching trees. Meanwhile, the average matching accuracy increased from 0.692 ± 0.175 (standard deviation) to 0.861 ± 0.152 ($p = 0.017$, t-test), demonstrating the effectiveness of the proposed matching approach in matching individual trees between multimodal images.

Supplementary Materials: The following supporting information can be downloaded at: https://www.mdpi.com/article/10.3390/rs15174128/s1. Figure S1: Aerial photographs respectively acquired in 2018 (a) and 2020 (b). Figure S2: The original offsets of individual trees between aerial photographs and CHMs derived from airborne LiDAR data in six landscapes.

Author Contributions: Conceptualization, methodology, programming, validation, visualization, writing—original draft preparation, Y.X.; formal analysis, investigation, Y.X. and T.W.; supervision, T.W. and A.K.S.; writing—review and editing, T.W., A.K.S. and T.W.G.; funding acquisition, Y.X. and A.K.S. All authors have read and agreed to the published version of the manuscript.

Funding: The China Scholarship Council (202008440522) and the ITC Research Fund co-funded this research. The work was also supported by the European Union's Horizon 2020 research and innovation program (834709), funded by the European Research Council (ERC).

Conflicts of Interest: The authors declare no conflict of interest.

References

1. Weinstein, B.G.; Marconi, S.; Bohlman, S.A.; Zare, A.; Singh, A.; Graves, S.J.; White, E.P. A remote sensing derived data set of 100 million individual tree crowns for the National Ecological Observatory Network. *eLife* **2021**, *10*, e62922.
2. Walther, G.-R.; Post, E.; Convey, P.; Menzel, A.; Parmesan, C.; Beebee, T.J.C.; Fromentin, J.-M.; Hoegh-Guldberg, O.; Bairlein, F. Ecological responses to recent climate change. *Nature* **2002**, *416*, 389–395.
3. Lassalle, G.; Ferreira, M.P.; La Rosa, L.E.C.; de Souza Filho, C.R. Deep learning-based individual tree crown delineation in mangrove forests using very-high-resolution satellite imagery. *ISPRS J. Photogramm. Remote Sens.* **2022**, *189*, 220–235.
4. Zheng, Z.; Zeng, Y.; Schuman, M.C.; Jiang, H.; Schmid, B.; Schaepman, M.E.; Morsdorf, F. Individual tree-based vs pixel-based approaches to mapping forest functional traits and diversity by remote sensing. *Int. J. Appl. Earth Obs. Geoinf.* **2022**, *114*, 103074.
5. Skidmore, A.K.; Coops, N.C.; Neinavaz, E.; Ali, A.; Schaepman, M.E.; Paganini, M.; Kissling, W.D.; Vihervaara, P.; Darvishzadeh, R.; Feilhauer, H.; et al. Priority list of biodiversity metrics to observe from space. *Nat. Ecol. Evol.* **2021**, *5*, 896–906.
6. Helfenstein, I.S.; Schneider, F.D.; Schaepman, M.E.; Morsdorf, F. Assessing biodiversity from space: Impact of spatial and spectral resolution on trait-based functional diversity. *Remote Sens. Environ.* **2022**, *275*, 113024.
7. Pettorelli, N.; Wegmann, M.; Skidmore, A.; Mücher, S.; Dawson, T.P.; Fernandez, M.; Lucas, R.; Schaepman, M.E.; Wang, T.; O'Connor, B.; et al. Framing the concept of satellite remote sensing essential biodiversity variables: Challenges and future directions. *Remote Sens. Ecol. Conserv.* **2016**, *2*, 122–131.
8. Bortolot, Z.J.; Wynne, R.H. Estimating forest biomass using small footprint LiDAR data: An individual tree-based approach that incorporates training data. *ISPRS J. Photogramm. Remote Sens.* **2005**, *59*, 342–360.
9. Jeronimo, S.M.A.; Kane, V.R.; Churchill, D.J.; McGaughey, R.J.; Franklin, J.F. Applying LiDAR Individual Tree Detection to Management of Structurally Diverse Forest Landscapes. *J. For.* **2018**, *116*, 336–346.
10. Uzoh, F.C.C.; Oliver, W.W. Individual tree diameter increment model for managed even-aged stands of ponderosa pine throughout the western United States using a multilevel linear mixed effects model. *For. Ecol. Manage.* **2008**, *256*, 438–445.
11. Ferreira, M.P.; Féret, J.-B.; Grau, E.; Gastellu-Etchegorry, J.-P.; Amaral, C.H.D.; Shimabukuro, Y.E.; de Souza Filho, C.R. Retrieving structural and chemical properties of individual tree crowns in a highly diverse tropical forest with 3D radiative transfer modeling and imaging spectroscopy. *Remote Sens. Environ.* **2018**, *211*, 276–291.
12. Weinhold, A.; Döll, S.; Liu, M.; Schedl, A.; Pöschl, Y.; Xu, X.; Neumann, S.; van Dam, N.M. Tree species richness differentially affects the chemical composition of leaves, roots and root exudates in four subtropical tree species. *J. Ecol.* **2022**, *110*, 97–116.
13. Zheng, Z.; Zeng, Y.; Schneider, F.D.; Zhao, Y.; Zhao, D.; Schmid, B.; Schaepman, M.E.; Morsdorf, F. Mapping functional diversity using individual tree-based morphological and physiological traits in a subtropical forest. *Remote Sens. Environ.* **2021**, *252*, 112170.
14. Schneider, F.D.; Morsdorf, F.; Schmid, B.; Petchey, O.L.; Hueni, A.; Schimel, D.S.; Schaepman, M.E. Mapping functional diversokity from remotely sensed morphological and physiological forest traits. *Nat. Commun.* **2017**, *8*, 1441.
15. Köhl, M.; Magnussen, S.; Marchetti, M. *Sampling Methods, Remote Sensing and GIS Multiresource Forest Inventory*; Springer: Berlin/Heidelberg, Germany, 2006; Volume 2.
16. Nielsen, A.B.; Östberg, J.; Delshammar, T. Review of urban tree inventory methods used to collect data at single-tree level. *Arboric. Urban For.* **2014**, *40*, 96–111. [CrossRef]
17. Saarinen, N.; Vastaranta, M.; Kankare, V.; Tanhuanpää, T.; Holopainen, M.; Hyyppä, J.; Hyyppä, H. Urban-Tree-Attribute Update Using Multisource Single-Tree Inventory. *Forests* **2014**, *5*, 1032–1052.
18. Wallace, L.; Sun, Q.; Hally, B.; Hillman, S.; Both, A.; Hurley, J.; Saldias, D.S.M. Linking urban tree inventories to remote sensing data for individual tree mapping. *Urban For. Urban Greening* **2021**, *61*, 127106. [CrossRef]
19. Alonzo, M.; Bookhagen, B.; Roberts, D.A. Urban tree species mapping using hyperspectral and lidar data fusion. *Remote Sens. Environ.* **2014**, *148*, 70–83.
20. Myeong, S.; Nowak, D.J.; Duggin, M.J. A temporal analysis of urban forest carbon storage using remote sensing. *Remote Sens. Environ.* **2006**, *101*, 277–282.
21. Lumnitz, S.; Devisscher, T.; Mayaud, J.R.; Radic, V.; Coops, N.C.; Griess, V.C. Mapping trees along urban street networks with deep learning and street-level imagery. *ISPRS J. Photogramm. Remote Sens.* **2021**, *175*, 144–157.
22. Ke, Y.; Quackenbush, L.J. A review of methods for automatic individual tree-crown detection and delineation from passive remote sensing. *Int. J. Remote Sens.* **2011**, *32*, 4725–4747. [CrossRef]
23. Brandtberg, T.; Walter, F. Automated delineation of individual tree crowns in high spatial resolution aerial images by multiple-scale analysis. *Mach. Vis. Appl.* **1998**, *11*, 64–73. [CrossRef]
24. Jurado, J.M.; López, A.; Pádua, L.; Sousa, J.J. Remote sensing image fusion on 3D scenarios: A review of applications for agriculture and forestry. *Int. J. Appl. Earth Obs. Geoinf.* **2022**, *112*, 102856. [CrossRef]

25. Dalponte, M.; Ørka, H.O.; Ene, L.T.; Gobakken, T.; Næsset, E. Tree crown delineation and tree species classification in boreal forests using hyperspectral and ALS data. *Remote Sens. Environ.* **2014**, *140*, 306–317. [CrossRef]
26. Kattenborn, T.; Leitloff, J.; Schiefer, F.; Hinz, S. Review on Convolutional Neural Networks (CNN) in vegetation remote sensing. *ISPRS J. Photogramm. Remote Sens.* **2021**, *173*, 24–49. [CrossRef]
27. Gougeon, F.A. A Crown-Following Approach to the Automatic Delineation of Individual Tree Crowns in High Spatial Resolution Aerial Images. *Can. J. Remote Sens.* **1995**, *21*, 274–284. [CrossRef]
28. Yang, J.; He, Y.; Caspersen, J. A multi-band watershed segmentation method for individual tree crown delineation from high resolution multispectral aerial image. In Proceedings of the 2014 IEEE Geoscience and Remote Sensing Symposium, Quebec City, QC, Canada, 13–18 July 2014.
29. Erikson, M. Segmentation of individual tree crowns in colour aerial photographs using region growing supported by fuzzy rules. *Can. J. For. Res.* **2003**, *33*, 1557–1563. [CrossRef]
30. Wagner, F.H.; Ferreira, M.P.; Sanchez, A.; Hirye, M.C.M.; Zortea, M.; Gloor, E.; Phillips, O.L.; de Souza Filho, C.R.; Shimabukuro, Y.E.; Aragão, L.E.O.C. Individual tree crown delineation in a highly diverse tropical forest using very high resolution satellite images. *ISPRS J. Photogramm. Remote Sens.* **2018**, *145*, 362–377. [CrossRef]
31. Tong, F.; Tong, H.; Mishra, R.; Zhang, Y. Delineation of Individual Tree Crowns Using High Spatial Resolution Multispectral WorldView-3 Satellite Imagery. *IEEE J. Sel. Top. Appl. Earth Obs. Remote Sens.* **2021**, *14*, 7751–7761. [CrossRef]
32. Gomes, M.F.; Maillard, P.; Deng, H. Individual tree crown detection in sub-meter satellite imagery using Marked Point Processes and a geometrical-optical model. *Remote Sens. Environ.* **2018**, *211*, 184–195. [CrossRef]
33. Huang, H.; Li, X.; Chen, C. Individual Tree Crown Detection and Delineation From Very-High-Resolution UAV Images Based on Bias Field and Marker-Controlled Watershed Segmentation Algorithms. *IEEE J. Sel. Top. Appl. Earth Obs. Remote Sens.* **2018**, *11*, 2253–2262. [CrossRef]
34. Miraki, M.; Sohrabi, H.; Fatehi, P.; Kneubuehler, M. Individual tree crown delineation from high-resolution UAV images in broadleaf forest. *Ecol. Inform.* **2021**, *61*, 101207. [CrossRef]
35. Safonova, A.; Hamad, Y.; Dmitriev, E.; Georgiev, G.; Trenkin, V.; Georgieva, M.; Dimitrov, S.; Iliev, M. Individual Tree Crown Delineation for the Species Classification and Assessment of Vital Status of Forest Stands from UAV Images. *Drones* **2021**, *5*, 77–94. [CrossRef]
36. Xi, X.; Xia, K.; Yang, Y.; Du, X.; Feng, H. Evaluation of dimensionality reduction methods for individual tree crown delineation using instance segmentation network and UAV multispectral imagery in urban forest. *Comput. Electron. Agric.* **2021**, *191*, 106506. [CrossRef]
37. Lei, L.; Yin, T.; Chai, G.; Li, Y.; Wang, Y.; Jia, X.; Zhang, X. A novel algorithm of individual tree crowns segmentation considering three-dimensional canopy attributes using UAV oblique photos. *Int. J. Appl. Earth Obs. Geoinf.* **2022**, *112*, 102893. [CrossRef]
38. Xu, X.; Iuricich, F.; Calders, K.; Armston, J.; De Floriani, L. Topology-based individual tree segmentation for automated processing of terrestrial laser scanning point clouds. *Int. J. Appl. Earth Obs. Geoinf.* **2023**, *116*, 103145. [CrossRef]
39. Moradi, F.; Javan, F.D.; Samadzadegan, F. Potential evaluation of visible-thermal UAV image fusion for individual tree detection based on convolutional neural network. *Int. J. Appl. Earth Obs. Geoinf.* **2022**, *113*, 103011. [CrossRef]
40. Leckie, D.; Gougeon, F.; Hill, D.; Quinn, R.; Armstrong, L.; Shreenan, R. Combined high-density lidar and multispectral imagery for individual tree crown analysis. *Can. J. Remote Sens.* **2003**, *29*, 633–649. [CrossRef]
41. Galvincio, J.D.; Popescu, S.C. Measuring individual tree height and crown diameter for mangrove trees with airborne LiDAR data. *Int. J. Adv. Eng. Manag. Sci.* **2016**, *2*, 239456.
42. Tang, H.; Dubayah, R.; Swatantran, A.; Hofton, M.; Sheldon, S.; Clark, D.B.; Blair, B. Retrieval of vertical LAI profiles over tropical rain forests using waveform lidar at La Selva, Costa Rica. *Remote Sens. Environ.* **2012**, *124*, 242–250. [CrossRef]
43. Shi, Y.; Skidmore, A.K.; Wang, T.; Holzwarth, S.; Heiden, U.; Pinnel, N.; Zhu, X.; Heurich, M. Tree species classification using plant functional traits from LiDAR and hyperspectral data. *Int. J. Appl. Earth Obs. Geoinf.* **2018**, *73*, 207–219. [CrossRef]
44. Khosravipour, A.; Skidmore, A.K.; Isenburg, M. Generating spike-free digital surface models using LiDAR raw point clouds: A new approach for forestry applications. *Int. J. Appl. Earth Obs. Geoinf.* **2016**, *52*, 104–114. [CrossRef]
45. Sankey, T.; Donager, J.; McVay, J.; Sankey, J.B. UAV lidar and hyperspectral fusion for forest monitoring in the southwestern USA. *Remote Sens. Environ.* **2017**, *195*, 30–43. [CrossRef]
46. Holmgren, J.; Persson, Å.; Söderman, U. Species identification of individual trees by combining high resolution LiDAR data with multi-spectral images. *Int. J. Remote Sens.* **2008**, *29*, 1537–1552. [CrossRef]
47. Zhen, Z.; Quackenbush, L.J.; Zhang, L. Trends in Automatic Individual Tree Crown Detection and Delineation—Evolution of LiDAR Data. *Remote Sens.* **2016**, *8*, 333. [CrossRef]
48. Li, H.; Hu, B.; Li, Q.; Jing, L. CNN-Based Individual Tree Species Classification Using High-Resolution Satellite Imagery and Airborne LiDAR Data. *Forests* **2021**, *12*, 1697–1718. [CrossRef]
49. Wu, B.; Yu, B.; Wu, Q.; Huang, Y.; Chen, Z.; Wu, J. Individual tree crown delineation using localized contour tree method and airborne LiDAR data in coniferous forests. *Int. J. Appl. Earth Obs. Geoinf.* **2016**, *52*, 82–94. [CrossRef]
50. Koch, B.; Heyder, U.; Weinacker, H. Detection of individual tree crowns in airborne lidar data. *Photogramm. Eng. Remote Sens.* **2006**, *72*, 357–363. [CrossRef]

51. Wan Mohd Jaafar, W.S.; Woodhouse, I.H.; Silva, C.A.; Omar, H.; Maulud, K.N.A.; Hudak, A.T.; Klauberg, C.; Cardil, A.; Mohan, M. Improving Individual Tree Crown Delineation and Attributes Estimation of Tropical Forests Using Airborne LiDAR Data. *Forests* **2018**, *9*, 759–781. [CrossRef]
52. Hamraz, H.; Contreras, M.A.; Zhang, J. A robust approach for tree segmentation in deciduous forests using small-footprint airborne LiDAR data. *Int. J. Appl. Earth Obs. Geoinf.* **2016**, *52*, 532–541. [CrossRef]
53. Torresan, C.; Carotenuto, F.; Chiavetta, U.; Miglietta, F.; Zaldei, A.; Gioli, B. Individual Tree Crown Segmentation in Two-Layered Dense Mixed Forests from UAV LiDAR Data. *Drones* **2020**, *4*, 10–29. [CrossRef]
54. Mohan, M.; Leite, R.V.; Broadbent, E.N.; Jaafar, W.S.W.M.; Srinivasan, S.; Bajaj, S.; Corte, A.P.D.; Amaral, C.H.D.; Gopan, G.; Saad, S.N.M.; et al. Individual tree detection using UAV-lidar and UAV-SfM data: A tutorial for beginners. *Open Geosci.* **2021**, *13*, 1028–1039. [CrossRef]
55. Rudge, M.L.M.; Levick, S.R.; Bartolo, R.E.; Erskine, P.D. Modelling the diameter distribution of savanna trees with drone-based LiDAR. *Remote Sens.* **2021**, *13*, 1266–1283. [CrossRef]
56. Allouis, T.; Durrieu, S.; Véga, C.; Couteron, P. Stem Volume and Above-Ground Biomass Estimation of Individual Pine Trees From LiDAR Data: Contribution of Full-Waveform Signals. *IEEE J. Sel. Top. Appl. Earth Obs. Remote Sens.* **2013**, *6*, 924–934. [CrossRef]
57. Shi, Y.; Wang, T.; Skidmore, A.K.; Heurich, M. Improving LiDAR-based tree species mapping in Central European mixed forests using multi-temporal digital aerial colour-infrared photographs. *Int. J. Appl. Earth Obs. Geoinf.* **2020**, *84*, 101970. [CrossRef]
58. Yang, J.; Kang, Z.; Cheng, S.; Yang, Z.; Akwensi, P.H. An Individual Tree Segmentation Method Based on Watershed Algorithm and Three-Dimensional Spatial Distribution Analysis From Airborne LiDAR Point Clouds. *IEEE J. Sel. Top. Appl. Earth Obs. Remote Sens.* **2020**, *13*, 1055–1067. [CrossRef]
59. Reitberger, J.; Schnörr, C.; Krzystek, P.; Stilla, U. 3D segmentation of single trees exploiting full waveform LIDAR data. *ISPRS J. Photogramm. Remote Sens.* **2009**, *64*, 561–574. [CrossRef]
60. Lee, J.; Cai, X.; Lellmann, J.; Dalponte, M.; Malhi, Y.; Butt, N.; Morecroft, M.; Schönlieb, C.B.; Coomes, D.A. Individual Tree Species Classification From Airborne Multisensor Imagery Using Robust PCA. *IEEE J. Sel. Top. Appl. Earth Obs. Remote Sens.* **2016**, *9*, 2554–2567. [CrossRef]
61. Hui, Z.; Cheng, P.; Yang, B.; Zhou, G. Multi-level self-adaptive individual tree detection for coniferous forest using airborne LiDAR. *Int. J. Appl. Earth Obs. Geoinf.* **2022**, *114*, 103028. [CrossRef]
62. Beese, L.; Dalponte, M.; Asner, G.P.; Coomes, D.A.; Jucker, T. Using repeat airborne LiDAR to map the growth of individual oil palms in Malaysian Borneo during the 2015–16 El Niño. *Int. J. Appl. Earth Obs. Geoinf.* **2022**, *115*, 103117. [CrossRef]
63. Qin, H.; Zhou, W.; Yao, Y.; Wang, W. Individual tree segmentation and tree species classification in subtropical broadleaf forests using UAV-based LiDAR, hyperspectral, and ultrahigh-resolution RGB data. *Remote Sens. Environ.* **2022**, *280*, 113143. [CrossRef]
64. Harikumar, A.; D'Odorico, P.; Ensminger, I. Combining Spectral, Spatial-Contextual, and Structural Information in Multispectral UAV Data for Spruce Crown Delineation. *Remote Sens.* **2022**, *14*, 2044–2062. [CrossRef]
65. Zhang, C.; Qiu, F. Mapping individual tree species in an urban forest using airborne lidar data and hyperspectral imagery. *Photogramm. Eng. Remote Sens.* **2012**, *78*, 1079–1087. [CrossRef]
66. Huang, R.; Zheng, S.; Hu, K. Registration of Aerial Optical Images with LiDAR Data Using the Closest Point Principle and Collinearity Equations. *Sensors* **2018**, *18*, 1770–1790. [CrossRef] [PubMed]
67. Bovolo, F.; Bruzzone, L.; Marchesi, S. Analysis and Adaptive Estimation of the Registration Noise Distribution in Multitemporal VHR Images. *IEEE Trans. Geosci. Remote Sens.* **2009**, *47*, 2658–2671. [CrossRef]
68. Wu, W.; Shao, Z.; Huang, X.; Teng, J.; Guo, S.; Li, D. Quantifying the sensitivity of SAR and optical images three-level fusions in land cover classification to registration errors. *Int. J. Appl. Earth Obs. Geoinf.* **2022**, *112*, 102868. [CrossRef]
69. Han, Y.; Bovolo, F.; Bruzzone, L. An Approach to Fine Coregistration Between Very High Resolution Multispectral Images Based on Registration Noise Distribution. *IEEE Trans. Geosci. Remote Sens.* **2015**, *53*, 6650–6662. [CrossRef]
70. Han, Y. Fine geometric alignment of very high resolution optical images using registration noise and quadtree structure. In Proceedings of the 2017 IEEE International Geoscience and Remote Sensing Symposium (IGARSS), Fort Worth, TX, USA, 23–28 July 2017.
71. Lee, J.; Cai, X.; Schönlieb, C.; Coomes, D.A. Nonparametric Image Registration of Airborne LiDAR, Hyperspectral and Photographic Imagery of Wooded Landscapes. *IEEE Trans. Geosci. Remote Sens.* **2015**, *53*, 6073–6084. [CrossRef]
72. Jiang, X.; Ma, J.; Xiao, G.; Shao, Z.; Guo, X. A review of multimodal image matching: Methods and applications. *Inf. Fusion* **2021**, *73*, 22–71. [CrossRef]
73. Xu, X.; Li, X.; Liu, X.; Shen, H.; Shi, Q. Multimodal registration of remotely sensed images based on Jeffrey's divergence. *ISPRS J. Photogramm. Remote Sens.* **2016**, *122*, 97–115. [CrossRef]
74. Näsi, R.; Honkavaara, E.; Blomqvist, M.; Lyytikäinen-Saarenmaa, P.; Hakala, T.; Viljanen, N.; Kantola, T.; Holopainen, M. Remote sensing of bark beetle damage in urban forests at individual tree level using a novel hyperspectral camera from UAV and aircraft. *Urban For. Urban Greening* **2018**, *30*, 72–83. [CrossRef]
75. Duncanson, L.; Dubayah, R. Monitoring individual tree-based change with airborne lidar. *Ecol. Evol.* **2018**, *8*, 5079–5089. [CrossRef]
76. Vastaranta, M.; Holopainen, M.; Yu, X.; Hyyppä, J.; Mäkinen, A.; Rasinmäki, J.; Melkas, T.; Kaartinen, H.; Hyyppä, H. Effects of Individual Tree Detection Error Sources on Forest Management Planning Calculations. *Remote Sens.* **2011**, *3*, 1614–1626. [CrossRef]

77. Mäyrä, J.; Keski-Saari, S.; Kivinen, S.; Tanhuanpää, T.; Hurskainen, P.; Kullberg, P.; Poikolainen, L.; Viinikka, A.; Tuominen, S.; Kumpula, T.; et al. Tree species classification from airborne hyperspectral and LiDAR data using 3D convolutional neural networks. *Remote Sens. Environ.* **2021**, *256*, 112322. [CrossRef]
78. Palenichka, R.M.; Zaremba, M.B. Automatic Extraction of Control Points for the Registration of Optical Satellite and LiDAR Images. *IEEE Trans. Geosci. Remote Sens.* **2010**, *48*, 2864–2879. [CrossRef]
79. Li, L.; Han, L.; Ding, M.; Cao, H.; Hu, H. A deep learning semantic template matching framework for remote sensing image registration. *ISPRS J. Photogramm. Remote Sens.* **2021**, *181*, 205–217. [CrossRef]
80. Mustaffar, M.; Mitchell, H.L. Improving area-based matching by using surface gradients in the pixel co-ordinate transformation. *ISPRS J. Photogramm. Remote Sens.* **2001**, *56*, 42–52. [CrossRef]
81. Ye, Y.; Shen, L. Hopc: A novel similarity metric based on geometric structural properties for multi-modal remote sensing image matching. *ISPRS Ann. Photogramm. Remote Sens. Spatial Inf. Sci.* **2016**, *3*, 9. [CrossRef]
82. Ye, Y.; Shan, J.; Bruzzone, L.; Shen, L. Robust registration of multimodal remote sensing images based on structural similarity. *IEEE Trans. Geosci. Remote Sens.* **2017**, *55*, 2941–2958. [CrossRef]
83. Wong, A.; Clausi, D.A. ARRSI: Automatic Registration of Remote-Sensing Images. *IEEE Trans. Geosci. Remote Sens.* **2007**, *45*, 1483–1493. [CrossRef]
84. Govindu, V.M.; Chellappa, R. Feature-based image to image registration. In *Image Registration for Remote Sensing*; Le Moigne, J., Netanyahu, N.S., Eastman, R.D., Eds.; Cambridge University Press: Cambridge, UK, 2011; pp. 215–239.
85. Lowe, D.G. Object recognition from local scale-invariant features. In Proceedings of the Seventh IEEE International Conference on Computer Vision, Kerkyra, Greece, 20–27 September 1999.
86. Paul, S.; Pati, U.C. Remote Sensing Optical Image Registration Using Modified Uniform Robust SIFT. *IEEE Geosci. Remote Sens. Lett.* **2016**, *13*, 1300–1304. [CrossRef]
87. Sedaghat, A.; Mohammadi, N. Uniform competency-based local feature extraction for remote sensing images. *ISPRS J. Photogramm. Remote Sens.* **2018**, *135*, 142–157. [CrossRef]
88. Goncalves, H.; Corte-Real, L.; Goncalves, J.A. Automatic Image Registration Through Image Segmentation and SIFT. *IEEE Trans. Geosci. Remote Sens.* **2011**, *49*, 2589–2600. [CrossRef]
89. Kupfer, B.; Netanyahu, N.S.; Shimshoni, I. An Efficient SIFT-Based Mode-Seeking Algorithm for Sub-Pixel Registration of Remotely Sensed Images. *IEEE Geosci. Remote Sens. Lett.* **2015**, *12*, 379–383. [CrossRef]
90. Hasan, M.; Jia, X.; Robles-Kelly, A.; Zhou, J.; Pickering, M.R. Multi-spectral remote sensing image registration via spatial relationship analysis on sift keypoints. In Proceedings of the 2010 IEEE International Geoscience and Remote Sensing Symposium, Honolulu, HI, USA, 25–30 July 2010.
91. Turner, D.; Lucieer, A.; Malenovský, Z.; King, D.H.; Robinson, S.A. Spatial Co-Registration of Ultra-High Resolution Visible, Multispectral and Thermal Images Acquired with a Micro-UAV over Antarctic Moss Beds. *Remote Sens.* **2014**, *6*, 4003–4024. [CrossRef]
92. Wong, A.; Orchard, J. Efficient FFT-accelerated approach to invariant optical–LIDAR registration. *IEEE Trans. Geosci. Remote Sens.* **2008**, *46*, 3917–3925. [CrossRef]
93. Dusmanu, M.; Rocco, I.; Pajdla, T.; Pollefeys, M.; Sivic, J.; Torii, A.; Sattler, T. D2-net: A trainable cnn for joint description and detection of local features. In Proceedings of the IEEE/CVF Conference on Computer Vision and Pattern Recognition (CVPR), Long Beach, CA, USA, 16–20 June 2019; pp. 8092–8101.
94. Blendowski, M.; Heinrich, M.P. Combining MRF-based deformable registration and deep binary 3D-CNN descriptors for large lung motion estimation in COPD patients. *Int. J. Comput. Assist. Radiol. Surg.* **2019**, *14*, 43–52. [CrossRef]
95. Cheng, X.; Zhang, L.; Zheng, Y. Deep similarity learning for multimodal medical images. *Comput. Methods Biomech. Biomed. Eng. Imaging Vis.* **2018**, *6*, 248–252. [CrossRef]
96. Wu, W.; Hu, S.; Shen, J.; Ma, L.; Han, J. Sensitization of 21% Cr Ferritic Stainless Steel Weld Joints Fabricated With/Without Austenitic Steel Foil as Interlayer. *J. Mater. Eng. Perform.* **2015**, *24*, 1505–1515. [CrossRef]
97. Wu, G.; Kim, M.; Wang, Q.; Gao, Y.; Liao, S.; Shen, D. Unsupervised Deep Feature Learning for Deformable Registration of MR Brain Images. In *Medical Image Computing and Computer-Assisted Intervention—MICCAI 2013*; Springer: Berlin/Heidelberg, Germany, 2013.
98. Simonovsky, M.; Gutiérrez-Becker, B.; Mateus, D.; Navab, N.; Komodakis, N. A Deep Metric for Multimodal Registration. In *Medical Image Computing and Computer-Assisted Intervention—MICCAI 2016*; Springer International Publishing: Cham, Switzerland, 2016.
99. Han, X.; Leung, T.; Jia, Y.; Sukthankar, R.; Berg, A.C. Matchnet: Unifying feature and metric learning for patch-based matching. In Proceedings of the IEEE Conference on Computer Vision and Pattern Recognition (CVPR), Boston, MA, USA, 7–12 June 2015; pp. 3279–3286.
100. Zagoruyko, S.; Komodakis, N. Learning to compare image patches via convolutional neural networks. In Proceedings of the IEEE Conference on Computer Vision and Pattern Recognition (CVPR), Boston, MA, USA, 7–12 June 2015; pp. 3279–3286.
101. Liao, R.; Miao, S.; de Tournemire, P.; Grbic, S.; Kamen, A.; Mansi, T.; Comaniciu, D. An artificial agent for robust image registration. In Proceedings of the Thirty-First AAAI Conference on Artificial Intelligence, San Francisco, CA, USA, 4–9 February 2017.
102. Haskins, G.; Kruger, U.; Yan, P. Deep learning in medical image registration: A survey. *Mach. Vis. Appl.* **2020**, *31*, 8. [CrossRef]

103. Zhang, J. Inverse-consistent deep networks for unsupervised deformable image registration. *arXiv* **2018**, arXiv:1809.03443.
104. Li, H.; Fan, Y. Non-rigid image registration using fully convolutional networks with deep self-supervision. *arXiv* **2017**, arXiv:1709.00799.
105. De Vos, B.D.; Berendsen, F.F.; Viergever, M.A.; Sokooti, H.; Staring, M.; Išgum, I. A deep learning framework for unsupervised affine and deformable image registration. *Med. Image Anal.* **2019**, *52*, 128–143. [CrossRef]
106. Zhu, X.X.; Tuia, D.; Mou, L.; Xia, G.-S.; Zhang, L.; Xu, F.; Fraundorfer, F. Deep learning in remote sensing: A comprehensive review and list of resources. *IEEE Geosci. Remote Sens. Mag.* **2017**, *5*, 8–36. [CrossRef]
107. Blaschke, T. Object based image analysis for remote sensing. *ISPRS J. Photogramm. Remote Sens.* **2010**, *65*, 2–16. [CrossRef]
108. Hossain, M.D.; Chen, D. Segmentation for Object-Based Image Analysis (OBIA): A review of algorithms and challenges from remote sensing perspective. *ISPRS J. Photogramm. Remote Sens.* **2019**, *150*, 115–134. [CrossRef]
109. Kotaridis, I.; Lazaridou, M. Remote sensing image segmentation advances: A meta-analysis. *ISPRS J. Photogramm. Remote Sens.* **2021**, *173*, 309–322. [CrossRef]
110. Blaschke, T.; Hay, G.J.; Kelly, M.; Lang, S.; Hofmann, P.; Addink, E.; Feitosa, R.Q.; van der Meer, F.; van der Werff, H.; van Coillie, F.; et al. Geographic Object-Based Image Analysis—Towards a new paradigm. *ISPRS J. Photogramm. Remote Sens.* **2014**, *87*, 180–191. [CrossRef]
111. Chandler, C.J.; van der Heijden, G.M.F.; Boyd, D.S.; Cutler, M.E.J.; Costa, H.; Nilus, R.; Foody, G.M. Remote sensing liana infestation in an aseasonal tropical forest: Addressing mismatch in spatial units of analyses. *Remote Sens. Ecol. Conserv.* **2021**, *7*, 397–410. [CrossRef]
112. Ma, K.; Chen, Z.; Fu, L.; Tian, W.; Jiang, F.; Yi, J.; Du, Z.; Sun, H. Performance and Sensitivity of Individual Tree Segmentation Methods for UAV-LiDAR in Multiple Forest Types. *Remote Sens.* **2022**, *14*, 298–317. [CrossRef]
113. Jin, X. *ENVI Automated Image Registration Solutions*; Internal Report; Harris Corporation: Melbourne, FL, USA, 2017; p. 26.
114. FAO. *Global Forest Resources Assessment 2020, Terms and Definitions*; FAO: Roma, Italy, 2020.
115. Kosztra, B.; Büttner, G.; Hazeu, G.; Arnold, S. *Updated CLC Illustrated Nomenclature Guidelines*; European Environment Agency: Wien, Austria, 2017; pp. 1–124.
116. Smart, N.; Eisenman, T.S.; Karvonen, A. Street Tree Density and Distribution: An International Analysis of Five Capital Cities. *Front. Ecol. Evol.* **2020**, *8*, 562646. [CrossRef]
117. Crowther, T.W.; Glick, H.B.; Covey, K.R.; Bettigole, C.; Maynard, D.S.; Thomas, S.M.; Smith, J.R.; Hintler, G.; Duguid, M.C.; Amatulli, G.; et al. Mapping tree density at a global scale. *Nature* **2015**, *525*, 201–205. [CrossRef] [PubMed]
118. Fekete, A.; Cserep, M. Tree segmentation and change detection of large urban areas based on airborne LiDAR. *Comput. Geosci.* **2021**, *156*, 104900. [CrossRef]
119. Mücher, C.A.; Roupioz, L.; Kramer, H.; Bogers, M.M.B.; Jongman, R.H.G.; Lucas, R.M.; Kosmidou, V.E.; Petrou, Z.; Manakos, I.; Padoa-Schioppa, E.; et al. Synergy of airborne LiDAR and Worldview-2 satellite imagery for land cover and habitat mapping: A BIO_SOS-EODHaM case study for the Netherlands. *Int. J. Appl. Earth Obs. Geoinf.* **2015**, *37*, 48–55. [CrossRef]
120. Hyyppä, J.; Hyyppä, H.; Leckie, D.; Gougeon, F.; Yu, X.; Maltamo, M. Review of methods of small-footprint airborne laser scanning for extracting forest inventory data in boreal forests. *Int. J. Remote Sens.* **2008**, *29*, 1339–1366. [CrossRef]
121. Khosravipour, A.; Skidmore, A.K.; Wang, T.; Isenburg, M.; Khoshelham, K. Effect of slope on treetop detection using a LiDAR Canopy Height Model. *ISPRS J. Photogramm. Remote Sens.* **2015**, *104*, 44–52. [CrossRef]
122. Alexander, C.; Korstjens, A.H.; Hill, R.A. Influence of micro-topography and crown characteristics on tree height estimations in tropical forests based on LiDAR canopy height models. *Int. J. Appl. Earth Obs. Geoinf.* **2018**, *65*, 105–113. [CrossRef]
123. Chen, Q.; Wang, X.; Hang, M.; Li, J. Research on the improvement of single tree segmentation algorithm based on airborne LiDAR point cloud. *Open Geosci.* **2021**, *13*, 705–716. [CrossRef]
124. Wang, X.; Chan, T.O.; Liu, K.; Pan, J.; Luo, M.; Li, W.; Wei, C. A robust segmentation framework for closely packed buildings from airborne LiDAR point clouds. *Int. J. Remote Sens.* **2020**, *41*, 5147–5165. [CrossRef]
125. Tobler, W.R. A computer movie simulating urban growth in the Detroit region. *Econ. Geogr.* **1970**, *46* (Suppl. 1), 234–240. [CrossRef]
126. Lu, G.Y.; Wong, D.W. An adaptive inverse-distance weighting spatial interpolation technique. *Comput. Geosci.* **2008**, *34*, 1044–1055. [CrossRef]
127. Feng, R.; Du, Q.; Li, X.; Shen, H. Robust registration for remote sensing images by combining and localizing feature- and area-based methods. *ISPRS J. Photogramm. Remote Sens.* **2019**, *151*, 15–26. [CrossRef]
128. Wang, S.; Quan, D.; Liang, X.; Ning, M.; Guo, Y.; Jiao, L. A deep learning framework for remote sensing image registration. *ISPRS J. Photogramm. Remote Sens.* **2018**, *145*, 148–164. [CrossRef]
129. Ye, Y.; Bruzzone, L.; Shan, J.; Bovolo, F.; Zhu, Q. Fast and Robust Matching for Multimodal Remote Sensing Image Registration. *IEEE Trans. Geosci. Remote Sens.* **2019**, *57*, 9059–9070. [CrossRef]
130. Zhu, B.; Ye, Y.; Zhou, L.; Li, Z.; Yin, G. Robust registration of aerial images and LiDAR data using spatial constraints and Gabor structural features. *ISPRS J. Photogramm. Remote Sens.* **2021**, *181*, 129–147. [CrossRef]
131. Xiao, Y.; Tian, Z.; Yu, J.; Zhang, Y.; Liu, S.; Du, S.; Lan, X. A review of object detection based on deep learning. *Multimed. Tools Appl.* **2020**, *79*, 23729–23791. [CrossRef]
132. Dhillon, A.; Verma, G.K. Convolutional neural network: A review of models, methodologies and applications to object detection. *Prog. Artif. Intell.* **2020**, *9*, 85–112. [CrossRef]

133. Wu, X.; Sahoo, D.; Hoi, S.C.H. Recent advances in deep learning for object detection. *Neurocomputing* **2020**, *396*, 39–64. [CrossRef]
134. Zhao, Z.Q.; Zheng, P.; Xu, S.T.; Wu, X. Object Detection With Deep Learning: A Review. *IEEE Trans. Neural Networks Learn. Syst.* **2019**, *30*, 3212–3232. [CrossRef]
135. Choi, K.; Lim, W.; Chang, B.; Jeong, J.; Kim, I.; Park, C.-R.; Ko, D.W. An automatic approach for tree species detection and profile estimation of urban street trees using deep learning and Google street view images. *ISPRS J. Photogramm. Remote Sens.* **2022**, *190*, 165–180. [CrossRef]
136. Liu, T.; Yao, L.; Qin, J.; Lu, N.; Jiang, H.; Zhang, F.; Zhou, C. Multi-scale attention integrated hierarchical networks for high-resolution building footprint extraction. *Int. J. Appl. Earth Obs. Geoinf.* **2022**, *109*, 102768. [CrossRef]
137. Feng, T.; Chen, S.; Feng, Z.; Shen, C.; Tian, Y. Effects of Canopy and Multi-Epoch Observations on Single-Point Positioning Errors of a GNSS in Coniferous and Broadleaved Forests. *Remote Sens.* **2021**, *13*, 2325–2339. [CrossRef]
138. Yuan, X.; Yuan, X.; Chen, J.; Wang, X. Large Aerial Image Tie Point Matching in Real and Difficult Survey Areas via Deep Learning Method. *Remote Sens.* **2022**, *14*, 3907–3924. [CrossRef]
139. Ørka, H.O.; Dalponte, M.; Gobakken, T.; Næsset, E.; Ene, L.T. Characterizing forest species composition using multiple remote sensing data sources and inventory approaches. *Scand. J. For. Res.* **2013**, *28*, 677–688. [CrossRef]
140. Hastings, J.H.; Ollinger, S.V.; Ouimette, A.P.; Sanders-DeMott, R.; Palace, M.W.; Ducey, M.J.; Sullivan, F.B.; Basler, D.; Orwig, D.A. Tree Species Traits Determine the Success of LiDAR-Based Crown Mapping in a Mixed Temperate Forest. *Remote Sens.* **2020**, *12*, 309–329. [CrossRef]
141. Jim, C.Y.; Liu, H.T. Species diversity of three major urban forest types in Guangzhou City, China. *For. Ecol. Manage.* **2001**, *146*, 99–114. [CrossRef]
142. Jim, C.Y.; Chen, W.Y. Diversity and distribution of landscape trees in the compact Asian city of Taipei. *Appl. Geogr.* **2009**, *29*, 577–587. [CrossRef]
143. Jin, E.J.; Yoon, J.H.; Bae, E.J.; Jeong, B.R.; Yong, S.H.; Choi, M.S. Particulate Matter Removal Ability of Ten Evergreen Trees Planted in Korea Urban Greening. *Forests* **2021**, *12*, 438–454. [CrossRef]
144. Źróbek-Sokolnik, A.; Dynowski, P.; Źróbek, S. Preservation and Restoration of Roadside Tree Alleys in Line with Sustainable Development Principles—Mission (Im)possible? *Sustainability* **2021**, *13*, 9635–9651. [CrossRef]
145. Bella, F. Driver perception of roadside configurations on two-lane rural roads: Effects on speed and lateral placement. *Accid. Anal. Prev.* **2013**, *50*, 251–262. [CrossRef]
146. Xu, X.; Sun, S.; Liu, W.; García, E.H.; He, L.; Cai, Q.; Xu, S.; Wang, J.; Zhu, J. The cooling and energy saving effect of landscape design parameters of urban park in summer: A case of Beijing, China. *Energy Build.* **2017**, *149*, 91–100. [CrossRef]
147. Goličnik, B.; Thompson, C.W. Emerging relationships between design and use of urban park spaces. *Landscape Urban Plann.* **2010**, *94*, 38–53. [CrossRef]
148. Belaire, J.A.; Whelan, C.J.; Minor, E.S. Having our yards and sharing them too: The collective effects of yards on native bird species in an urban landscape. *Ecol. Appl.* **2014**, *24*, 2132–2143. [CrossRef] [PubMed]
149. Vila-Ruiz, C.P.; Meléndez-Ackerman, E.; Santiago-Bartolomei, R.; Garcia-Montiel, D.; Lastra, L.; Figuerola, C.E.; Fumero-Caban, J. Plant species richness and abundance in residential yards across a tropical watershed: Implications for urban sustainability. *Ecol. Soc.* **2014**, *19*, 22–32. [CrossRef]
150. Avolio, M.; Blanchette, A.; Sonti, N.F.; Locke, D.H. Time Is Not Money: Income Is More Important Than Lifestage for Explaining Patterns of Residential Yard Plant Community Structure and Diversity in Baltimore. *Front. Ecol. Evol.* **2020**, *8*, 85–98. [CrossRef]
151. Kirkpatrick, J.B.; Daniels, G.D.; Davison, A. Temporal and spatial variation in garden and street trees in six eastern Australian cities. *Landscape Urban Plann.* **2011**, *101*, 244–252. [CrossRef]
152. Shen, Z.; Li, J.; Sheng, Y.; Warner, T.A.; Zhao, L. A Multitemporal Remote Sensing Image Registration Method Based on Water Bodies for the Lake-Rich Region. *IEEE J. Sel. Top. Appl. Earth Obs. Remote Sens.* **2019**, *12*, 4327–4341. [CrossRef]
153. Sima, A.A.; Buckley, S.J.; Kurz, T.H.; Schneider, D. Semi-Automated Registration Of Close-Range Hyperspectral Scans Using Oriented Digital Camera Imagery And A 3d Model. *Photogramm. Rec.* **2014**, *29*, 10–29. [CrossRef]
154. Liu, D.; Gong, P.; Kelly, M.; Guo, Q. Automatic registration of airborne images with complex local distortion. *Photogramm. Eng. Remote Sens.* **2006**, *72*, 1049–1059. [CrossRef]
155. Ip, A.; El-Sheimy, N.; Mostafa, M. Performance analysis of integrated sensor orientation. *Photogramm. Eng. Remote Sens.* **2007**, *73*, 89. [CrossRef]
156. Ye, Y.; Zhu, B.; Tang, T.; Yang, C.; Xu, Q.; Zhang, G. A robust multimodal remote sensing image registration method and system using steerable filters with first- and second-order gradients. *ISPRS J. Photogramm. Remote Sens.* **2022**, *188*, 331–350. [CrossRef]
157. Ye, Y.; Yang, C.; Zhu, B.; Zhou, L.; He, Y.; Jia, H. Improving Co-Registration for Sentinel-1 SAR and Sentinel-2 Optical Images. *Remote Sens.* **2021**, *13*, 928. [CrossRef]
158. Mielcarek, M.; Stereńczak, K.; Khosravipour, A. Testing and evaluating different LiDAR-derived canopy height model generation methods for tree height estimation. *Int. J. Appl. Earth Obs. Geoinf.* **2018**, *71*, 132–143. [CrossRef]
159. Brandtberg, T.; Warner, T.A.; Landenberger, R.E.; McGraw, J.B. Detection and analysis of individual leaf-off tree crowns in small footprint, high sampling density lidar data from the eastern deciduous forest in North America. *Remote Sens. Environ.* **2003**, *85*, 290–303. [CrossRef]

160. Liu, H.; Dong, P. A new method for generating canopy height models from discrete-return LiDAR point clouds. *Remote Sens. Lett.* **2014**, *5*, 575–582. [CrossRef]
161. Weinstein, B.G.; Marconi, S.; Aubry-Kientz, M.; Vincent, G.; Senyondo, H.; White, E.P. DeepForest: A Python package for RGB deep learning tree crown delineation. *Methods Ecol. Evol.* **2020**, *11*, 1743–1751. [CrossRef]

remote sensing

MDPI

Article

Exploring Influences of Leaves on Urban Species Identification Using Handheld Laser Scanning Point Cloud: A Case Study in Hong Kong

Meilian Wang [1] and Man Sing Wong [1,2,*]

[1] Department of Land Surveying and Geo-Informatics, The Hong Kong Polytechnic University, Hong Kong, China; meilianp.wang@polyu.edu.hk

[2] Research Institute of Land and Space, The Hong Kong Polytechnic University, Hong Kong, China

* Correspondence: ls.charles@polyu.edu.hk; Tel.: +852-3400-8959

Abstract: Urban tree species identification performs a significant role in tree management and the sustainable development of cities. Conventionally, multispectral or hyperspectral remote sensing images are applied to identify species. However, spectral profiles of trees on images are easily affected by surroundings and illuminations, resulting in urban trees of different species possibly having similar spectral features. The handheld laser scanning (HLS) technique can capture 3D structural information of trees and be confirmed to be effective in reducing the problem of spectral similarity through tree structural properties (TSP). TSP usually varies in different leaf conditions, especially TSP of tropical tree species. In this study, we investigated the effects of leaves on urban tropical tree species identification using HLS. A total of 89 metrics that characterized the TSP were evaluated, including 19 branches, 12 stems, 45 crowns, and 13 entire tree metrics. All metrics were derived under different leaf conditions. The correlation and importance of these metrics were further evaluated. Our results demonstrated that crown metrics perform the most important role in urban species identification in leaf-on and leaf-off conditions and that the combination of metrics derived in different leaf conditions can improve the identification accuracy. Furthermore, we discovered 9 robust metrics that perform well in all leaf conditions, including 3 crowns, 2 branches, 2 stems, and 2 entire tree metrics. These metrics give a deep understanding of numerous structural properties and provide a significant reference for the relevant structure-based classification of other tropical species. This study also illustrated that HLS could help to overcome the spectrum-related limitations and improve the efficiency of species identification and sustainable forest management.

Keywords: handheld laser scanning; structural properties; metric importance; optimal metric set; tropical species classification

1. Introduction

Information about species distribution in urban areas performs a significant role in tree management and conservation. Studies of tree species classification were commonly implemented by identifying spectral features from remote sensing images, such as multispectral and hyperspectral images [1,2]. However, the performance of spectral features is limited by the similarity problem, i.e., the spectral features of the same species may be different due to some factors, such as varying shapes in details and the surrounding environment [3], or the spectral features of different species may be similar [4]. This problem can be reduced by considering tree structures that vary between species because of the different branching patterns and foliage distributions [5]. The light detection and ranging (LiDAR) technology that measures distances using roundtrip time of pulsed laser energy between targets and sensors [6] can capture and represent tree structure information via three-dimensional (3D) point clouds.

The conventional terrestrial LiDAR (TLS) emits laser pulses from a sensor standing on the ground. To capture as complete as possible the structural information, time-consuming preparations and point cloud fusion are required [7]. HLS technology, which places a laser scanning sensor on a handheld moving platform, is able to capture dense and complete tree point clouds economically and efficiently [8]. Its flexible mobility not only reduces the occlusion effect (i.e., the trees far away from the scanner may be occluded by trees near the scanner, causing an incomplete point cloud of trees away from the scanner) but also simplifies the complex preparation and co-registration procedures. Additionally, performance of HLS for forest inventory evaluation has been validated. For instance, Chen et al. [9] achieved an estimation of the diameter at the breast height (DBH) of an arbor forest in the Haidian District in Beijing with an RMSE of 1.58 cm using ZEB-REVO-RT. Oveland et al. [10] and Su et al. [11] accurately estimated the DBH of boreal forest in the southeastern part of Norway with an RMSE of 14.3% using GeoSLAM ZEB1. However, many studies used HLS to estimate the inventory of trees in forests, rarely focusing on inventory surveys of urban trees. To bridge this gap, this study aims to investigate the potential of HLS for urban tree inventory.

Species identification from 3D point clouds usually relies on tree structural properties (TSP) characterized by structural metrics, such as tree height, crown diameter, and DBH [12,13]. Over decades, many structural metrics have been developed and applied to classify species, such as explicit structural parameters [3], quantitative structural features [14], and salient geometric features [15]. These metrics can be divided into two categories according to their extraction methods. One category is metrics that are extracted directly from individual tree point clouds. Explicit structural parameters and salient geometric features both belong to this category. Another category is metrics that are extracted from the reconstructed 3D tree model, which is hierarchically generated by cylinder fitting [14]. Metrics extracted from point clouds mainly represent external geometric characteristics, while metrics extracted from 3D models can describe internal and external geometric characteristics of trees. Most studies concentrate on optimization and improvement of identification accuracy by combining diverse types of data or developing optimized algorithms [16,17]. A comprehensive and thorough understanding of the contribution of structural metrics for species identification is rarely studied [18,19], such as the relationship between derived metrics and specific structural properties of different species and the importance of different types of structural metrics for species identification [20]. Therefore, we proposed to estimate the performance and importance of existing structural metrics for tree species identification using HLS in an urban setting.

A challenge for urban tree species identification based on structural metrics is that structural metrics would change with leaf conditions [21]. Different leaf conditions result in different tree structures [22], thus influencing structural metric values and species identification accuracy. However, related studies demonstrated that leaf condition shows varied influences on the identification of different species. Hamraz et al. [23] demonstrated that leaf-off data could provide more useful information for the identification of mixed temperate species in southeastern Kentucky, while Shi et al. [5] indicated no greatly different performance between leaf-on and leaf-off conditions for the identification of species in Central Europe but the combination of metrics derived under two leaf conditions could improve identification accuracy. The influence of leaf conditions on tropical species identification was rarely studied. Thus, we propose to evaluate certain influences of leaf conditions on tropical species identification in an urban setting.

In this study, we collected 89 structural metrics from previous studies, including 19 in branches, 12 in stems, 45 in crowns, and 13 of entire tree metrics. Under the assumption that leaf conditions may affect the extraction of structural metrics and urban tropical species identification results, an experiment evaluating the specific influence of leaf conditions on species identification was conducted. We removed leaf points by combining the TLSeparation [24] algorithm and manual refinement. TLSeparation is a method developed for the separation of wood and leaf points from individual tree point clouds based on 3D geometric features and has been set as a Python library. All metrics were extracted under two leaf conditions (leaf-on and leaf-off).

2. Materials and Methods

2.1. Study Site and Research Data

Hong Kong (HK) is located south of China (22°08′–22°35′N, 113°49′–114°31′E). The certain climate, namely, long rainy summer and tiny temperate differences in four seasons, creates a great environment for vegetation, resulting in considerable tropical and subtropical tree species diversity. To improve the sustainable development of HK, the government invests enormous efforts in managing urban trees. Information on most trees has been collected and kept, including location, species, and basic structural parameters. Compared to forests, urban trees tend to have low density, and their structural characteristics are less affected by neighboring trees [25]. These make HK urban trees good study objects.

For the selection of tree species, we proposed three rules, i.e., separability, availability, and invariability. Separability means target samples can be greatly separated spatially to ensure the completion of individual tree point cloud and structural metrics extraction. There should be clear boundaries between target trees and neighbors. Availability indicates that enough trees of species are available to be measured. Invariability indicates that the general shape of the target species is not easily affected by neighbors. Following these three rules, four species were finally selected from the dominant species in Hong Kong [26–28], i.e., *Aleurites moluccana* (L.) Willd (AM), *Ficus altissima* Blume (FA), *Delonix regia* (Boj. ex Hook.) Raf. (DR), and *Hibiscus tiliaceus* L. (HT). AM and HT have spherical crowns with simple and complex structures, respectively, while FA and DR have upright funnel crowns with complex and simple structures, respectively. The simple and complex structure stands for the number of main branches of a tree. The number of main branches of AM are relatively smaller than the number of main branches of HT and FA. A total of 85 AM, 69 FA, 65 DR, and 68 HT trees were chosen as research samples (Figure 1). Basic structural parameters, i.e., tree height (TH), DBH, and crown width (CS) of all trees, were simultaneously measured in the field (Table 1). TH and CS were measured by a laser rangefinder. DBH was measured by a tree caliper. The height difference between the lowest and highest points of a tree is defined as TH. The maximum horizontal difference of a crown is defined as CS. The trunk diameter at a tree height of 1.3 m is defined as DBH.

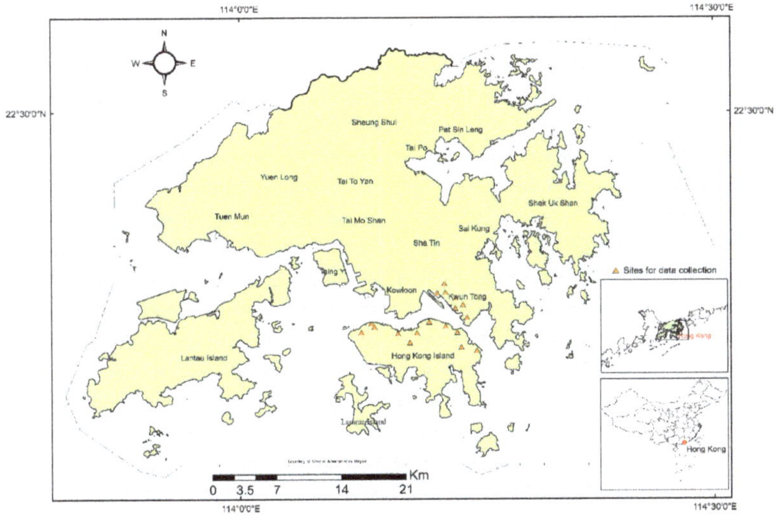

Figure 1. Research site. The yellow icons are the site for the tree point cloud collection. The collected trees are mainly in the park or along the roadside in Hong Kong Island and Kwun Tong District.

Table 1. Basic structural information of each species. Subscript "avg", "median", and "sd" indicate the average, median, and standard deviation values, respectively. The unit of TH and CS is m and the unit of DBH is mm.

Species	TH			DBH			CS		
	TH_{avg}	TH_{median}	TH_{sd}	DBH_{avg}	DBH_{median}	DBH_{sd}	CS_{avg}	CS_{median}	CS_{sd}
AM	13.27	12.69	3.31	353.0	335.0	12.64	9.02	8.83	3.79
FA	12.96	13.33	3.53	311.6	311.2	32.54	10.79	9.99	3.81
DR	10.76	10.28	3.63	238.6	284.0	19.28	11.47	11.81	4.61
HT	9.55	9.70	1.56	284.6	346.6	22.21	7.33	7.37	2.22

Tree point clouds (Figure 2) were collected during May and December in the year 2020 by ZEB Horizon (https://geoslam.com/solutions/zeb-horizon/) accessed on 10 May 2023, which is a lightweight personal HLS developed by GeoSLAM Ltd., consisting of 16 sensors with a wavelength of 903 nm. This equipment can capture approximately 300,000 points per second with a relative accuracy of up to 6 mm. The scanning range is up to 100 m. Its field of view is 360° horizontally and 270° vertically. A detailed description of ZEB Horizon is listed in Table 2.

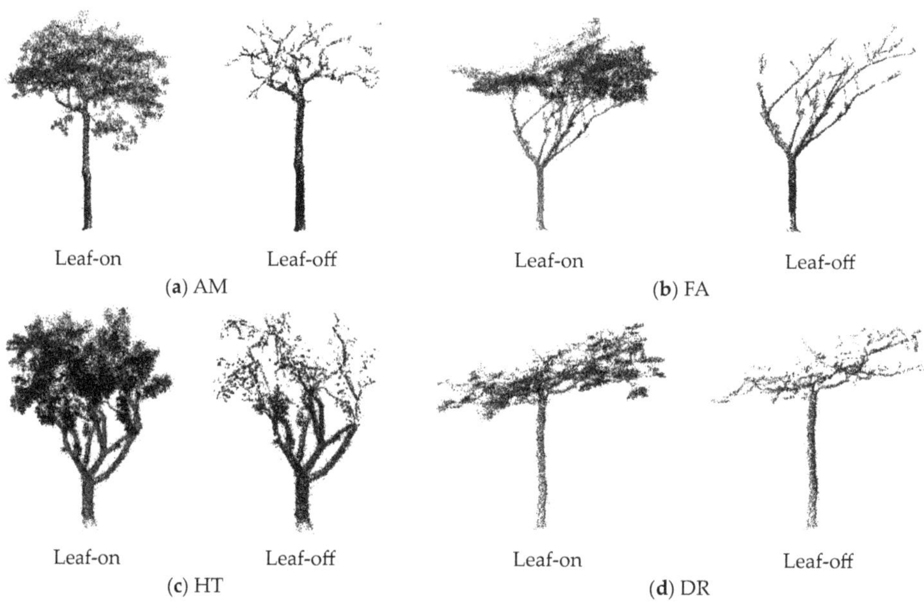

Figure 2. Point cloud examples of four species. The four species are (**a**) *Aleurites moluccana*, (**b**) *Ficus altissima*, (**c**) *Hibiscus tiliaceus*, (**d**) *Delonix regia*.

To capture as complete as possible information about trees, we held the scanner and walked around the target trees twice. The captured point cloud was automatically processed using GeoSLAM Hub software which utilizes simultaneous localization and mapping (SLAM) algorithm to locate the scanner and register 3D point cloud [29]. The accuracy of the SLAM algorithm used in GeoSLAM Hub is less than 3 cm [30]. Processed point clouds are outputted in LAS 1.2 format.

A statistical outlier removal approach was applied to reduce noise points. The average distances of each point to neighbors were computed. Then, points farther than the sum value of average distance and standard deviation of point distances were regarded as noise [31]. Subsequently, ground points were removed using a cloth simulation filtering algorithm, which simulates the drop procedure of cloth to the ground surface [32]. The noise removal and ground point removal were conducted with CloudCompare 2.0 software [33].

Table 2. Datasheet of ZEB Horizon.

Range	100 m	Intensity	yes
Laser	Class 1/λ 903 nm	Real-time processing	yes
FOV	360 × 270 degree	Scanner points per second	300,000
Protection class	IP54	No. of sensors	16
Processing	Post	Vertical angular resolution	2°
Scanner weight	1.45 kg	Horizontal angular resolution	0.2°
Datalogger weight (including battery)	1.4 kg	Relative accuracy	up to 6 mm
Colorized point cloud	yes	Raw data file size	25–50 MB/min

2.2. Individual Tree Point Cloud Segmentation

A four-phased approach was employed to segment individual trees (Figure 3). In phases 1–3, Canopy Height Model (CHM) was first created by rasterizing elevation differences. Based on CHM, we employed Marker-controlled Watershed Segmentation (MCWS) algorithm to coarsely segment individual trees. The marker was able to alleviate over-segmentation by determining segmented object locations [34]. We used tree locations, which were detected through fitting cylinders to stem using Density-based Spatial Clustering of Applications with Noise Algorithm (DBSCAN) at tree height of 1.3–1.4 m [35], as markers of the MCWS algorithm.

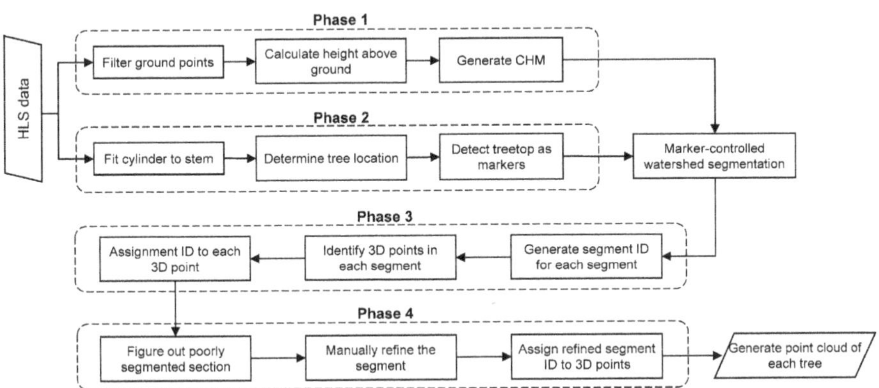

Figure 3. Procedure of individual tree segmentation.

After coarse segmentation, each 2D segment was assigned a segment ID, and the segmentation of the 3D point cloud was conducted (i.e., the points located within a segment were assigned the corresponding segment ID). In phase 4, unsuccessfully segmented trees were filtered. Subsequently, manual refinement was executed to further separate individual tree points clouds.

2.3. Structural Metric Derivation

Many structural metrics have been developed for species identification. We divided existing metrics into four categories based on the part of a tree that structural metrics are related to, i.e., branch, stem, crown, and entire tree metrics. Eighty-nine metrics were derived in this study (Tables 3 and 4). Branch and stem metrics were extracted based on the construction of TreeQSM which was developed by Raumonen et al. [36]. Before QSM construction, several parameters (i.e., PatchDiam1, PatchDiam2Min, and PatchDiam2Max) were optimized by grid search based on four tree point clouds (a tree for each species).

PatchDiam1 was used to control patch size during the first cover sets generation. PatchDiam2Min and PatchDiam2Max controlled minimum and maximum patch size during the second cover sets generation, respectively. Testing ranges of three parameters were [0.06, 0.08, 0.12, 0.14] (PatchDiam1), [0.02, 0.03, 0.04, 0.05] (PatchDiam2Min), and [0.05, 0.07, 0.11, 0.13] (PatchDiam2Max).

Table 3. Definition of stem, branch, and entire tree metrics.

Type	No.	Definition	No.	Definition
Stem	S1	DBH	S7	Stem length
	S2	DBH/TH	S8	Stem length/TH
	S3	DBH/tree volume	S9	Stem direction at axis x
	S4	DBH/min stem radius	S10	Stem direction at axis y
	S5	Stem radial irregularity	S11	Stem direction at axis z
	S6	Stem taper	S12	Stem volume
Branch	B1	Branch symmetry	B11	Overall branch length/branch volume
	B2	Median 1st branch angle	B12	Mean top 10 branch length
	B3	Sum 1st branch angle	B13	Mean 1st branch radius
	B4	Average 1st branch angle	B14	B13/TH
	B5	Branch angle ratio	B15	Mean top 10 branch radius
	B6	SD of 1st branch angle	B16	Average branch distance
	B7	Sum of 1st branch length	B17	B16/DBH
	B8	Mean 1st branch length	B18	Branch volume below 55% of the tree
	B9	B8/TH	B19	Branch density
	B10	B8/DBH		
Entire tree	T1	TH	T8	Volume below 55% of TH
	T2	Mean grid height/TH	T9	Total cylinder length/tree volume
	T3	Mean grid height	T10	Tree volume/crown cover area;
	T4	Relative coverage ratio	T11	Tree volume/TH
	T5	Stem grid ratio	T12	Tree volume/crown diameter
	T6	Tree volume	T13	Total surface area of entire tree
	T7	Volume distribution		

Where TH is tree height, mean grid height means grid representing a tree with an edge length of 0.1 m.

Table 4. Definition of crown metrics.

No.	Definition	No.	Definition
C1	Highest crown bottom	C24	Height of max CS/TH
C2	Lowest crown bottom	C25	Height of max CS/C5
C3	Crown start height	C26	Crown $area_{xy}$/TH
C4	Crown start height/TH	C27	Crown $area_{xy}$/crown vertical area
C5	Largest crown length	C28	Average equivalent center height/C2
C6	Crown length ratio	C29	Average crown grid height/C2
C7	Mean crown length	C30	C34/C18
C8	C5/crown diameter	C31	Crown symmetry
C9	C5/TH	C32	Alpha volume/convex hull volume
C10	CD_{xy}/TH	C33	Crown top evenness
C11	CD_{xy}/the shortest crown length	C34	Standard deviation of C28
C12	CD_{xy}/CCD_{xy}	C35	Crown bottom evenness
C13	CD_{xy}	C36	Crown volume
C14	CCD_{xy}	C37	C36/crown projection area
C15	CD_{yz}/CCD_{yz}	C38	C36/number of crown points
C16	CD_{xz}/CCD_{xz}	C39	Volume of crown grids with max point density/corresponding grid volume
C17	Equivalent centers radius/crown radius	C40	Standard deviation of grid number of all crown layers
C18	Standard deviation of C17 of crown profiles	C41	Crown difference ratio
C19	Semi-axis x of paraboloid fitted to crown	C42	Coefficient variation of crown
C20	Semi-axis y of paraboloid fitted to crown	C43	Total area of voxel one-side covers/convex hull area from two perpendicular side views
C21	Semi-axis z of paraboloid fitted to crown	C44	Residual sum of squared errors of ellipsoid fitting
C22	Crown equivalent diameter	C45	shape signature index
C23	Diameter of crown minimum circumscribed circle		

Where CD_{xy}, CD_{yz} and CD_{xz} and mean maximum diameter of projected crown on x-y, y-z, x-z plane, respectively. CCD_{xy}, CCD_{yz} and CCD_{xz} mean maximum distance at the opposite direction of crown diameter of the projected crown on x-y, y-z, x-z plane, respectively.

Then, model precision was calculated for each QSM by comparing the average difference of TH, DBH, and CS between derived and field-measured parameters. The smaller differences, the more precise the QSM. Optimized parameters were finally selected according to model precision. Afterwards, QSM was built five times for each tree using same input parameters to reduce the uncertainty, unreliability, and stochasticity. The appropriate model was estimated by calculating mean distance between cylinder models and points of the trunk and all branches by the select-optimum function of TreeQSM. The optimized structural parameters were output by this function. In addition, a part of the crown and entire tree metrics were extracted directly from the point cloud. Other metrics were extracted from gridding tree point clouds (tree points were gridded using an edge with a length of 0.02). Derivation of all metrics was conducted using MATLAB software. The detailed definition and format of eighty-nine metrics were listed in Supplementary File.

All metrics were derived under two leaf conditions, respectively, to investigate the effect of leaf points on tree structures and species identification. Leaf-on condition means individual tree point clouds contain leaf points, and leaf-off condition means leaf points were not contained in individual tree point clouds. Leaf-off condition was achieved by removing leaf points using the TLSeparation algorithm developed by Vicari et al. [24] and manual refinement. TLSeparation is likely to introduce significant underestimation of crown structural complexity [25]. Manual refinement thus was used to improve separation results of TLSeparation and to obtain better leaf-off individual tree point clouds. According to the work of Demol et al. [37] and Lau et al. [38], TreeQSM has limitations on the 3D model reconstruction of small branches, especially branches whose diameters are thicker than 20 cm. Therefore, this study mainly focuses on the extraction of structural metrics of first- and second-order branches.

Removal of highly related metrics not only saves computing power but also boosts the performance of identification models. We used Pearson's correlation coefficient, a measure of the correlation between two sets of data [39], to analyze the relationship between all metrics. The threshold of r was used to split highly correlated and lowly correlated metrics [40]. We used a popular threshold of 0.7 to distinguish the correlation coefficient (i.e., $|r| > 0.7$ demonstrates a high correlation, otherwise demonstrating a low correlation) [41,42].

2.4. Metric Importance Assessment and Metric Selection

A hybrid method was proposed to evaluate the importance of individual metrics (i.e., incorporate the importance assessment with identification procedure). During classifier construction, the metric that made the split, the corresponding changes of Gini impurity and the number of affected samples were all tracked once a split was done. The times a metric was used to split a node can be counted by Gini impurity (Equation (1)), which is a measurement used to determine how the features of a dataset should split nodes to form the tree [43]. The more frequently a metric is used during a classifier construction, the more important it is [43]. After classifier construction, the total gain that measures the decrease in node impurity of each metric can be computed. Accordingly, the importance ranking of all metrics can be generated. The importance assessment procedure was achieved using the Python library skicit-learn [44].

$$Gini = 1 - \sum_{i=1}^{k} p_i^2 \quad (1)$$

where k is the number of classes. p is the probability of samples belonging to class i at a given node.

Redundant and irrelevant metrics adversely affect the performance of identification models with the increase in metric dimensions [45,46]. For example, highly related or insignificant metrics increase dimensions without improving identification accuracy [16,47]. Metric selection seeks to find appropriate metrics for improving identification efficiency by maximizing the performance of identification models and minimizing metric numbers [48]. In this study, structural metrics were selected based on correlation analysis and importance assessment. In each pair, the metric that had lower importance values ($|r| < 0.7$) was removed. When the importance of the two metrics was the same, their total $|r|$ values were compared. Then, metrics that have relatively higher total $|r|$ values were removed. In addition, an accumulated importance was introduced to evaluate the importance of metric types for tropical species identification. The accumulated importance equals to the total sum of importance values of all metrics of the same type.

2.5. Tropical Species Classification

Machine learning approaches have been commonly applied to solve identification problems. Among various machine learning models, XGBoost has shown great ability on a metric-based identification problem in recent years [49]. It is an optimized and scalable end-to-end tree-boosting system based on a gradient-decision tree and provides a parallel tree-boosting method which fits many large or small trees to the reweighted version of the training data [50]. We employed XGBoost to identify species using selected structural metrics. Overall, trees were divided into two datasets. About 20% of trees were randomly selected as dataset 2 for evaluating the robustness and reliability of identification models and selected structural metrics. The rest 80% of the trees were used as dataset 1. Approximately 70% of trees of dataset 1 were used as training data and 30% as testing data to show the performance of the identification model. The model trained using training data in dataset 1 was implemented to identify species of dataset 2. Training and testing of identification models were implemented under two leaf conditions, respectively, to explore the effect of leaves on species identification.

The performance of XGBoost model is controlled by a number of parameters. Obtaining the performance of each parameter value is time-consuming and computationally intensive. We integrated a grid search that can exhaustively search subsets from parameter space and cross-validation to find optimal parameter sets. A five-step strategy was proposed for parameter optimization: (1) Determine learning rate and an optimum number of trees; (2) Tune parameters of each tree; (3) Optimize regularization parameters; (4) Decrease learning rate and repeat the above-mentioned steps; and (5) Finish parameter optimization until model accuracy does not increase. The experiment was carried out with the "scikit-learn" package [44].

The performance of selected metrics and identification model was evaluated by three assessment approaches (Equations (2)–(4)), namely, overall accuracy (O_a), user's accuracy (U_a), and procedure's accuracy (P_a). O_a illustrated the general performance of the identification model on four species. U_a and P_a were used for each species to reduce the influence of imbalanced samples between species on result assessment. U_a evaluates identification results from the point of view of species, and P_a evaluates identification results from perspective of model. To each species, U_a demonstrated rate of correct prediction made by identification model compared to all predictions. P_a demonstrated rate of correct predictions compared to true sample.

$$O_a = \frac{N_{CorrPred\_all}}{N_{total}}, \quad (2)$$

$$U_a = \frac{N_{CorrPred}}{N_{pred}}, \quad (3)$$

$$P_a = \frac{N_{CorrPred}}{N_{true}}, \quad (4)$$

where $N_{CorrPred\_all}$ is the number of precisely predicted samples, N_{total} is the count of all samples, N_{pred} is the number of predicted samples of a species, $N_{Corrpred}$ is the number of samples predicted as that species, and N_{true} is the number of true samples of a species.

3. Results

3.1. Validation of HLS-Based Tree Structure Characterization

The differences between manually measured and HLS-derived metrics, including mean, standard deviation (SD), and median of difference values, are listed in Table 5. Overall mean differences of DBH, TH, and CS are 4.35 mm, 0.69 m, and 0.82 m, respectively, demonstrating that HLS can effectively capture basic structural information of trees. In terms of specific species, FA has relatively larger errors than others in TH and DBH. This may be because some aerial roots of FA influenced the automated measurement of structural metrics. For CS, the possible reason for the large differences between manually measured and HLS-derived metrics is human vision error.

Table 5. Difference between manual measured and HLS-derived metrics. Sd means standard deviation. MPE means the mean percentage error (unit is %).

Species	TH (m)				DBH (mm)				CS (m)			
	Mean	Median	Sd	MPE	Mean	Median	Sd	MPE	Mean	Median	Sd	MPE
AM	0.67	0.65	0.45	6.14	3.27	4.03	2.01	3.80	0.83	0.84	0.43	4.34
FA	0.71	0.79	0.44	5.97	4.88	4.54	2.79	4.17	0.80	0.81	0.39	4.91
DR	0.57	0.51	0.35	4.78	3.62	3.77	1.94	3.89	0.84	0.91	0.55	5.36
HT	0.79	0.58	0.45	5.56	5.01	5.12	2.88	3.78	0.82	0.86	0.52	5.63
Overall	0.69	0.65	0.44	5.52	4.35	4.73	2.42	4.30	0.82	0.84	0.47	4.18

3.2. Correlation of Structural Metrics and Metric Selection

The number of highly correlated metrics in leaf-on conditions is larger than in leaf-off conditions (Figures 4 and 5), demonstrating that the diversity of inner structures represented by metrics derived under leaf-off conditions is more remarkable than external structures represented by metrics derived under leaf-on conditions. With regard to metric types, the correlation coefficient of metrics, especially branch and crown metrics, tends to be larger in leaf-on than in leaf-off conditions. About half of crown metrics have high correlation coefficient values under two-leaf conditions. This notion may be explained by two possible situations: (1) existing crown metrics represent the same or similar characteristics of a crown using various formats; and (2) some metrics may not be appropriate to tropical species. Metrics with high correlation coefficients under two leaf conditions predominantly relate to branch length, CS, crown cover, crown length, stem length, TH, and tree volume.

The metric selection was conducted on the basis of assessment of the correlation coefficient and the importance of all metrics in the first calculation. Under the leaf-on condition, 42 metrics were removed, including 6 branch metrics, 24 crown metrics, 4 stem metrics, and 8 entire tree metrics. Under leaf-off condition, 38 metrics were removed, including 4 branch metrics, 22 crown metrics, 4 stem metrics, and 8 entire tree metrics. Although most crown metrics were developed in previous studies, many of them are similar or have relatively low importance for species classification. The same situation was observed in entire tree-related metrics. The remaining metrics under two leaf conditions are listed in Table 6.

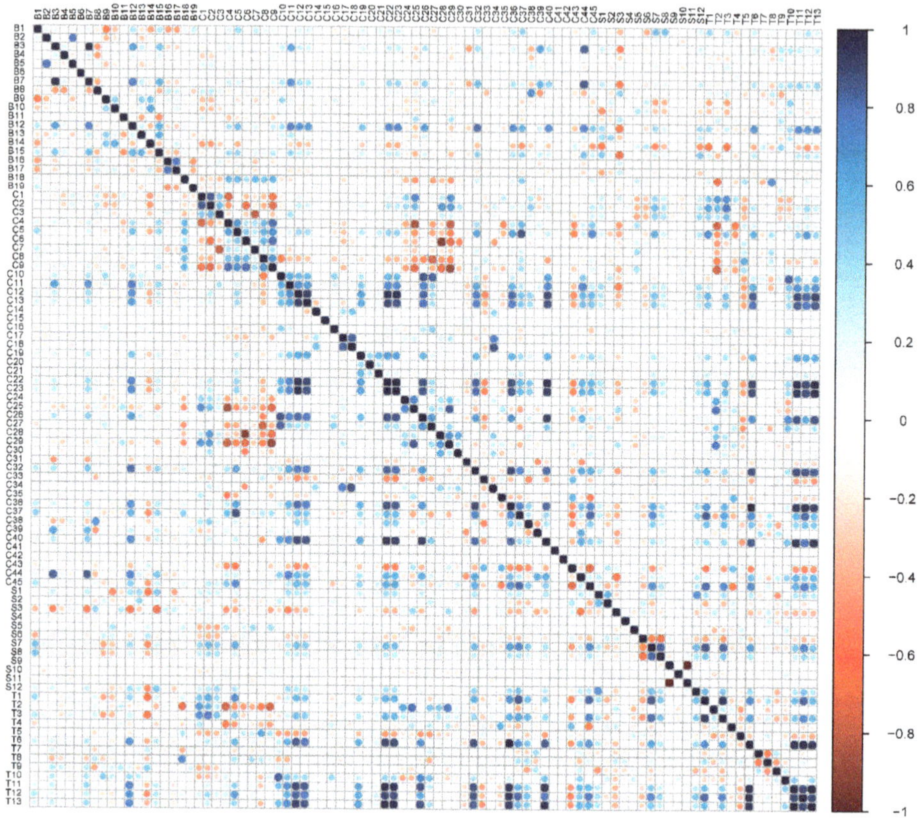

Figure 4. Correlation coefficient of all metrics under leaf-on condition. The higher the positive correlation, the blue is darker. The higher the negative correlation, the red is darker.

Table 6. Remaining metrics under leaf-on and leaf-off conditions after removing highly correlated metrics.

Type	Leaf Condition	Remaining Metrics
Branch	Leaf-on	B1, B2, B3, B4, B6, B9, B10, B11, B13, B14, B16, B17, B18, B19
	Leaf-off	B1, B2, B3, B5, B6, B8, B9, B11, B13, B14, B15, B16, B17, B18, B19
Crown	Leaf-on	C3, C6, C8, C10, C14, C15, C16, C17, C19, C20, C21, C24, C30, C31, C33, C35, C38, C39, C41, C42, C43
	Leaf-off	C2, C3, C5, C6, C8, C10, C14, C15, C16, C17, C19, C20, C22, C25, C30, C31, C33, C35, C38, C41, C42, C43, C44
Stem	Leaf-on	S1, S2, S3, S4, S5, S6, S9, S10
	Leaf-off	S1, S2, S3, S4, S5, S6, S9, S10
Entire tree	Leaf-on	T3, T4, T5, T7, T9
	Leaf-off	T2, T3, T4, T5, T7, T8, T9

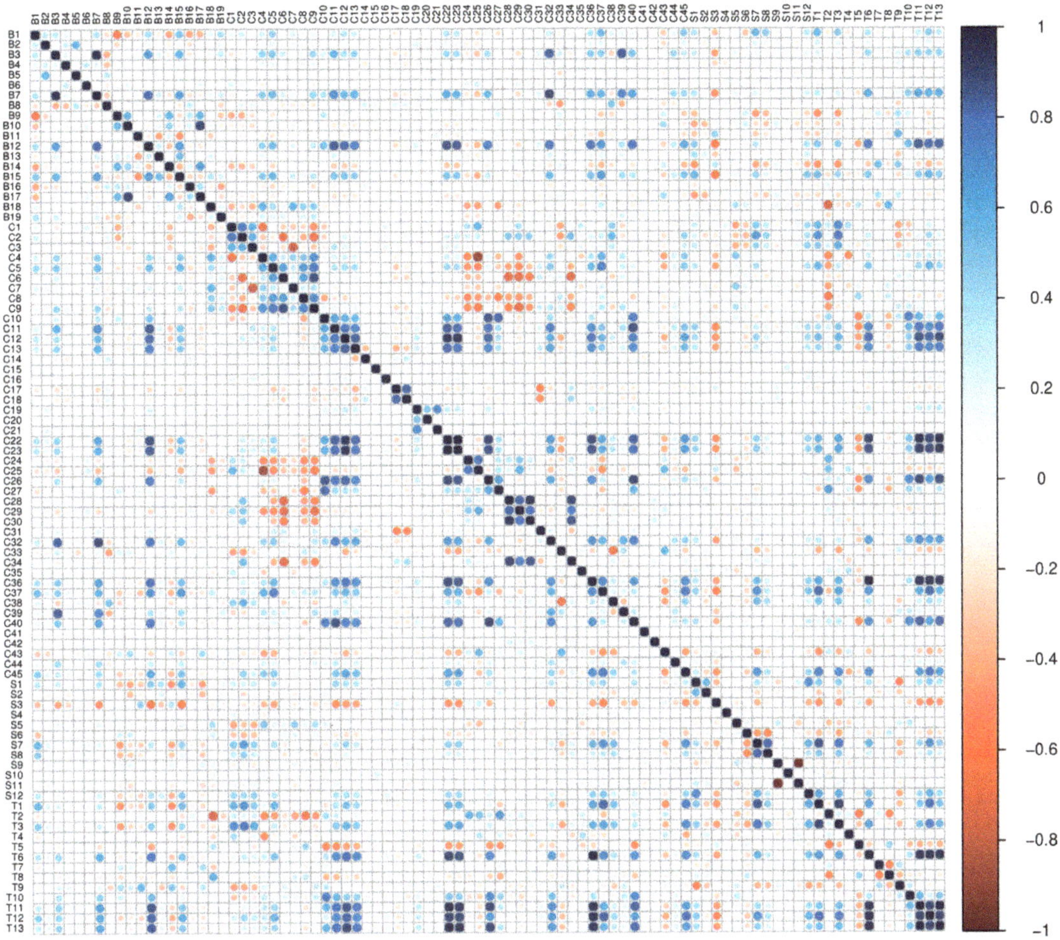

Figure 5. Correlation coefficient of all metrics under leaf-off condition. The higher the positive correlation, the darker the blue is. The higher the negative correlation, the darker the red is.

3.3. Importance of Structural Metrics

Among all structural metrics, crown structural metrics were found to be the predominant metrics for differentiating four tropical species in HK under two leaf conditions according to accumulated importance values of each type (Table 7). Their accumulated importance values remain the largest in the four types of metrics, and no significant difference was observed between the two leaf conditions. This finding agrees that crown structure is a fundamental attribute of a tree [51]. The aforementioned situation may also be the reason many previous studies predominantly used crown metrics to classify species from point cloud [5,52,53]. The performance of branch and stem metrics significantly varies under different leaf conditions. Branch metrics perform increasingly important roles from leaf-on to leaf-off conditions, while stem metrics show an opposite trend. This situation may be because the internal structural and morphological characteristics of trees were shown clearer under leaf-off conditions. Thus, branch metrics more explicitly express differences between species.

Table 7. Overall accumulated importance of each type of metric.

Type	Leaf-On		Type	Leaf-Off	
Branch	14	0.1633	Branch	15	0.2010
Crown	21	0.3961	Crown	23	0.3844
Stem	8	0.2735	Stem	8	0.2496
Entire	5	0.1676	Entire	7	0.1649

From the importance of individual metrics, the stem shape metric performs the most important role in tropical species identification under all leaf conditions (Figure 6). This notion agrees with the demonstration of Stal et al. [54], wherein stem properties are important parameters of a tree. However, unlike commonly used stem metrics such as DBH for species identification, the most important stem metric in this study is S5. Two possible reasons for this situation are as follows: (1) in comparison with DBH, which only measures the diameter of a stem, stem radial irregularity the takes size and shape of a stem into consideration. Taking HT and AM as an example, S5 can still identify them when stem diameters are the same because the trunk of AM tends to grow straighter, and the shape of AM trunk is closer to the perfect circle. However, the HT trunk has greater irregularity. (2) S5 has lower relationships with other metrics, making identification models more sensitive to its variations between species.

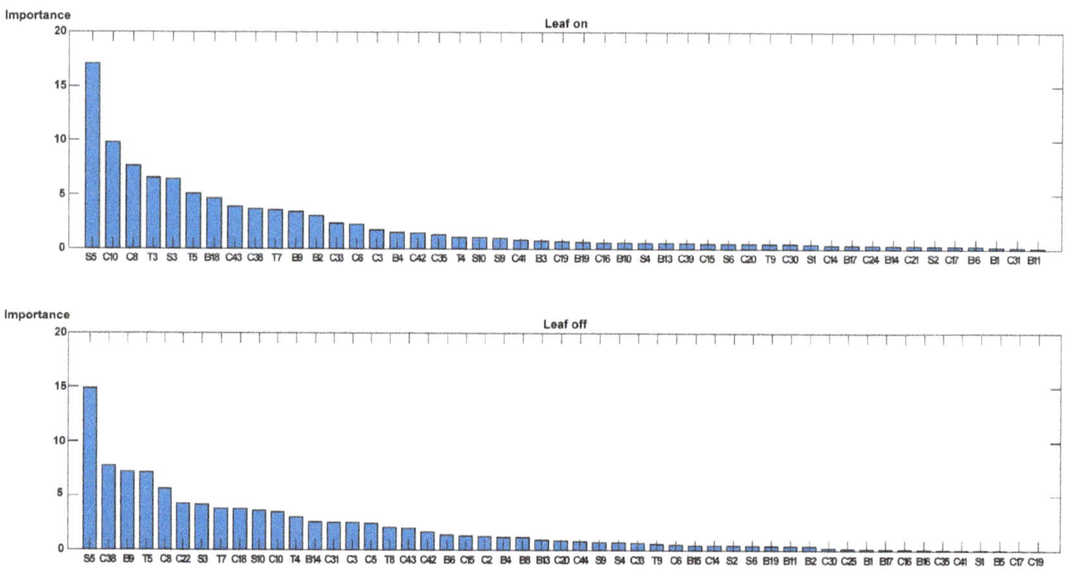

Figure 6. Importance ranking of the structural metrics under leaf-on and leaf-off conditions.

In addition, some structural metrics perform different roles in tropical tree species identification under different leaf conditions. When leaves exist, C10 and C8 are identified as important metrics for species identification (ranking as second and third). However, under leaf-off conditions, C10 and C8 become less important. Meanwhile, B9 and C38 become the second and third most important metrics. Unlike commonly used metrics in previous studies, such as TH, CS, and DBH [18,19], the first 15 important metrics under two-leaf conditions are more concrete and tangible, especially the first five metrics. In the first five metrics under leaf-on conditions, the crown shape is described through the ratio between CS and TH and the ratio between the horizontal and vertical spread. The ratio between metrics and TH is also proven to perform better because it can reduce the influence of tree ages [3]. Stem shape is described through S5, and TH is the mean height of all tree grids. Moreover, the first five metrics under leaf-off condition characterize properties of stem diameter, crown volume, branch length, stem volume, and CS by stem radial irregularity, the ratio of crown volume to the number of crown points, the mean length of first-order branches, the ratio between the number stem grids and entire tree grids, and the ratio of the vertical length of the crown to the horizontal length of the crown, respectively.

We discovered nine robust metrics that consistently show large importance values under two leaf conditions, including three crown metrics, two branch metrics, two stem metrics, and two entire tree metrics, namely, S5, C10, C8, S3, T5, B18, T7, B9, and C3. These metrics appear in the first 15 metrics under two leaf conditions but with different rankings. The frequently employed branch length and branch volume were also verified to be effective and efficient for tropical species classification. Three robust crown metrics are more appropriate to tropical species from the aspect of ecosystem and plant morphology compared with the commonly used crown metrics, such as crown diameter, crown cover area, and crown base height [55–57]. Shenkin et al. [51], who studied the role of ecosystem and phylogeny on crown size and shape, demonstrated that crown vertical length and crown volume of tropical species vary more across gradients than crown diameter and projected crown area.

The crown depth and crown base height can be affected by structures and leaves of the upper crown [58]. Specifically, the sparse structure and compound leaves of the upper crown could allow lateral light to effectively penetrate deeper into the canopy, allowing the crown bottom to maintain a positive carbon balance and grow deeper [49,59]. These phenomena are reasons why crown metrics C3, C8, and C38, which describe the characteristics of the crown bottom, crown vertical length (also called crown depth), and crown volume, respectively, outperformed other commonly used metrics in tropical species classification. The two entire tree metrics relate to tree volume, confirming again the effectiveness of tree volume for the classification of species, not only for boreal and temperate species [60] but also for tropical species.

3.4. Results of Species Identification

Identification results under different leaf conditions are shown in Table 8. Overall accuracy values indicated that the selected first 15 important metrics could identify tropical species with an accuracy of approximately 70%. The overall accuracy obtained under the leaf-on condition is slightly better than that under the leaf-off condition in the two datasets. In terms of performance on each species, noticeable differences can be observed. The P_a and U_a values of AM and HT are generally higher than those of FA and DR under two leaf conditions in two datasets. This may be explained from the aspect of plant morphology, i.e., AM and HT have a spherical crown, while FA and DR trees have an upright funnel crown. Structural metrics may be more sensitive to spherical shapes than to upright funnel shapes. There are also differences in specific classification performance on two species that have similar shapes. To the spherical crown, leaf condition has smaller influences on simple structure, while to the upright funnel crown, the leaf condition has a larger influence on the simple structure than on the complex tree structure.

Table 8. Identification results of four tropical species under leaf-on and leaf-off conditions.

Classification Results	Dataset 1				Dataset 2			
	Leaf-On		Leaf-Off		Leaf-On		Leaf-Off	
	P_a	U_a	P_a	U_a	P_a	U_a	P_a	U_a
AM	85.71	75.00	92.85	81.25	86.67	76.47	86.67	76.47
FA	66.67	54.55	55.56	45.45	75.00	60.00	66.67	66.67
DR	37.5	100.00	50.00	80.00	37.50	60.00	66.67	66.67
HT	90.00	81.82	70.00	77.77	75.00	100.00	25.00	50.00
O_a	73.17		70.73		71.43		70.27	

The combination of metrics derived under two leaf conditions improved overall classification accuracy from 70.27% to 81.7% (Table 9). However, the performance of specific species significantly varies. The combination of metrics derived under two leaf conditions improves classification results of FA and DR, with the procedure's accuracy increasing from 55.56% to 75.00% and from 37.5% to 75.00%, respectively, showing that the combination of metrics derived under two leaf conditions could improve classification results of upright funnel crown, while the combination of metrics derived under two leaf condition shows different performance on AM and HT in testing data (i.e., datasets 1 and 2). In dataset 1, P_a of AM achieved using the combination of metrics derived under two leaf conditions is slightly larger than that achieved using metrics derived under the leaf-on condition but is smaller than that achieved under the leaf-off condition. However, in dataset 2, P_a of AM achieved using the combination of metrics derived under two-leaf condition is less than that achieved using metrics derived under leaf-on and leaf-off conditions. P_a value of HT achieved using the combination of metrics derived from two leaf conditions is less than that achieved using metrics derived under leaf-on but larger than that achieved using metrics derived under the leaf-off condition in dataset 1. However, the identification result of HT using the combination of metrics derived under two leaf conditions is larger than that using metrics derived under single leaf conditions in dataset 2.

Table 9. Confusion matrix of classification results of four tropical species under combination of structural metrics derived under two leaf conditions.

		Dataset 1					**Dataset 2**				
		Prediction					**Prediction**				
		AM	**FA**	**DR**	**HT**	P_a	**AM**	**FA**	**DR**	**HT**	P_a
Referenced	AM	13	1	0	1	86.67	12	2	1	0	75.00
	FA	1	9	1	1	75.00	2	9	0	1	75.00
	DR	1	1	7	0	75.00	0	3	6	0	66.67
	HT	1	0	0	7	87.50	1	0	0	7	87.50
	U_a	81.25	81.81	87.50	77.78		80.00	64.29	85.71	87.5	
O_a				81.71					75.56		

4. Discussion

4.1. Performance Analysis

Several studies have identified tropical species based on structural properties characterized by structural metrics from point clouds in recent years. However, the importance of these metrics was rarely compared. In this study, we extracted 89 structural metrics that represent traits of different parts of a tree from the HLS point cloud, including branch, crown, stem, and entire tree metrics, and assessed their correlations and importance for tropical species identification under different leaf conditions. The identification performance of these metrics was also evaluated.

Our correlation coefficient analysis illustrated that about half of existing structural metrics are highly correlated. Among them, approximately two-thirds of branch and stem metrics have high correlation coefficient values, and more than half of crown and entire tree metrics have high correlation coefficient values. This notion may be explained by two possible situations: (1) existing crown metrics represent the same or similar characteristics of a crown using various formats; and (2) some metrics may not be appropriate to tropical species. Thus, it is essential to find effective and efficient metrics for tropical species identification. According to important assessment results, the structural properties of the crown and stem were identified as the most important components for the identification of tropical species. This finding may also be the reason many studies mainly used crown metrics to classify species. In addition, stem metrics perform important roles. However, different from other studies which use DBH, the most important stem metric is stem radial irregularity. There are two possible reasons: (1) in comparison with DBH that only measures the diameter of a stem, stem radial irregularity takes the size and shape of a stem into consideration. (2) stem radial irregularity has a lower relationship with other metrics, making the identification model more sensitive to its variations between species.

Several robust and more concrete metrics were discovered, providing reference significance to further studies. Identification results demonstrated that leaf condition affects species identification. Based on overall accuracy, the leaf-off condition could slightly improve the performance of structural metrics, while combination of metrics derived under two leaf conditions could significantly improve identification accuracy. This could be due to the fact that metrics derived under the leaf-off condition can better represent inner crown and branch structures compared to metrics derived under leaf-on condition. The combination of metrics derived under two leaf conditions integrated the internal and external structural properties of a tree, thus obtaining better identification results. However, specific influence of leaf conditions on the identification of tropical trees depends on the species. This may relate to the surrounding environment, plant morphology, and ecology.

4.2. Influence Factors

This study obtained positive importance assessment and species identification results. Each species has at least two misidentified trees. This situation may be due to the effects of the surrounding environment and their growth-defense tradeoffs [61]. Tomé and Burkhart [62] illustrated that the growth of individual trees on particular sites could be affected by local neighbors and competition status. Park trees in Hong Kong usually have high species diversity and a complex growth environment [63]. To adapt to various external stress conditions and maximize survival efficiency, plants evolve a complex and sophisticated regulatory mechanism to mediate the balance of growth and external stress [64,65]. This adaptation leads to the convergence of tree traits, resulting in the structure differs from the common pattern of their corresponding species, lowering variation between species [66,67]. In addition, to maintain the neatness and beauty, trees in the park are sometimes pruned, resulting that the precision of their external shape and structural characteristics are affected. Samples used in this study are selected from parks which away from residential areas and near forests to minimize the influence of pruning and maintenance activities. It is difficult to completely reduce the impact of pruning activities. This study was conducted on four tropical species growing in a park in Hong Kong. The limited number of species, certain environments, and ecosystems may also affect the assessment of the importance and applicability of selected important metrics. In addition, tree point clouds were obtained by holding the scanner and walking around trees twice. During the generation of tree point clouds, the point clouds need to be registered, and repeated points need to be removed by GeoSLAM Hub software. This procedure may be affected by system errors of GeoSLAM Hub software, thus influencing the quality of tree point clouds and the extraction of structural metrics.

Furthermore, four species were selected as examples of tropical species for assessment of structural metrics. Although some inefficient and highly correlated metrics were detected and optimal metric sets were proposed in this study, it is not enough to be used as a standard guideline for the identification of all tropical species. We will try to explore the capability and applicability of structural metrics on more species and samples in diverse environments. Branch and stem metrics derived under all leaf conditions are extracted on the basis of the construction of 3D tree models. The 3D model was constructed by hierarchically fitting cylinders to stems and branches. However, branch points, particularly high-order branch points within the crown, may be misidentified under leaf-on condition. Accordingly, metrics related to traits of high-order branches may be inaccurate. This is the reason the importance of branch metrics is relatively low.

4.3. Application Analysis

Our study affirmed that HLS could be used to reduce obscuration effects and obtain structural measurements. Meanwhile, HLS can easily handle forest inventory and structural heterogeneity, tree growth modeling, and tree structural health monitoring. The application in other fields, such as 3D model construction of buildings and survey of individual trees and forests, can be performed using HLS. This study explored the importance and performance of numerous existing structural metrics for species identification, evaluated their correlation, and discussed the association of structural metrics, identification, and structural properties. Results of our capability test showed that good identification results could be obtained with several important and optimal metrics instead of many metric sets. This method not only decreases identification time but also saves computing power, providing a significant reference for later work.

4.4. Potential Improvement

Our assessment of structural metrics indicated that identifying species based on the structural characteristics of a species is practical and feasible, but there is still space for improvement. Many structural metrics were included and evaluated in this study, and optimal metrics were verified to be effective for tropical species identification. Some properties of trees, such as texture traits of foliage, crown, and stem, cannot be resolved. According to crown criteria illustrated by González-Orozco et al. [68] and Trichon [69], the division within a crown remarkably varies between species. Some crowns do not have clear divisions, while others have two or more divisions with each component. Foliage texture has many types, for example, smooth, mottled, granular, grainy, and spotted [68]. If these texture features can be taken into consideration, species identification using structural metrics derived from the HLS point cloud will be improved. Traditionally, species identification is conducted based on multispectral or hyperspectral photographs, which can represent spectral information of crowns and leaves. If structural characteristics and spectral information can be integrated as features for species identification, more details and knowledge of trees may enhance identification precision. In addition, structural metrics and the influence of leaves were explored based on standalone trees. Their performance on trees in dense forests is still unknown. Therefore, further experiments are worth being designed and carried out to exploit and improve the robustness, practicality, and applicability.

5. Conclusions

This study evaluated the correlation coefficient between 89 existing structural metrics, including crown, stem, branch, and entire tree metrics, and assessed the use of these metrics for tropical species classification under different leaf conditions using HLS point cloud. Approximately two-thirds of branch and stem metrics are highly correlated, and more than half of crown and entire tree metrics are highly correlated. In terms of metric importance, crown, and stem metrics were identified as the most important components. Leaf conditions (i.e., leaf-on and leaf-off) were found to have an influence on tropical species classification. The combination of metrics derived under leaf-on, and leaf-off conditions can significantly improve the identification accuracy of four tropical species. Furthermore, we investigated fifteen optimal metrics based on correlation analysis and importance metrics, and nine robust structural metric sets were proposed and validated. The most important structural metrics discovered in this study are more concrete compared to commonly used structural metrics. For example, we found CS characterized by the ratio between the horizontal and vertical maximum spread of a crown is more significant than horizontal spread and spread area for species identification. Although plenty of structural metrics were developed, many of them are identical. It is essential to investigate efficient structural metrics for the identification of more species, such as boreal and temperate species. Our exploration of the connection between metrics and structural properties and assessment of the importance of identification of four tropical species provide a significant reference for further research, not only studies on the effectiveness of structural metrics but also studies on the identification of other tropical species.

Supplementary Materials: The following supporting information can be downloaded at: https://www.mdpi.com/article/10.3390/rs15112826/s1, Figure S1: Illustration of stem metric derivation. Figure S2: Illustration of crown metrics. Figure S3: Illustration of crown profile and relevant metrics. Table S1: Detailed definition and format of stem structural metrics. Table S2: Detailed definition and format of branch structural metrics. Table S3: Detailed definition and format of entire tree structural metrics. Table S4: Detailed definition and format of crown structural metrics. References [70–74] are cited in the supplementary materials.

Author Contributions: Conceptualization, M.S.W.; methodology, M.S.W. and M.W.; validation, M.W.; formal analysis and investigation, M.W.; writing—original draft preparation, M.W.; writing—review and editing, M.S.W.; visualization, M.W. All authors have read and agreed to the published version of the manuscript.

Funding: This research was funded by General Research Fund (Grant Nos. 15603920 and 15609421) and the Collaborative Research Fund (Grant No. C5062-21GF) from the Research Grants Council, Hong Kong, China. M.S. Wong also acknowledged the support from the project 1-CD81, Research Institute for Land and Space, the Hong Kong Polytechnic University, Hong Kong, China.

Data Availability Statement: The data that support the findings of this study are available from the corresponding author upon reasonable request.

Acknowledgments: We express our gratitude to the anonymous reviewers and the editor for their valuable comments and suggestions to improve the manuscript.

Conflicts of Interest: The authors declare no conflict of interest. The funders had no role in the design of the study; in the collection, analyses, or interpretation of data; in the writing of the manuscript; or in the decision to publish the results.

References

1. Field, C.B.; Barros, V.R. *Climate Change 2014–Impacts, Adaptation and Vulnerability: Regional Aspects*; Cambridge University Press: Cambridge, UK, 2014.
2. Cao, K.; Zhang, X. An improved Res-UNet model for tree species classification using airborne high-resolution images. *Remote Sens.* **2020**, *12*, 1128. [CrossRef]
3. Lin, Y.; Herold, M. Tree species classification based on explicit tree structure feature parameters derived from static terrestrial laser scanning data. *Agric. For. Meteorol.* **2016**, *216*, 105–114. [CrossRef]
4. Ghiyamat, A.; Shafri, H.Z. A review on hyperspectral remote sensing for homogeneous and heterogeneous forest biodiversity assessment. *Int. J. Remote Sens.* **2010**, *31*, 1837–1856. [CrossRef]
5. Shi, Y.; Skidmore, A.K.; Wang, T.; Holzwarth, S.; Heiden, U.; Pinnel, N.; Zhu, X.; Heurich, M. Tree species classification using plant functional traits from LiDAR and hyperspectral data. *Int. J. Appl. Earth Obs. Geoinf.* **2018**, *73*, 207–219. [CrossRef]
6. Alonzo, M.; Bookhagen, B.; Roberts, D.A. Urban tree species mapping using hyperspectral and lidar data fusion. *Remote Sens. Environ.* **2014**, *148*, 70–83. [CrossRef]
7. Bauwens, S.; Bartholomeus, H.; Calders, K.; Lejeune, P. Forest inventory with terrestrial LiDAR: A comparison of static and hand-held mobile laser scanning. *Forests* **2016**, *7*, 127. [CrossRef]
8. Balenović, I.; Liang, X.; Jurjević, L.; Hyyppä, J.; Seletković, A.; Kukko, A. Hand-held personal laser scanning–current status and perspectives for forest inventory application. *Croat. J. For. Eng.* **2021**, *42*, 165–183. [CrossRef]
9. Chen, S.; Liu, H.; Feng, Z.; Shen, C.; Chen, P. Applicability of personal laser scanning in forestry inventory. *PLoS ONE* **2019**, *14*, e0211392. [CrossRef] [PubMed]
10. Oveland, I.; Hauglin, M.; Giannetti, F.; Schipper Kjørsvik, N.; Gobakken, T. Comparing three different ground based laser scanning methods for tree stem detection. *Remote Sens.* **2018**, *10*, 538. [CrossRef]
11. Su, Y.; Guo, Q.; Jin, S.; Guan, H.; Sun, X.; Ma, Q.; Hu, T.; Wang, R.; Li, Y. The development and evaluation of a backpack LiDAR system for accurate and efficient forest inventory. *IEEE Geosci. Remote Sens. Lett.* **2020**, *18*, 1660–1664. [CrossRef]
12. Budei, B.C.; St-Onge, B.; Hopkinson, C.; Audet, F.-A. Identifying the genus or species of individual trees using a three-wavelength airborne lidar system. *Remote Sens. Environ.* **2018**, *204*, 632–647. [CrossRef]
13. Shi, Y.; Wang, T.; Skidmore, A.K.; Heurich, M. Important LiDAR metrics for discriminating forest tree species in Central Europe. *J. Photogramm. Remote Sens.* **2018**, *137*, 163–174. [CrossRef]
14. Åkerblom, M.; Raumonen, P.; Mäkipää, R.; Kaasalainen, M. Automatic tree species recognition with quantitative structure models. *Remote Sens. Environ.* **2017**, *191*, 1–12. [CrossRef]
15. Yao, W.; Krzystek, P.; Heurich, M. Tree species classification and estimation of stem volume and DBH based on single tree extraction by exploiting airborne full-waveform LiDAR data. *Remote Sens. Environ.* **2012**, *123*, 368–380. [CrossRef]
16. Fassnacht, F.E.; Latifi, H.; Stereńczak, K.; Modzelewska, A.; Lefsky, M.; Waser, L.T.; Straub, C.; Ghosh, A. Review of studies on tree species classification from remotely sensed data. *Remote Sens. Environ.* **2016**, *186*, 64–87. [CrossRef]

17. Wang, M.; Wong, M.S.; Abbas, S. Tropical Species Classification with Structural Traits Using Handheld Laser Scanning Data. *Remote Sens.* **2022**, *14*, 1948. [CrossRef]
18. Terryn, L.; Calders, K.; Disney, M.; Origo, N.; Malhi, Y.; Newnham, G.; Raumonen, P.; Verbeeck, H. Tree species classification using structural features derived from terrestrial laser scanning. *J. Photogramm. Remote Sens.* **2020**, *168*, 170–181. [CrossRef]
19. Xi, Z.; Hopkinson, C.; Rood, S.B.; Peddle, D.R. See the forest and the trees: Effective machine and deep learning algorithms for wood filtering and tree species classification from terrestrial laser scanning. *J. Photogramm. Remote Sens.* **2020**, *168*, 1–16. [CrossRef]
20. Koenig, K.; Höfle, B. Full-waveform airborne laser scanning in vegetation studies—A review of point cloud and waveform features for tree species classification. *Forests* **2016**, *7*, 198. [CrossRef]
21. Arseniou, G.; MacFarlane, D.W.; Seidel, D. Measuring the contribution of leaves to the structural complexity of urban tree crowns with terrestrial laser scanning. *Remote Sens.* **2021**, *13*, 2773. [CrossRef]
22. Sumnall, M.J.; Hill, R.A.; Hinsley, S.A. Comparison of small-footprint discrete return and full waveform airborne lidar data for estimating multiple forest variables. *Remote Sens. Environ.* **2016**, *173*, 214–223. [CrossRef]
23. Hamraz, H.; Contreras, M.A.; Zhang, J. A robust approach for tree segmentation in deciduous forests using small-footprint airborne LiDAR data. *Int. J. Appl. Earth Obs. Geoinf.* **2016**, *52*, 532–541. [CrossRef]
24. Vicari, M.B.; Disney, M.; Wilkes, P.; Burt, A.; Calders, K.; Woodgate, W. Leaf and wood classification framework for terrestrial LiDAR point clouds. *Methods Ecol. Evol.* **2019**, *10*, 680–694. [CrossRef]
25. Arseniou, G.; MacFarlane, D.W. Fractal dimension of tree crowns explains species functional-trait responses to urban environments at different scales. *Ecol. Appl.* **2021**, *31*, e02297. [CrossRef] [PubMed]
26. Government of Hong Kong. *The Natural Environment, Plants & Animals in Hong Kong*; Conservation, N., Ed.; Government of Hong Kong: Hong Kong, 2021.
27. Jim, C.Y. Roadside trees in urban Hong Kong: Part I census methodology. *Arboric. J.* **1996**, *20*, 221–237. [CrossRef]
28. Jim, C.Y. Tree–habitat relationships in urban Hong Kong. *Environ. Conserv.* **1992**, *19*, 209–218. [CrossRef]
29. Ford, K.R.; Breckheimer, I.K.; Franklin, J.F.; Freund, J.; Kroiss, S.; Larson, A.J.; Theobald, E.; HilleRisLambers, J. Competition alters tree growth responses to climate at individual and stand scales. *Can. J. For. Res.* **2017**, *47*, 53–62. [CrossRef]
30. Sammartano, G.; Spanò, A. Point clouds by SLAM-based mobile mapping systems: Accuracy and geometric content validation in multisensor survey and stand-alone acquisition. *Appl. Geomat.* **2018**, *10*, 317–339. [CrossRef]
31. Walfish, S. A review of statistical outlier methods. *Pharm. Technol.* **2006**, *30*, 82.
32. Zhang, W.; Qi, J.; Wan, P.; Wang, H.; Xie, D.; Wang, X.; Yan, G. An easy-to-use airborne LiDAR data filtering method based on cloth simulation. *Remote Sens.* **2016**, *8*, 501. [CrossRef]
33. Girardeau-Montaut, D. CloudCompare. France: EDF R&D Telecom ParisTech, 11. Available online: https://pcp2019.ifp.uni-stuttgart.de/presentations/04-CloudCompare_PCP_2019_public.pdf (accessed on 10 May 2023).
34. Yang, J.; Kang, Z.; Cheng, S.; Yang, Z.; Akwensi, P.H. An Individual Tree Segmentation Method Based on Watershed Algorithm and Three-Dimensional Spatial Distribution Analysis from Airborne LiDAR Point Clouds. *IEEE J. Sel. Top. Appl. Earth Obs. Remote Sens.* **2020**, *13*, 1055–1067. [CrossRef]
35. Ester, M.; Kriegel, H.-P.; Sander, J.; Xu, X. A density-based algorithm for discovering clusters in large spatial databases with noise. *kdd* **1996**, *96*, 226–231.
36. Raumonen, P.; Åkerblom, M.; Kaasalainen, M.; Casella, E.; Calders, K.; Murphy, S. Massive-scale tree modelling from TLS data. *ISPRS Ann. Photogramm. Remote Sens. Spat. Inf. Sci.* **2015**, *2*, 189. [CrossRef]
37. Demol, M.; Wilkes, P.; Raumonen, P.; Krishna Moorthy Parvathi, S.; Calders, K.; Gielen, B.; Verbeeck, H. Volumetric overestimation of small branches in 3D reconstructions of Fraxinus excelsior. *Silva Fenn.* **2022**, *56*, 10550. [CrossRef]
38. Lau, A.; Bentley, L.P.; Martius, C.; Shenkin, A.; Bartholomeus, H.; Raumonen, P.; Malhi, Y.; Jackson, T.; Herold, M. Quantifying branch architecture of tropical trees using terrestrial LiDAR and 3D modelling. *Trees* **2018**, *32*, 1219–1231. [CrossRef]
39. Benesty, J.; Chen, J.; Huang, Y.; Cohen, I. *Pearson Correlation Coefficient, in Noise Reduction in Speech Processing*; Springer: Berlin/Heidelberg, Germany, 2009; pp. 1–4.
40. Dormann, C.F.; Elith, J.; Bacher, S.; Buchmann, C.; Carl, G.; Carré, G.; Marquéz, J.R.G.; Gruber, B.; Lafourcade, B.; Leitão, P.J. Collinearity: A review of methods to deal with it and a simulation study evaluating their performance. *Ecography* **2013**, *36*, 27–46. [CrossRef]
41. Schober, P.; Boer, C.; Schwarte, L.A. Correlation coefficients: Appropriate use and interpretation. *Anesth. Analg.* **2018**, *126*, 1763–1768. [CrossRef] [PubMed]
42. Obilor, E.I.; Amadi, E.C. Test for significance of Pearson's correlation coefficient. *Int. J. Innov. Math. Stat. Energy Policies* **2018**, *6*, 11–23.
43. Shi, X.; Wong, Y.; Li, M.Z.-F.; Palanisamy, C.; Chai, C. A feature learning approach based on XGBoost for driving assessment and risk prediction. *Accid. Anal. Prev.* **2019**, *129*, 170–179. [CrossRef] [PubMed]
44. Pedregosa, F.; Varoquaux, G.; Gramfort, A.; Michel, V.; Thirion, B.; Grisel, O.; Blondel, M.; Prettenhofer, P.; Weiss, R.; Dubourg, V. Scikit-learn: Machine learning in Python. *J. Mach. Learn. Res.* **2011**, *12*, 2825–2830.45.
45. Al-Tashi, Q.; Abdulkadir, S.J.; Rais, H.M.; Mirjalili, S.; Alhussian, H. Approaches to multi-objective feature selection: A systematic literature review. *IEEE Access* **2020**, *8*, 125076–125096. [CrossRef]

46. Liu, H.; Motoda, H. *Feature Selection for Knowledge Discovery and Data Mining*; Springer Science & Business Media: New York, NY, USA, 2012; Volume 454.
47. Amiri, N.; Heurich, M.; Krzystek, P.; Skidmore, A. Feature Relevance Assessment of Multispectral Airborne Lidar Data for Tree Species Classification. *Int. Arch. Photogramm. Remote Sens. Spat. Inf. Sci.* **2018**, *42*, 31–34. [CrossRef]
48. Cai, J.; Luo, J.; Wang, S.; Yang, S. Feature selection in machine learning: A new perspective. *Neurocomputing* **2018**, *300*, 70–79. [CrossRef]
49. Manju, N.; Harish, B.; Prajwal, V. Ensemble feature selection and classification of internet traffic using XGBoost classifier. *Int. J. Comput. Netw. Inf. Secur.* **2019**, *11*, 37. [CrossRef]
50. Chen, T.; Guestrin, C. Xgboost: A scalable tree boosting system. In Proceedings of the 22nd ACM Sigkdd International Conference on Knowledge Discovery and Data Mining, San Francisco, CA, USA, 13–17 August 2016.
51. Shenkin, A.; Bentley, L.P.; Oliveras, I.; Salinas, N.; Adu-Bredu, S.; Marimon-Junior, B.H.; Marimon, B.S.; Peprah, T.; Choque, E.L.; Trujillo Rodriguez, L. The influence of ecosystem and phylogeny on tropical tree crown size and shape. *Front. For. Glob. Chang.* **2020**, *3*, 501757. [CrossRef]
52. Harikumar, A.; Bovolo, F.; Bruzzone, L. An internal crown geometric model for conifer species classification with high-density lidar data. *IEEE Trans. Geosci. Remote Sens.* **2017**, *55*, 2924–2940. [CrossRef]
53. Li, J.; Hu, B.; Noland, T.L. Classification of tree species based on structural features derived from high density LiDAR data. *Agric. For. Meteorol.* **2013**, *171*, 104–114. [CrossRef]
54. Stal, C.; Verbeurgt, J.; De Sloover, L.; De Wulf, A. Assessment of handheld mobile terrestrial laser scanning for estimating tree parameters. *J. For. Res.* **2021**, *32*, 1503–1513. [CrossRef]
55. Du Toit, F.; Coops, N.C.; Goodbody, T.R.; Stoehr, M.; El-Kassaby, Y.A. Deriving internal crown geometric features of Douglas-fir from airborne laser scanning in a realized-gain trial. *For. Int. J. For. Res.* **2021**, *94*, 442–454. [CrossRef]
56. Kim, S. *Individual Tree Species Identification Using LIDAR-Derived Crown Structures and Intensity Data*; Citeseer: Princeton, NJ, USA, 2008.
57. Maschler, J.; Atzberger, C.; Immitzer, M. Individual tree crown segmentation and classification of 13 tree species using airborne hyperspectral data. *Remote Sens.* **2018**, *10*, 1218. [CrossRef]
58. King, D.A. Allometry and life history of tropical trees. *J. Trop. Ecol.* **1996**, *12*, 25–44. [CrossRef]
59. Purves, D.W.; Lichstein, J.W.; Pacala, S.W. Crown plasticity and competition for canopy space: A new spatially implicit model parameterized for 250 North American tree species. *PLoS ONE* **2007**, *2*, e870. [CrossRef]
60. Torabzadeh, H.; Leiterer, R.; Hueni, A.; Schaepman, M.E.; Morsdorf, F. Tree species classification in a temperate mixed forest using a combination of imaging spectroscopy and airborne laser scanning. *Agric. For. Meteorol.* **2019**, *279*, 107744. [CrossRef]
61. Huot, B.; Yao, J.; Montgomery, B.L.; He, S.Y. Growth–defense tradeoffs in plants: A balancing act to optimize fitness. *Mol. Plant* **2014**, *7*, 1267–1287. [CrossRef]
62. Tomé, M.; Burkhart, H.E. Distance-dependent competition measures for predicting growth of individual trees. *For. Sci.* **1989**, *35*, 816–831.
63. Jim, C.Y. The urban forestry programme in the heavily built-up milieu of Hong Kong. *Cities* **2000**, *17*, 271–283. [CrossRef]
64. Simonovsky, M.; Komodakis, N. Dynamic edge-conditioned filters in convolutional neural networks on graphs. In Proceedings of the IEEE Conference on Computer Vision and Pattern Recognition, Honolulu, HI, USA, 26 June 2017.
65. Xie, Z.; Nolan, T.M.; Jiang, H.; Yin, Y. AP2/ERF transcription factor regulatory networks in hormone and abiotic stress responses in Arabidopsis. *Front. Plant Sci.* **2019**, *10*, 228. [CrossRef] [PubMed]
66. MacFarlane, D.W.; Kane, B. Neighbour effects on tree architecture: Functional trade-offs balancing crown competitiveness with wind resistance. *Funct. Ecol.* **2017**, *31*, 1624–1636. [CrossRef]
67. Poorter, L.; Bongers, L.; Bongers, F. Architecture of 54 moist-forest tree species: Traits, trade-offs, and functional groups. *Ecology* **2006**, *87*, 1289–1301. [CrossRef] [PubMed]
68. González-Orozco, C.E.; Mulligan, M.; Trichon, V.; Jarvis, A. Taxonomic identification of Amazonian tree crowns from aerial photography. *Appl. Veg. Sci.* **2010**, *13*, 510–519. [CrossRef]
69. Trichon, V. Crown typology and the identification of rain forest trees on large-scale aerial photographs. In *Tropical Forest Canopies: Ecology and Management: Proceedings of ESF Conference, Oxford University, 12–16 December 1998*; Springer: Berlin/Heidelberg, Germany, 2001.
70. Colaço, A.F.; Trevisan, R.G.; Molin, J.P.; Rosell-Polo, J.R. A method to obtain orange crop geometry information using a mobile terrestrial laser scanner and 3D modeling. *Remote Sens.* **2017**, *9*, 763. [CrossRef]
71. Cupec, R.; Vidović, I.; Filko, D.; Đurović, P. Object recognition based on convex hull alignment. *Pattern Recognit.* **2020**, *102*, 107199. [CrossRef]
72. Di Gennaro, S.F.; Matese, A. Evaluation of novel precision viticulture tool for canopy biomass estimation and missing plant detection based on 2.5 D and 3D approaches using RGB images acquired by UAV platform. *Plant Methods* **2020**, *16*, 91. [CrossRef] [PubMed]

73. Lin, Y.; Hyyppä, J. A comprehensive but efficient framework of proposing and validating feature parameters from airborne LiDAR data for tree species classification. *Int. J. Appl. Earth Obs. Geoinf.* **2016**, *46*, 45–55. [CrossRef]
74. Liu, H.; Wu, C. Incorporating crown shape information for identifying ash tree species. *Photogramm. Eng. Remote Sens.* **2018**, *84*, 495–503. [CrossRef]

remote sensing

MDPI

Article

Mapping Seasonal Leaf Nutrients of Mangrove with Sentinel-2 Images and XGBoost Method

Jing Miao [1,2], Jianing Zhen [3], Junjie Wang [1,4,*], Demei Zhao [1,2], Xiapeng Jiang [1,2], Zhen Shen [1,2], Changjun Gao [5] and Guofeng Wu [1,2]

1 MNR Key Laboratory for Geo-Environmental Monitoring of Great Bay Area, Guangdong Key Laboratory of Urban Informatics, Shenzhen Key Laboratory of Spatial Smart Sensing and Services, Shenzhen University, Shenzhen 518060, China; 2060325008@email.szu.edu.cn (J.M.); 2050322003@email.szu.edu.cn (D.Z.); 1900325011@email.szu.edu.cn (X.J.); 2100325026@email.szu.edu.cn (Z.S.); guofeng.wu@szu.edu.cn (G.W.)
2 School of Architecture and Urban Planning, Shenzhen University, Shenzhen 518060, China
3 Guangdong Laboratory of Artificial Intelligence and Digital Economy (SZ), Shenzhen 518000, China; zhenjn@radi.ac.cn
4 College of Life Sciences and Oceanography, Shenzhen University, Shenzhen 518060, China
5 Guangdong Provincial Key Laboratory of Silviculture, Protection and Utilization, Guangdong Academy of Forestry, Guangzhou 510520, China; gaochangjun015@163.com
* Correspondence: wang_2015@szu.edu.cn

Abstract: Monitoring the seasonal leaf nutrients of mangrove forests helps one to understand the dynamics of carbon (C) sequestration and to diagnose the availability and limitation of nitrogen (N) and phosphorus (P). To date, very little attention has been paid to mapping the seasonal leaf C, N, and P of mangrove forests with remote sensing techniques. Based on Sentinel-2 images taken in spring, summer, and winter, this study aimed to compare three machine learning models (XGBoost, extreme gradient boosting; RF, random forest; LightGBM, light gradient boosting machine) in estimating the three leaf nutrients and further to apply the best-performing model to map the leaf nutrients of 15 seasons from 2017 to 2021. The results showed that there were significant differences in leaf nutrients ($p < 0.05$) across the three seasons. Among the three machine learning models, XGBoost with sensitive spectral features of Sentinel-2 images was optimal for estimating the leaf C ($R^2 = 0.655$, 0.799, and 0.829 in spring, summer, and winter, respectively), N ($R^2 = 0.668$, 0.743, and 0.704) and P ($R^2 = 0.539$, 0.622, and 0.596) over the three seasons. Moreover, the red-edge (especially B6) and near-infrared bands (B8 and B8a) of Sentinel-2 images were efficient estimators of mangrove leaf nutrients. The information of species, elevation, and canopy structure (leaf area index [LAI] and canopy height) would be incorporated into the present model to improve the model accuracy and transferability in future studies.

Keywords: mangrove; Sentinel-2 image; seasonal leaf nutrients; XGBoost; red edge

1. Introduction

Carbon (C) is the most abundant nutrient element in the dry matter of leaves [1], and nitrogen (N) and phosphorus (P) are essential nutrients in the construction of nucleic acid and proteins in plants [2]. Mangrove forest is one of the most species-diverse and productive marine ecosystems [3], which are the main contributors to blue C in coastal zones [4]. Leaf nutrients often vary between seasons to adapt to the growth process and seasonal climatic cycles [5]. Seasonal nutrient monitoring helps to reflect the dynamics of C sequestration and diagnose the availability and limitation of N and P, which is important to understanding the growth status and nutrient utilization strategies of the mangrove ecosystem [6].

In the last two decades, airborne and satellite-based remote sensing techniques have been key methods in the dynamic monitoring of mangrove growth and health [7]. At the landscape scale, most studies have focused on the mapping of the extent, species composition, biomass, leaf area index (LAI), and chlorophyll content of mangrove forests using medium- (e.g., Landsat-7/8 and Sentinel-2) and high-spatial-resolution satellite (e.g., Worldview-2/3 and Pléiades-1) and Unmanned Aerial Vehicle (UAV-based) multispectral and hyperspectral images [8–10]. In the visible and near-infrared bands, vegetation has obvious reflection and absorption features which are formed by the electronic transition of electromagnetic radiation, vibration, and harmonic generation of chemical bonds (e.g., C-H and N-H bond) through the chemical components of plant leaves [11]. Based on the spectral mechanism, leaf nutrients (C, N, and P) are widely estimated in crop and grassland ecosystems [12,13], however, such estimations have rarely been investigated in mangrove forests.

Due to the abundant spectral details, proximal and UAV-based hyperspectral data are mainly used in the remote estimation of leaf nutrients, and several studies claimed that red-edge wavelengths are sensitive to leaf N and P estimation [13]. Satellite-based hyperspectral imagery is rarely used, because there are far fewer satellites in orbit than multispectral satellites. Among the commonly used multispectral satellites, only Sentinel-2 contains three red-edge bands sensitive to plant growth and health, which could provide great potential for mapping mangrove seasonal leaf nutrients. Yet, little attention has been paid to the long-term monitoring of mangrove seasonal leaf nutrients using Sentinel-2 imagery.

Since leaf C, N, and P are not input parameters to physical models, these nutrients have only been directly estimated with empirical models, including simple regression with one vegetation index [14,15], multiple linear regression (e.g., partial least squares regression) [16,17], and machine learning models (e.g., random forest [RF] and support vector machine [SVM]) [18]. Moreover, the machine learning model generally obtains higher accuracy in estimating leaf nutrients [19]. Recently, due to good accuracy and rapid computational speed, gradient boosting algorithms (e.g., XGBoost [extreme gradient boosting] and LightGBM [light gradient boosting machine]) have been successfully used to remotely assess the parameters of ecological environments (e.g., biomass, soil organic carbon, and leaf chlorophyll content) [18,20,21]. However, the performances of the XGBoost and LightGBM models have not been evaluated in mapping mangrove leaf nutrients.

Using seasonal Sentinel-2 images of mangrove forests in Gaoqiao Mangrove Reserve, China, this study aimed to explore the seasonal response of mangrove leaf nutrients (C, N, and P), compare three machine learning models (XGBoost, RF, and LightGBM) in estimating leaf nutrients, and further to extend the best-performing model to map the leaf nutrients of 15 seasons from 2017 to 2021. The results could facilitate our understanding of seasonal nutrient cycling and limitations in mangrove ecosystems.

2. Materials and Methods

2.1. Field Sampling

Three field surveys were conducted in three seasons (spring of 2018, winter of 2019, and summer of 2020) in Gaoqiao Mangrove Reserve (Figure 1), Guangdong Province, China. The dominant mangrove species of the reserve are *Aegiceras corniculatum, Bruguiear gymnorrhiza, Avicennia marina, Rhizophora stylosa, Sonneratia apetala*, and *Kandelia candel*. With a random sampling strategy and field survey accessibility, a total of 53, 62, and 57 plots (15 m × 15 m) containing one single species (Table 1) were randomly set in 2018, 2019, and 2020, respectively. For each plot, five trees were randomly selected with a distance of 2–5 m between trees, and five mature and healthy leaves were randomly collected from the top canopy of each tree. The geographical location of the center of each plot was recorded by a differential GPS with a positional accuracy less than 20 cm to avoid position mismatch between the image and plot, hence, the corresponding pixel (10 m × 10 m) of geometrically corrected Sentinel-2 imagery could represent the spectral information of the plot.

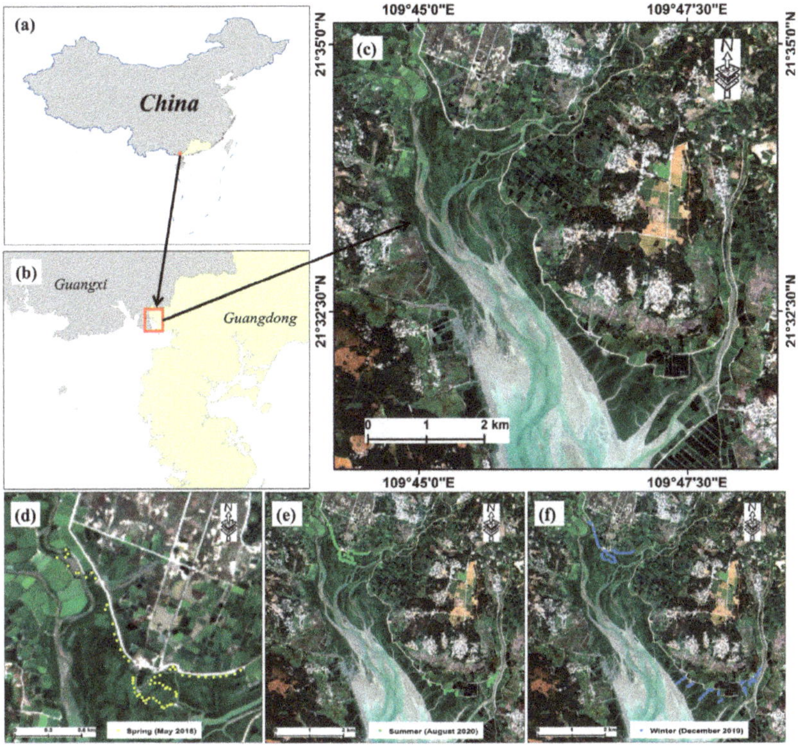

Figure 1. Study area (**a–c**) and spatial distribution of sampling plots in three seasons (**d–f**) (spring: 22–26 May 2018; winter: 20–28 December 2019; summer: 6–12 August 2020). The true-color image is based on the Sentinel-2 image (date: 24 November 2019) with a color combination of band 4 (red), band 3 (green), and band 2 (blue).

Table 1. Number of plot samples collected in three seasons.

Species \ Season	Spring in 2018	Summer in 2020	Winter in 2019
Aegiceras corniculatum	35	16	23
Bruguiear gymnorrhiza	10	13	22
Avicennia marina	0	16	6
Rhizophora stylosa	3	6	8
Sonneratia apetala	1	3	0
Kandelia candel	4	3	3

2.2. Chemical Analysis of Leaf Nutrients

All the fresh leaf samples were dried in an oven at 65 °C for 72 h. Afterwards, C, N, and P concentrations (unit: g/kg) were determined (Table 2) by the high temperature external thermal potassium dichromate oxidation-volumetric method [22], Kjeldahl method including digestion, distillation, and titration [23], and vanadate–molybdate yellow colorimetric method using sulfuric acid–hydrogen peroxide digestion [24], respectively.

Table 2. Statistics of leaf nutrients (unit: g/kg) in three seasons.

Season (Number of Samples)	Nutrient	Min	Max	Mean	CV (%)
Spring (53)	C	408.38	481.02	449.07	3.54
	N	8.05	15.61	10.58	15.49
	P	0.70	1.85	0.90	19.21
Summer (62)	C	424.73	542.08	492.75	5.93
	N	7.23	19.28	10.59	22.64
	P	0.62	2.00	0.86	28.24
Winter (57)	C	403.48	501.87	456.66	6.49
	N	9.77	22.86	14.94	20.98
	P	0.83	2.68	1.57	33.99

CV: coefficient of variation.

2.3. Sentinel-2 Images Pre-Processing and Vegetation Indices Extraction

Three cloudless Sentinel-2 images (Level 1C, acquisition date: 23 May 2018, 24 November 2019, and 23 August 2020), acquired as consistently with the sampling date as possible, were downloaded from the USGS official website (https://earthexplorer.usgs.gov/, accessed on 20 July 2021). The Sen2Cor module from the Sentinel Application Platform (SNAP) was used for atmospheric correction, converting the image top-level reflectance to canopy reflectance (12 spectral bands in the range of 443–2190 nm, Table 3). The bands (B5, B6, B7, B8, B11, and B12) with a spatial resolution of 20 m and 60 m were then resampled to 10 m using the Sen2Res module in SNAP (http://step.esa.int/main/snap-supported-plugins/sen2res/, accessed on 20 July 2021). The Sen2Res algorithm was proposed by Brodu [25], and it employs the general geometric information to unmix low-resolution pixels to realize super-resolution reconstruction of low-resolution bands while keeping the spectral characteristics unchanged.

Table 3. Band information of the Sentinel-2 satellite.

Band	Center Wavelength/nm	Bandwidth/nm	Spatial Resolution/m
B1 (Coastal aerosol)	443	20	60
B2 (Blue)	490	65	10
B3 (Green)	560	35	10
B4 (Red)	665	30	10
B5 (Red-edge1)	705	15	20
B6 (Red-edge2)	740	15	20
B7 (Red-edge3)	783	20	20
B8 (NIR)	842	115	10
B8a (Narrow NIR)	865	20	20
B9 (Water Vapor)	945	20	60
B10 (Cirrus)	1380	30	60
B11 (SWIR1)	1610	90	20
B12 (SWIR2)	2190	180	20

With the 12 bands of 10 m spatial resolution, a total of 30 vegetation indices (VIs, Table 4) were extracted for each image. Some studies claimed that leaf N and P had a close relationship with leaf pigments [12,26], and red-edge bands were sensitive to leaf N and P [13], hence, 28 selected VIs were related to leaf chlorophyll, carotenoid, anthocyanin, and red-edge bands. Moreover, two mangrove-related VIs (MI and MFI) were chosen. The Pearson's correlations of mangrove leaf C, N, and P concentrations against the spectral features of Sentinel-2 images (12 bands + 30 VIs) were calculated and compared for the three seasons. The images were then geometrically corrected with a UAV-based digital orthophoto (spatial resolution: 0.2 m).

Table 4. Vegetation indices used in this study.

Vegetation Index	Formula	Sentinel-2 Bands	Reference
Normalized Difference Vegetation Index (NDVI)	$(R_{NIR} - R_{red})/(R_{NIR} + R_{red})$	B5, B8	[27]
Green NDVI (gNDVI)	$(R_{750} - R_{550})/(R_{750} + R_{550})$	B3, B6	
Optimized Soil-Adjusted Vegetation Index (OSAVI)	$1.16 * (R_{800} - R_{670})/(R_{800} + R_{670} + 0.16)$	B4, B8	[28]
Red-Edge Inflection Point (REIP)	$R_{700} + 40 * \frac{0.5*(R_{670} - R_{780}) - R_{700}}{R_{740} - R_{700}}$	B4, B5, B6, B7	[29]
Simple Ratio Index (SR_{705})	R_{750}/R_{705}	B5, B6	[30]
Enhanced Vegetation Index (EVI)	$2.5 * (R_{NIR} - R_{red})/(R_{NIR} + 6 * R_{red} - 7.5 * R_{blue} + 1)$	B2, B4, B8	[31]
$SR_{Chl\ a}$	$R_{672}/(R_{550} * R_{708})$	B3, B4, B5	[27]
$SR_{Chl\ b}$	R_{672}/R_{550}	B3, B4	
SR_{chl}	$R_{860}/(R_{550} * R_{708})$	B3, B5, B8	
Modified Cab Absorption in Reflectance Index (MCARI)	$[(R_{700} - R_{670}) - 0.2 * (R_{700} - R_{550})] * (R_{700}/R_{670})$	B3, B4, B5	[32]
Modified Chlorophyll Absorption in Reflectance Index (MCARI1)	$1.2 * [2.5 * (R_{800} - R_{670}) - 1.3 * (R_{800} - R_{550})]$	B3, B4, B8	
Transformed CARI (TCARI)	$3 * [(R_{700} - R_{670}) - 0.2 * (R_{700} - R_{550}) * (R_{700}/R_{670})]$	B3, B4, B5	[33]
MCARI/OSAVI	$\frac{[(R_{700} - R_{670}) - 0.2*(R_{700} - R_{550})]*(R_{700}/R_{670})}{[1.16*(R_{800} - R_{670})/(R_{800} + R_{670} + 0.16)]}$	B3, B4, B5, B8	[34]
TCARI/OSAVI	$\frac{3*[(R_{700} - R_{670}) - 0.2*(R_{700} - R_{550})]*(R_{700}/R_{670})}{[1.16*(R_{800} - R_{670})/(R_{800} + R_{670} + 0.16)]}$	B3, B4, B5, B8	
Red-Edge Position (REP)	$705 + 35 * [0.5 * (R_{665} + R_{783}) - R_{705}]/(R_{740} - R_{705})$	B4, B5, B6, B7	[35]
Pigment Specific Simple Ratio for Chla (PSSRa)	R_{800}/R_{675}	B4, B8	[36]
Green chlorophyll index (CI_{green})	$(R_{RE3}/R_{green}) - 1$	B3, B7	[37]
Green chlorophyll index ($CI_{red\text{-}edge}$)	$\left(R_{RE3}/R_{red\ edge}\right) - 1$	B5, B7	
Disease Water Stress Index (DSWI)	$(R_{803} + R_{549})/(R_{1659} + R_{681})$	B3, B4, B8, B11	[38]
Moisture Stress Index (MSI)	R_{SWIR}/R_{NIR}	B8, B11	[39]
Red and Green Pigment Indices (RGI)	R_{690}/R_{550}	B3, B5	[40]
Anthocyanin Reflectance Index (ARI)	$(1/R_{green}) * \left(1/R_{red\ edge}\right)$	B3, B8a	[30]
Carotenoid Reflectance (CRI)	$(R_{510})^{-1} - (R_{550})^{-1}$	B2, B3	
Carotenoid Reflectance (CRI2)	$(R_{510})^{-1} - (R_{700})^{-1}$	B2, B5	
Visible Atmospherically Resistant Index ($VARI_{green}$)	$(R_{green} - R_{red})/(R_{green} + R_{red} - R_{blue})$	B2, B3, B4	
Cater Stress Index (CSI2)	R_{695}/R_{760}	B5, B7	[41]
Apparent Clumping Index (ACI)	R_{green}/R_{NIR}	B3, B8	[42]
Red-Edge Normalized Difference Vegetation Index (NDRE1)	$(R_{RE2} - R_{RE1})/(R_{RE2} + R_{RE1})$	B5, B6	[43]
Mangrove Index (MI)	$(R_{NIR} - R_{SWIR}/R_{NIR} * R_{SWIR}) * 10{,}000$	B8, B12	[44]
Mangrove Forest Index (MFI)	$[(R_{RE1} - R_{B\lambda 1}) + (R_{RE2} - R_{B\lambda 2}) + (R_{RE3} - R_{B\lambda 3}) + (R_{RE4} - R_{B\lambda 4})]/4$ $R_{B\lambda i} = R_{2190} + (R_{665} - R_{2190}) * (2190 - \lambda i)/(2190 - 665)$	B4, B5, B6, B7, B8a, B12	[45]

2.4. Estimation of Mangrove Seasonal Leaf Nutrients with Machine Learning Method

2.4.1. Three Machine Learning Models

In this study, three machine learning models were employed to estimate mangrove leaf nutrients with seasonal Sentinel-2 images: XGBoost (extreme gradient boosting), random forest (RF), and light gradient boosting machine (LightGBM). All three methods are ensemble learning-based algorithms, which could improve the generalization ability and robustness of basic learners by combining the prediction results of multiple base learners.

XGBoost is proposed based on the gradient-boosting decision tree (GBDT) [46]. Compared to the traditional GBDT algorithm, XGBoost adopts second-order Taylor expansion into the cost function to avoid model overfitting [46]. Moreover, XGBoost uses a sparse-aware split lookup method to process sparse data [47], which is practical for dealing with the limited number of observation samples in the field of quantitative remote sensing [48]. In the training stage, one decision tree is incremented during each iteration, gradually forming a strong evaluator with a combination of multiple trees. The objective function for the tth iteration is defined as following:

$$obj^{(t)} = \sum_{i=1}^{n}\left[g_i f_t(x_i) + \frac{1}{2}\, h_i f_t^2(x_i)\right] + \omega(f_t) \tag{1}$$

$$\begin{aligned} g_i &= \partial_{\hat{y}^{(t-1)}} l\left(y_i, \hat{y}^{(t-1)}\right) \\ h_i &= \partial^2_{\hat{y}^{(t-1)}} l\left(y_i, \hat{y}^{(t-1)}\right) \end{aligned} \tag{2}$$

where y_i is the field-measured value; $\hat{y}_i{}^{(t-1)}$ is the predicted value of $(t-1)$-th iteration; f_t represents the kth decision tree; x_i represents the feature vector of the ith sample; $l\left(y_i, \hat{y}^{(t-1)}\right)$ is the prediction error of the learning model consisting of the previous $t-1$ trees; g_i and h_i are the first and second derivative of the prediction model and the current model, respectively; and $\omega(f_t)$ is the regular term of the objective function in each iteration.

RF is proposed based on the bagging method [49]; it establishes multiple decision trees and mutually independent weak estimators at a time, and it votes on all independent weak estimators. RF selects the optimal estimators with the highest votes as the final model-prediction result. The algorithm generates a new training sample set by randomly extracting k samples from the original training sample set, and then generates k decision trees according to the bootstrapping sample set to form a random forest. Several studies claimed that XGBoost was superior to the popular machine learning method (RF) in the remote estimation of the biomass of mangrove and the chlorophyll of pepper leaf [10,49].

LightGBM is proposed based on GBDT with a different splitting strategy of the leaf nodes [50]. Unlike the undifferentiated level-wise strategy of XGBoost, LightGBM uses a leaf-wise strategy of leaf node splitting which selects the nodes that benefit most from splitting and reduces computational effort [50]. The algorithm employs gradient-based one-side sampling (GOSS) and exclusive feature bundling (EFB) for faster training. GOSS picks data with a larger gradient from the samples to increase their contribution to the calculated information gain. EFB reduces the data dimensionality by merging similar data features. The objective function depending on the leaf-wise strategy is defined as follows:

$$G = \frac{1}{2}\left(\frac{\left(\sum_{i\in I_L} g_i\right)^2}{\sum_{i\in I_L} h_i + \lambda} + \frac{\left(\sum_{i\in I_R} g_i\right)^2}{\sum_{i\in I_R} h_i + \lambda} - \frac{\left(\sum_{i\in I} g_i\right)^2}{\sum_{i\in I} h_i + \lambda}\right) \tag{3}$$

where g_i and h_i indicate the first and second derivative statistics of the loss function, and I_L and I_R are the sample sets of the leaf and right branches, respectively.

All three methods have the efficient ability of feature selection based on the importance score, which can be quantified using several metrics, such as the gain, weights, total gain, and total coverage of a node. We chose the metric value of gain to explain the relative importance of each corresponding feature. The higher score values indicate a greater contribution to the model performance. The three models were built using Python version 3.8, and the settings of the main parameters are shown in Table 5.

Table 5. Main parameter settings of the three machine learning methods.

Algorithm	Algorithm Library	Main Parameter Settings
XGBoost	https://github.com/dmlc/xgboost/ (accessed on 20 September 2021), version 1.5.0	max_depth = 5, learning_rate = 0.1, n_estimators = 200, min_child_weight = 1
RF	https://github.com/kjw0612/awesome-random-forest (accessed on 20 September 2021), version 1.2.2	n_estimators = 200, criterion = 'mse', max_depth = None, min_samples_split = 2, min_samples_leaf = 1
LightGBM	https://github.com/microsoft/LightGBM (accessed on 20 September 2021), version 3.3.1	learning_rate = 0.1, num_leaves = 31, min_data_in_leaf = 20, n_estimators = 200

2.4.2. Two Modeling Strategies

In this study, two modeling strategies were used to compare the performances of the three machine learning models across the three seasons: model development with the dataset of a single season and the pooled dataset of three seasons. Due to the possible data redundancy of the spectral features of Sentinel-2 images, feature selection was required to choose the sensitive features for modeling. Based on the importance score derived from each machine learning method, the features with an importance score greater than the standard deviation of all score values were selected for further modeling.

For the first modeling strategy, a total of nine regression models (three machine learning methods × three seasons) were established and cross-validated by the leave-one-out cross-validation procedure in estimating each leaf nutrient. For the second modeling strategy, all the samples were combined, a total of three regression models were established and cross-validated by estimating each leaf nutrient, and the samples in each season were predicted by the combined model.

2.4.3. Model Evaluation

The coefficient of determination (R^2), relative root mean square error (RRMSE), and residual prediction deviation (RPD) were calculated to evaluate the performance of each model:

$$R^2 = 1 - \frac{\sum_{i=1}^{n} (y_i - \hat{y}_i)^2}{\sum_{i=1}^{n} (y_i - \overline{y_i})^2} \tag{4}$$

$$\text{RMSE} = \sqrt{\frac{\sum_{i=1}^{n} (y_i - \hat{y}_i)^2}{n}} \tag{5}$$

$$\text{RRMSE} = \text{RMSE}/\overline{y_i} \tag{6}$$

$$\text{RPD} = \text{SD}/\text{RMSE} \tag{7}$$

$$\text{MAE} = \frac{1}{n} \times \sum_{i=1}^{n} \left(\frac{|y_i - \hat{y}_i|}{y_i} \times 100\%\right) \tag{8}$$

where y_i and $\hat{y}_i$ are the measured and predicted value of the ith sample, respectively; $\overline{y_i}$ is the mean value of measured leaf nutrient, n is the number of leaf samples, and SD is the standard deviation. A higher R^2 and RPD and a lower RRMSE (unit: %) indicate a better model performance.

Based on the best-performing model, all the combined samples of the three seasons were classified into four groups according to the quartile values of the leaf nutrients, and the number of samples and mean absolute error (MAE, unit: %) of each group were calculated.

2.5. Mapping of Seasonal Mangrove Leaf Nutrients

For each nutrient, the optimal machine learning model was chosen to map its spatial distribution in each season. To investigate the spatial distribution of leaf nutrients, hotspot analysis was further conducted in ArcGIS 10.8 using the Getis-Ord Gi* parameters to identify statistically significant clusters of hot and cold spots. The p value and z score were obtained to judge whether to reject the null hypothesis. The p value indicates the probability that the spatial pattern of the leaf nutrient concentration was created by a random process. The Z score indicates the multiples of the standard deviation, and a higher Z score (>0) with a lower p value indicates greater clustering of higher value (hot spot), while a lower Z score (<0) with a lower p value indicates greater clustering of lower value (cold spot). Moreover, there is no significant spatial clustering when the Z score is close to 0.

On the other hand, to understand the changes of seasonal leaf nutrients in mangrove forests across different years, we obtained another 12 cloudless Sentinel-2 images (Table 6) from 2017 to 2021 to map the concentrations of leaf C, N, and P based on the corresponding best-performing model.

Table 6. The acquisition time of 12 Sentinel-2 images from 2017 to 2021.

Year	Season	Acquisition Date
2017	Spring	8 April 2017
	Summer	27 June 2017
	Winter	19 December 2017
2018	Summer	31 August 2018
	Winter	23 January 2019
2019	Spring	23 May 2019
	Summer	2 July 2019
2020	Spring	8 March 2020
	Winter	8 December 2020
2021	Spring	17 May 2021
	Summer	4 September 2021
	Winter	28 November 2021

3. Results

3.1. Seasonal Variation of Mangrove Leaf Nutrients

Among the three seasons (Table 2), the leaf samples from summer showed higher leaf C concentration (mean = 492.75 g/kg) than those from spring (mean = 449.07 g/kg) and winter (mean = 457.16 g/kg). However, higher leaf N (mean = 14.94 g/kg) and P (mean = 1.58 g/kg) concentrations were observed in winter than in spring (mean = 10.58 and 0.90 g/kg) and summer (mean = 10.59 and 0.86 g/kg). Moreover, considering any two seasons, the one-way analysis of variation (ANOVA) results (Table 7) showed that there was no significant difference in leaf C between spring and winter ($p > 0.05$), and no significant difference in leaf N or P was observed between spring and summer. However, considering all the three seasons, there were significant differences in leaf C, N, and P ($p < 0.05$).

Based on all the leaf samples from the three seasons (Figure 2), leaf N was positively correlated with leaf P, while leaf C was negatively correlated with leaf N and P ($p < 0.01$), and the N–P correlation was stronger than the C–N and C–P correlation. Based on the leaf samples from a single season, the N–P correlation (0.880, 0.848, and 0.686 in winter, summer, and spring, respectively; $p < 0.01$) was also stronger than the C–P (−0.716, −0.631, and −0.420; $p < 0.01$) and C-N correlation (−0.612, −0.526, and −0.334; $p < 0.01$).

Table 7. ANOVA of leaf nutrients between seasons.

Nutrient	Season	Spring	Summer	Winter
C	Spring			
	Summer	0.000 *		
	Winter	0.078	0.000 *	
N	Spring			
	Summer	0.978		
	Winter	0.000 *	0.000 *	
P	Spring			
	Summer	0.586		
	Winter	0.000 *	0.000 *	

* Significant ($p < 0.05$).

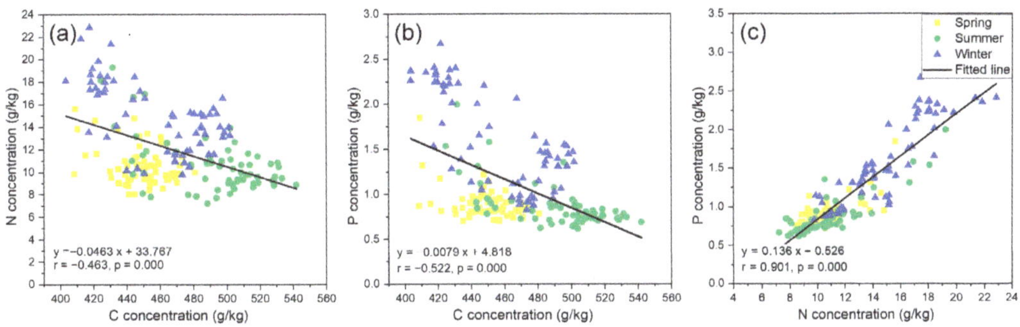

Figure 2. Intercorrelation of leaf C–N concentration (**a**), leaf C–P concentration (**b**), and leaf N–P concentration (**c**) with all the samples ($n = 172$), respectively.

3.2. Correlation of Leaf Nutrients against Spectral Features of Seasonal Sentinel-2 Images

Leaf C had the highest correlation with DSWI index, B11, and MCARI/OSAVI (r = 0.693, 0.751, and 0.743, $p < 0.01$) in spring, summer, and winter, respectively (Figure 3); leaf N had the highest correlation with CIgreen, B11, and B6 (r = −0.351, −0.450, and −0.633, $p < 0.01$) in spring, summer, and winter, respectively; leaf P had the highest correlation with B3, B11, and B6 (r = 0.490, −0.561, and −0.730, $p < 0.01$) in spring, summer, and winter, respectively. Considering the mean absolute correlation coefficient of the 42 spectral features against leaf nutrients, the three leaf nutrients reported stronger correlation (mean $|r|$ = 0.51, 0.44, and 0.49 for C, N, and P, respectively) in winter than summer (mean $|r|$ = 0.50, 0.19, and 0.23) and spring (mean $|r|$ = 0.39, 0.28, and 0.36).

3.3. Comparison of Three Machine Learning Models in Estimating Leaf Nutrients

When using the dataset of a single season to estimate the three leaf nutrients (Table 8), XGBoost model with sensitive features (Figure 4) reported higher accuracy (R^2 = 0.655–0.829, RRMSE = 1.687–2.408%, RPD = 1.703–2.418 for leaf C estimation; R^2 = 0.668–0.743, RRMSE = 6.090–9.668%, RPD = 1.736–1.973 for leaf N estimation; R^2 = 0.539–0.622, RRMSE = 4.659–19.560%, RPD = 1.473–1.627 for leaf P estimation) than RF and LightGBM in each season, except the cases of leaf C estimation in summer and leaf P estimation in summer and winter. Considering the mean value of the performance parameters of the three machine learning models, leaf C was estimated with higher accuracy in winter (mean R^2 = 0.829, mean RRMSE = 2.401%, mean RPD = 2.418) than in summer and spring; leaf N and P were estimated with higher accuracy in summer (mean R^2 = 0.743 and 0.622, mean RRMSE = 9.668% and 16.251%, mean RPD = 1.973 and 1.627) than spring and winter.

When using the pooled dataset of three seasons to estimate leaf nutrients (Table 9), all three models reported very poor performance in spring ($R^2 < 0.3$), and the XGBoost model with sensitive spectral features (Figure 4) showed a stronger performance (mean R^2 = 0.513, 0.347, and 0.389 for C, N, and P estimation) than RF (mean R^2 = 0.477, 0.262, and 0.376) and LightGBM (mean R^2 = 0.453, 0.232, and 0.33) considering the mean value of R^2 in summer and winter. Considering the mean value of the performance parameters of the three models, leaf C was estimated with higher accuracy in summer (mean R^2 = 0.788, mean RRMSE = 1.604%, mean RPD = 2.171) than in winter and spring; leaf N and P were estimated with higher accuracy in winter (mean R^2 = 0.504 and 0.439, mean RRMSE = 9.321% and 12.993%, mean RPD = 1.419 and 1.335) than in summer and spring.

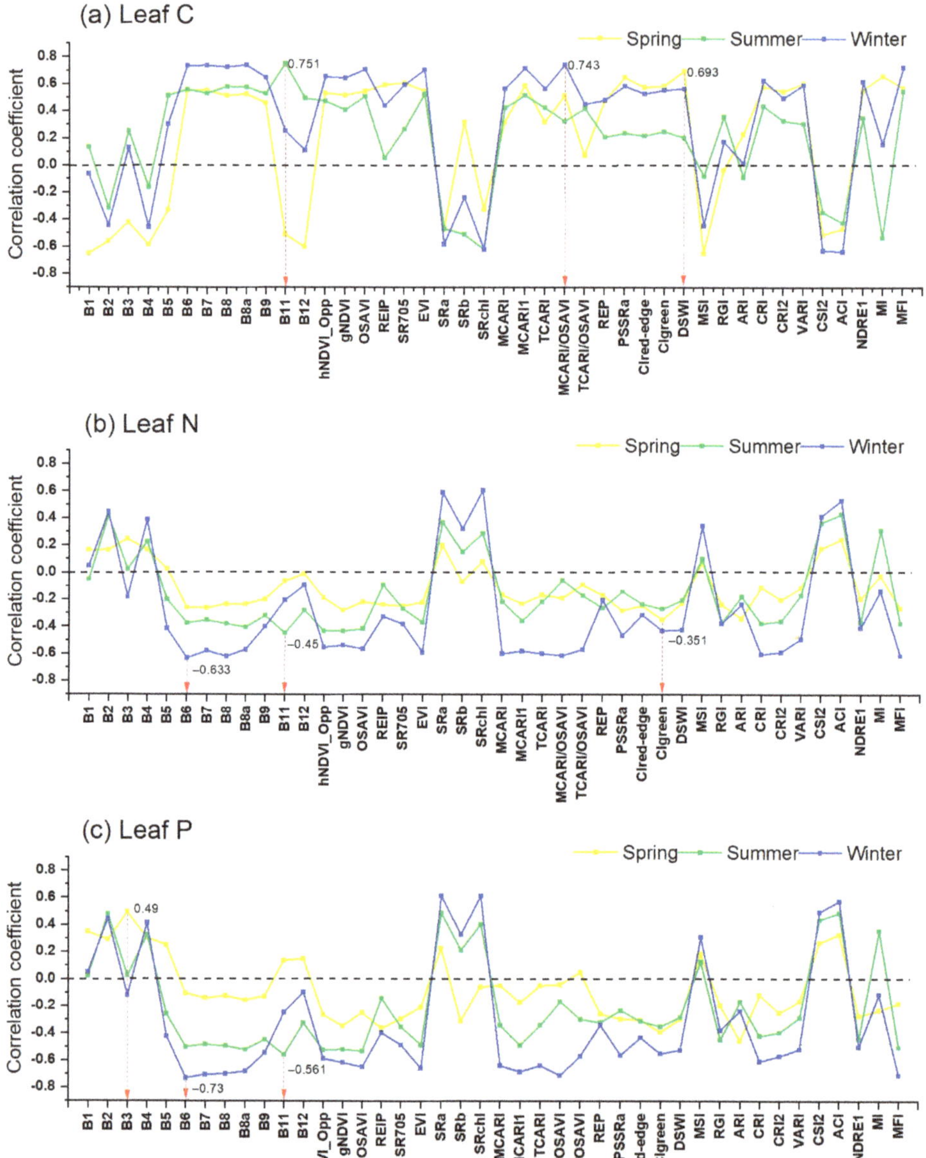

Figure 3. Correlations of leaf C (**a**), N (**b**), and P (**c**) against spectral features (12 spectral bands + 30 VIs) of Sentinel-2 images in spring, summer, and winter. The red arrow indicates the feature with the highest correlation in each season.

Table 8. Model performance in estimating leaf C, N, and P using the dataset of a single season.

Model	Nutrient	Spring			Summer			Winter		
		R^2	RRMSE(%)	RPD	R^2	RRMSE(%)	RPD	R^2	RRMSE(%)	RPD
XGBoost	C	0.655	1.687	1.703	0.799	2.408	2.230	0.829	2.401	2.418
	N	0.668	6.090	1.736	0.743	9.668	1.973	0.704	8.998	1.838
	P	0.539	4.659	1.473	0.622	16.251	1.627	0.596	19.560	1.573
RF	C	0.549	1.717	1.489	0.811	2.314	2.300	0.824	2.420	2.383
	N	0.629	13.699	1.642	0.684	10.063	1.779	0.637	9.944	1.660
	P	0.415	6.168	1.309	0.652	14.662	1.700	0.613	18.156	1.607
LightGBM	C	0.415	1.952	1.308	0.401	3.014	1.292	0.803	2.538	2.253
	N	0.654	5.512	1.700	0.133	9.813	1.074	0.015	3.835	1.008
	P	0.273	6.755	1.173	0.207	14.292	1.122	0.627	17.596	1.637

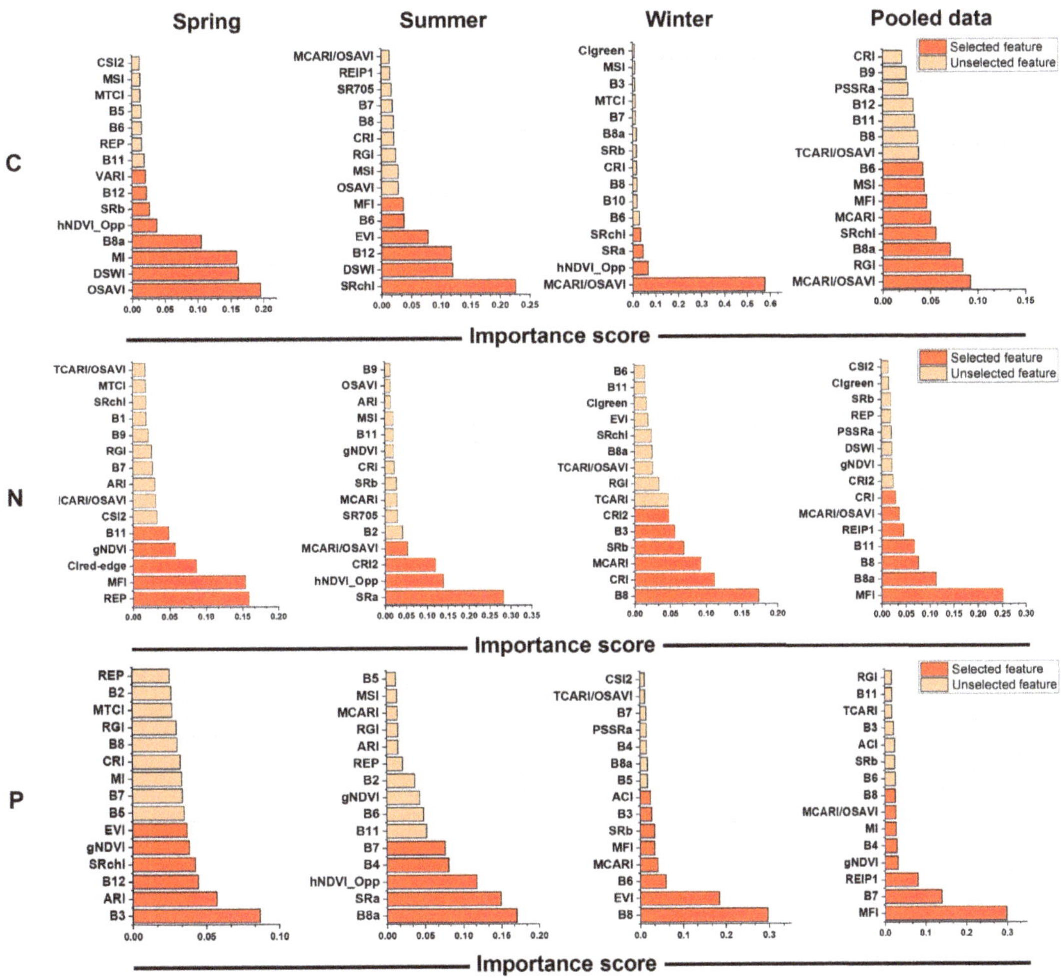

Figure 4. The ranking of the importance scores of the spectral features (bands + VIs) of Sentinel-2 images in correlating with leaf C, N, and P using the dataset of a single season and the pooled dataset of the three seasons. The importance scores were derived from the XGBoost algorithm, and the selected features were marked in red.

Table 9. Model performance in estimating leaf C, N, and P using the pooled dataset of three seasons.

Model	Nutrient	Spring			Summer			Winter		
		R^2	RRMSE(%)	RPD	R^2	RRMSE(%)	RPD	R^2	RRMSE(%)	RPD
XGBoost	C	0.218	1.842	1.131	0.788	1.604	2.171	0.534	2.599	1.464
	N	0.021	9.757	1.011	0.504	9.321	1.419	0.516	10.481	1.438
	P	0.057	15.218	1.030	0.434	13.635	1.329	0.677	14.180	1.759
RF	C	0.133	1.987	1.074	0.730	1.818	1.924	0.569	2.543	1.523
	N	0.038	11.251	1.020	0.294	11.894	1.190	0.453	9.933	1.352
	P	0.079	16.392	1.042	0.439	12.993	1.335	0.611	15.016	1.603
LightGBM	C	0.169	2.094	1.097	0.758	1.714	2.035	0.432	2.882	1.327
	N	0.000	12.883	1.000	0.405	10.167	1.297	0.290	12.965	1.186
	P	0.000	19.315	1.000	0.417	15.544	1.309	0.573	16.101	1.530

Overall, the three models performed better using the dataset of a single season than using the pooled dataset of three seasons in estimating the three leaf nutrients in each season. Moreover, the XGBoost model always provided higher accuracy than RF and LightGBM in estimating the three leaf nutrients. Hence, an XGBoost model with sensitive spectral features using a single dataset was further used for mapping the spatial distribution of seasonal leaf nutrients from 2017 to 2021.

Based on the XGBoost method (Figure 4), MCARI/OSAVI using the bands of B3, B4, B5, and B8; OSAVI using the bands of B4 and B8, and SRchl using the bands of B3, B5, and B8 were the most sensitive to leaf C; REP using the bands of B4, B5, B6, and B7, SRa using the bands of B3, B4, B5, and B8, and MFI using the bands of B4, B5, B6, B7, B8a, and B12 were the most sensitive to leaf N; and B3, B8a, B8, and MFI were the most sensitive to leaf P.

On the other hand, the scatter plots of field-measured versus estimated leaf nutrients (Figure 5) showed that the estimated concentrations of leaf nutrients in spring were lower than in summer and winter. Due to the absence of samples with measured leaf C concentrations of 460–470 g/kg, two-point clusters of leaf C with abnormal distribution were observed in summer and winter. Moreover, the samples with medium C concentrations (443.63–473.10 g/kg) and lower N (8.52–11.72 g/kg) and lower P concentrations (0.71–1.12 g/kg) tended to have lower estimation errors (MAE < 10%, Table 10), while the samples with lower C concentrations (414.16–443.63 g/kg) and medium N (14.92–18.12 g/kg) and medium P concentrations (1.54–1.95 g/kg) tended to have higher estimation errors.

Table 10. The mean absolute error (MAE) of measured versus estimated leaf nutrients in different data groups based on the scatter points of Figure 5.

Leaf Nutrient	Data Group	Spring		Summer		Winter	
		n	MAE(%)	*n*	MAE(%)	*n*	MAE(%)
C	414.16–443.63	14	1.98	4	3.06	24	2.37
	443.63–473.10	39	1.28	7	1.59	10	2.07
	473.10–502.57	0	–	22	2.52	28	1.98
	502.57–532.04	0	–	24	1.84	0	–
N	8.52–11.72	46	6.68	50	8.55	7	6.87
	11.72–14.92	7	5.53	3	9.53	22	6.30
	14.92–18.12	0	–	4	12.22	27	9.14
	18.12–21.33	0	–	0	–	6	7.81
P	0.71–1.12	53	7.62	49	9.50	14	9.02
	1.12–1.54	0	–	6	21.41	19	22.09
	1.54–1.95	0	–	2	25.52	7	22.59
	1.95–2.37	0	–	0	–	22	15.36

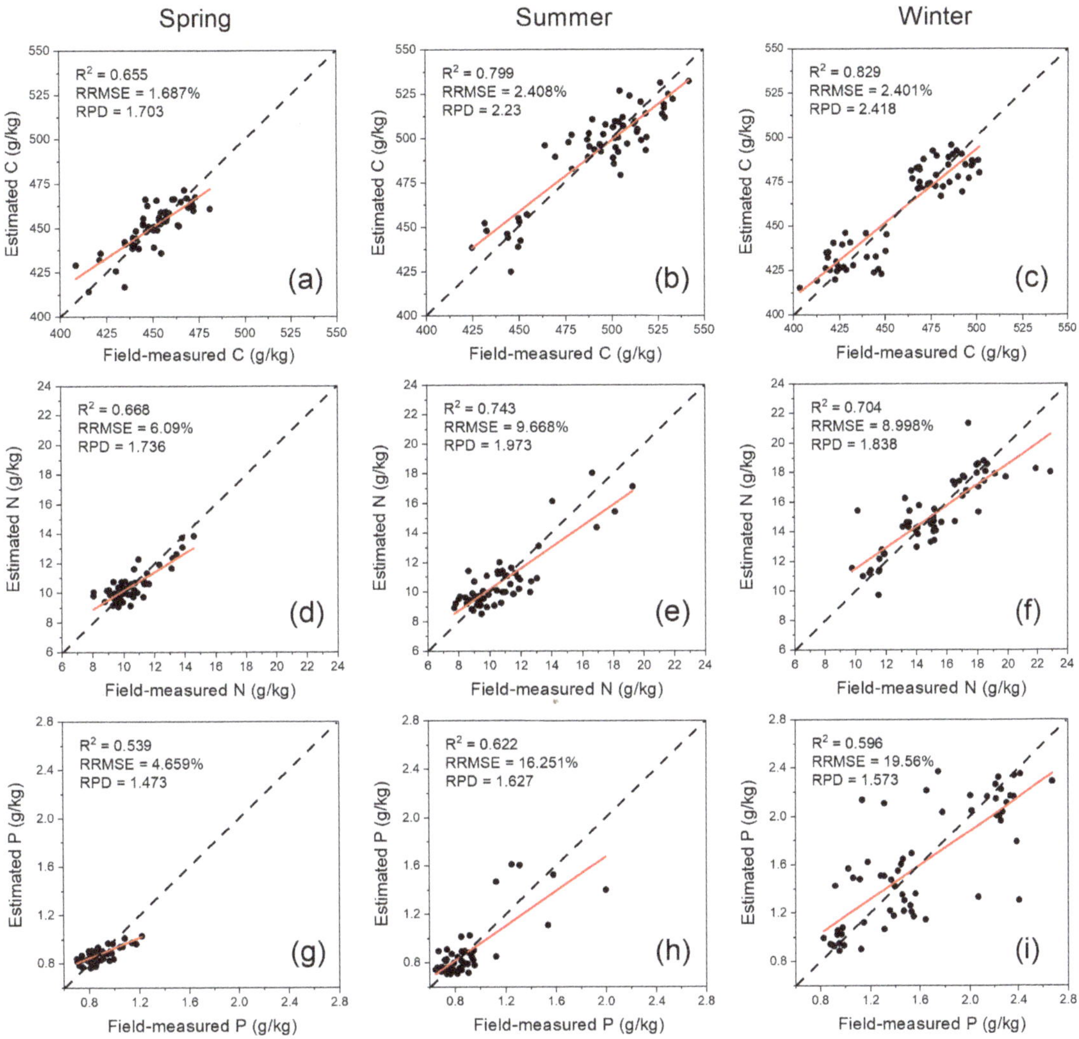

Figure 5. Scatter plots of field-measured versus estimated leaf C (**a–c**), N (**d–f**), and P (**g–i**) using the XGBoost model with the dataset of a single season (the fitted line showed in red).

3.4. Mapping Seasonal Leaf Nutrients with XGBoost Model

3.4.1. Mapping Leaf C, N, and P Concentrations in Three Seasons

Based on the XGBoost model using the sensitive spectral features of the dataset of a single season, leaf C, N, and P concentrations were mapped in spring, summer, and winter, respectively (Figure 6). The mapped C concentrations in summer (mean = 499.564 g/kg, range = 454.886–530.845 g/kg) were higher than those in winter (mean = 481.287 g/kg, range = 435.731–510.978 g/kg) and spring (mean = 449.757 g/kg, range = 406.960–481.936 g/kg). Leaf C concentrations were higher in the northwestern part of the study area, while lower C concentrations were mostly observed in the estuary and nearshore areas with low vegetation coverage.

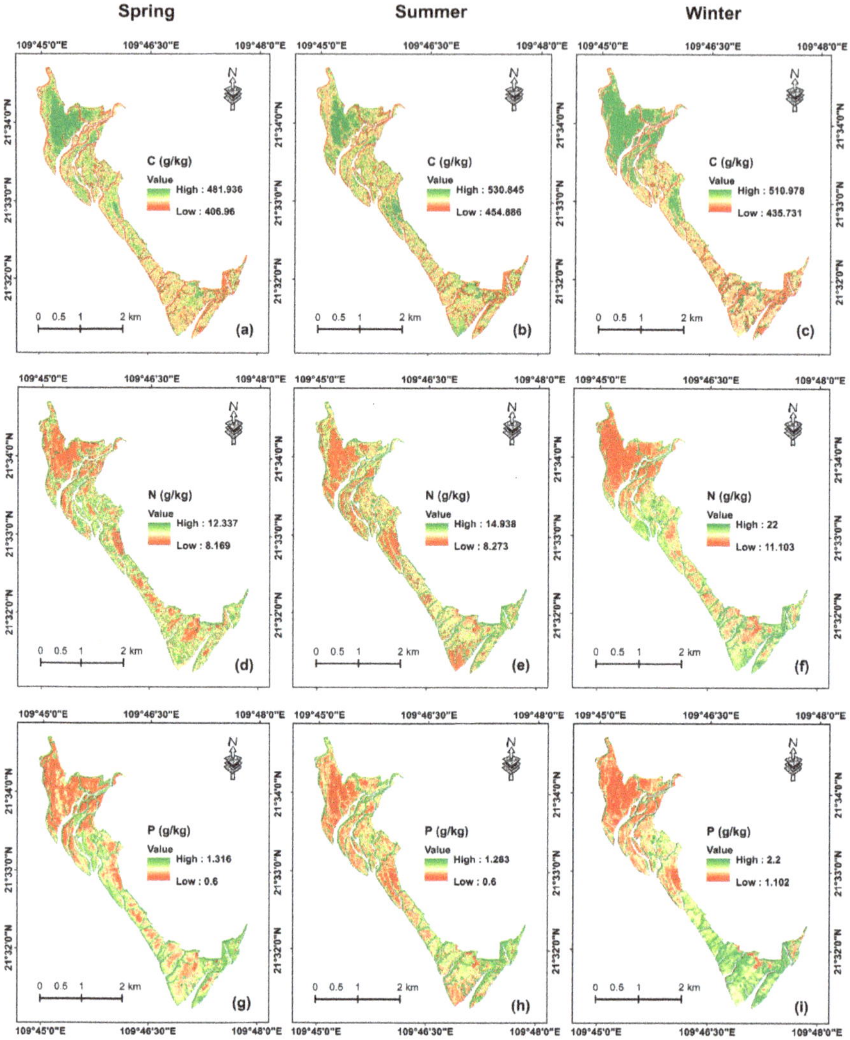

Figure 6. Maps of leaf C (**a–c**), N (**d–f**), and P (**g–i**) concentrations in spring, summer, and winter derived from the XGBoost model using sensitive spectral features of Sentinel-2 images.

The mapped N concentrations (mean = 16.937 g/kg, range = 11.103–22.000 g/kg) in winter were higher than those in spring (mean = 9.184 g/kg, range = 8.169–12.337 g/kg) and summer (mean = 9.059 g/kg, range = 8.273–14.938 g/kg). Moreover, the area with higher C was found to have lower N. The mapped leaf P concentrations showed similar spatial distribution characteristics to the mapped N concentrations, and leaf P showed higher concentrations in winter (mean = 1.513 g/kg, range = 1.102–2.200 g/kg) than in spring (mean = 0.832 g/kg, range = 0.600–1.316 g/kg) and summer (mean = 0.773, range = 0.600–1.283 g/kg).

The hotspot analysis of mapped leaf nutrients (Figure 7) showed that there were different spatial patterns of aggregation across different seasons. The significant hot spot areas of leaf C were mainly located in the northwest and middle part of the study area, while the significant cold spot areas of leaf C were mainly located near the rivers. In contrast, the aggregations of leaf N and P in most of the study area were not significant; the significant hot spot areas of leaf N and P were mainly located near the rivers and the boundary, while the significant cold spot areas were mainly located in the northwest of the study area.

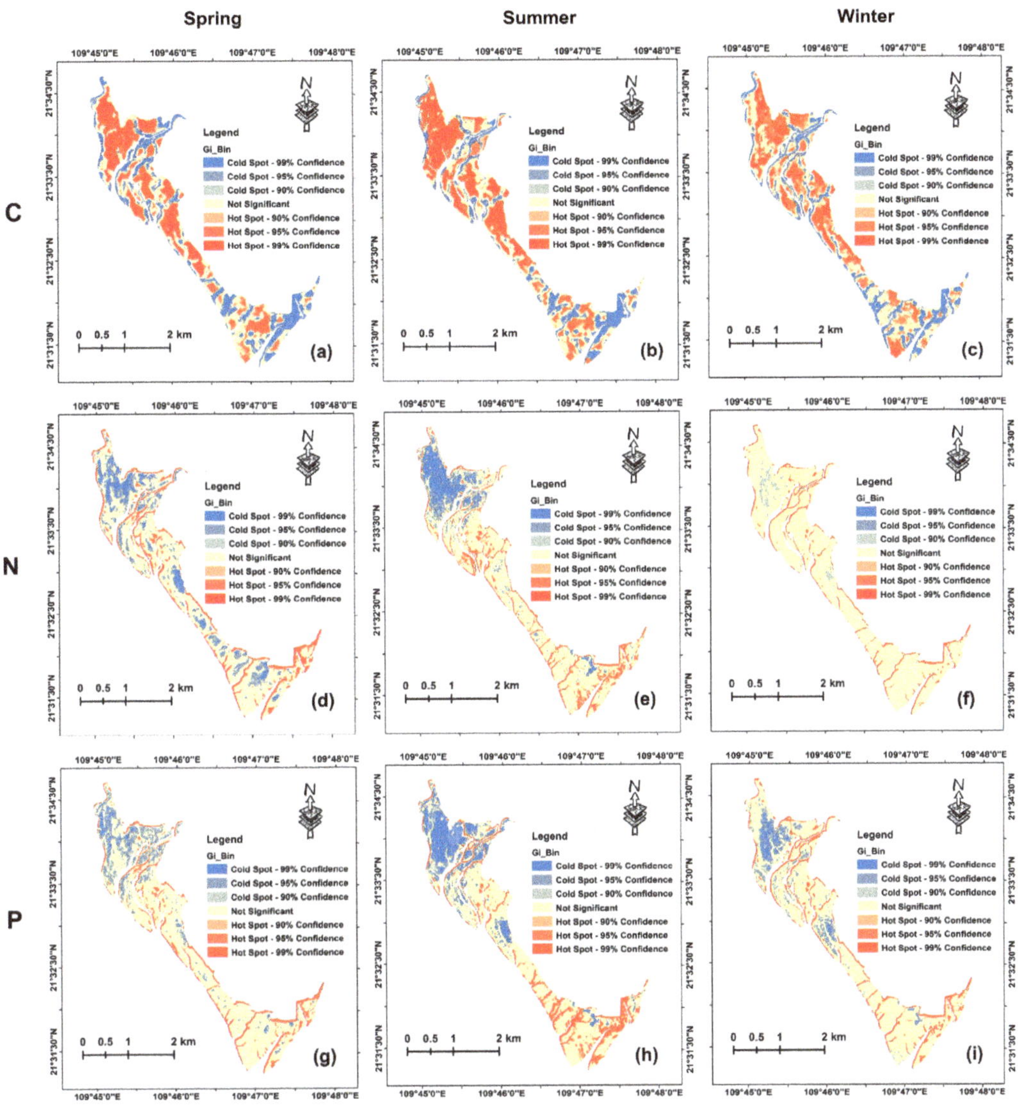

Figure 7. Hotspot analysis of mapped leaf C (**a–c**), N (**d–f**), and P (**g–i**) concentrations in spring, summer, and winter.

3.4.2. Mapping Seasonal Leaf C, N and P Concentrations from 2017 to 2021

During 2017–2021, the mean mapped leaf C concentration in summer (range = 463.003–499.564 g/kg) was higher with wider data variation than in winter (range = 459.631–481.287 g/kg) and in spring (range = 449.757–465.282 g/kg) (Figure 8); the mean mapped leaf N and P concentrations in winter were much higher than those in spring and summer (Figures 9 and 10). Moreover, the mean leaf N and P concentrations were stable with minor variations across the 15 seasons of the five years.

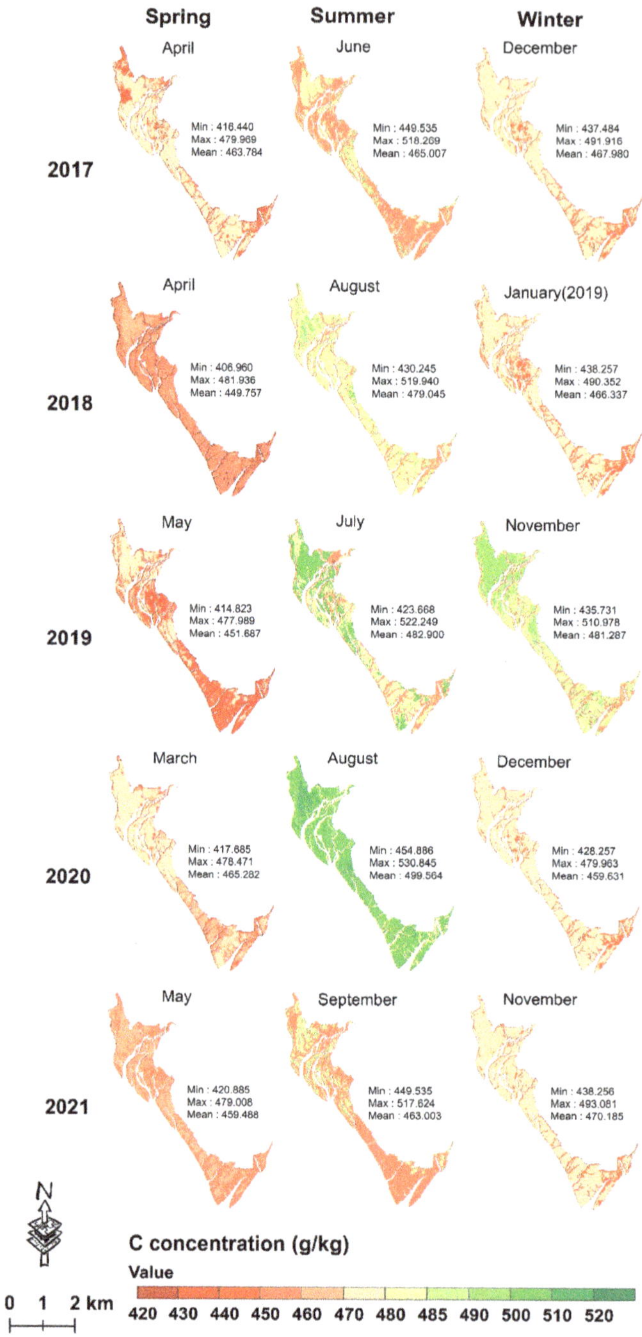

Figure 8. Seasonal leaf C mapping from 2017 to 2021 using the XGBoost model with the dataset of a single season.

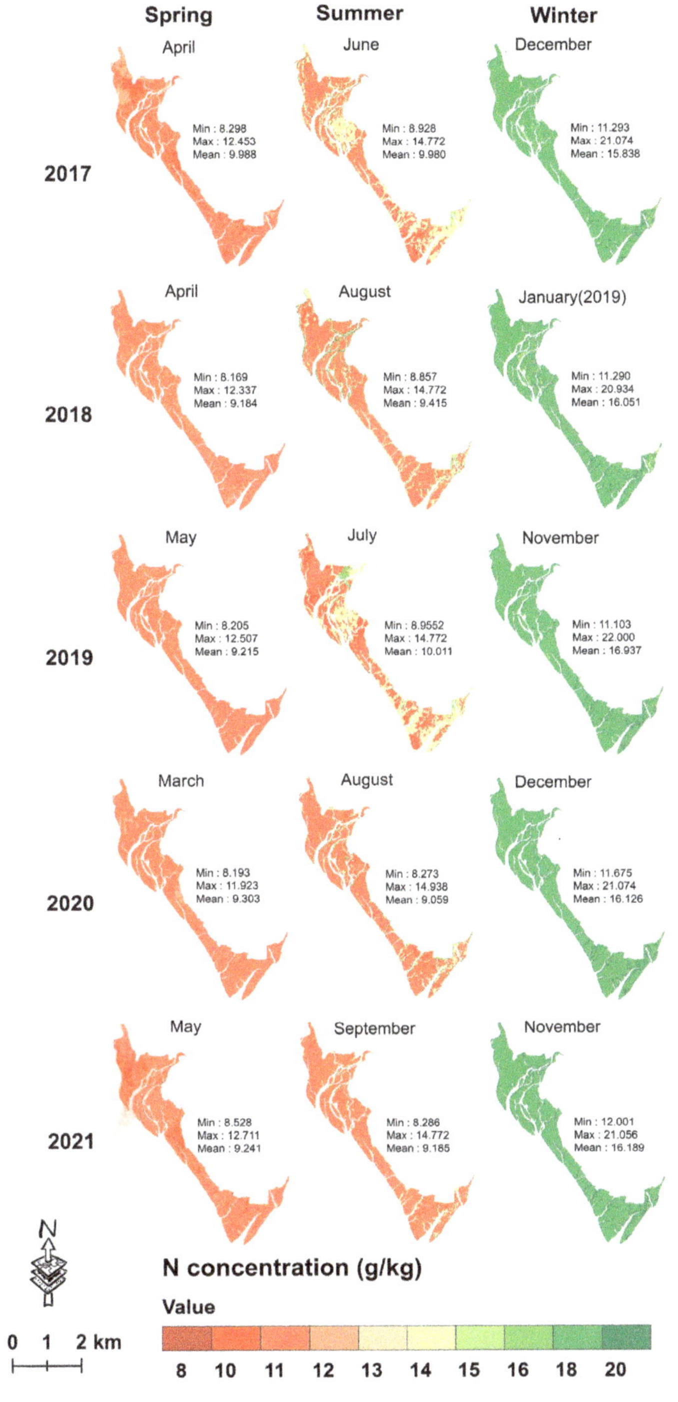

Figure 9. Seasonal leaf N mapping from 2017 to 2021 using the XGBoost model with the dataset of a single season.

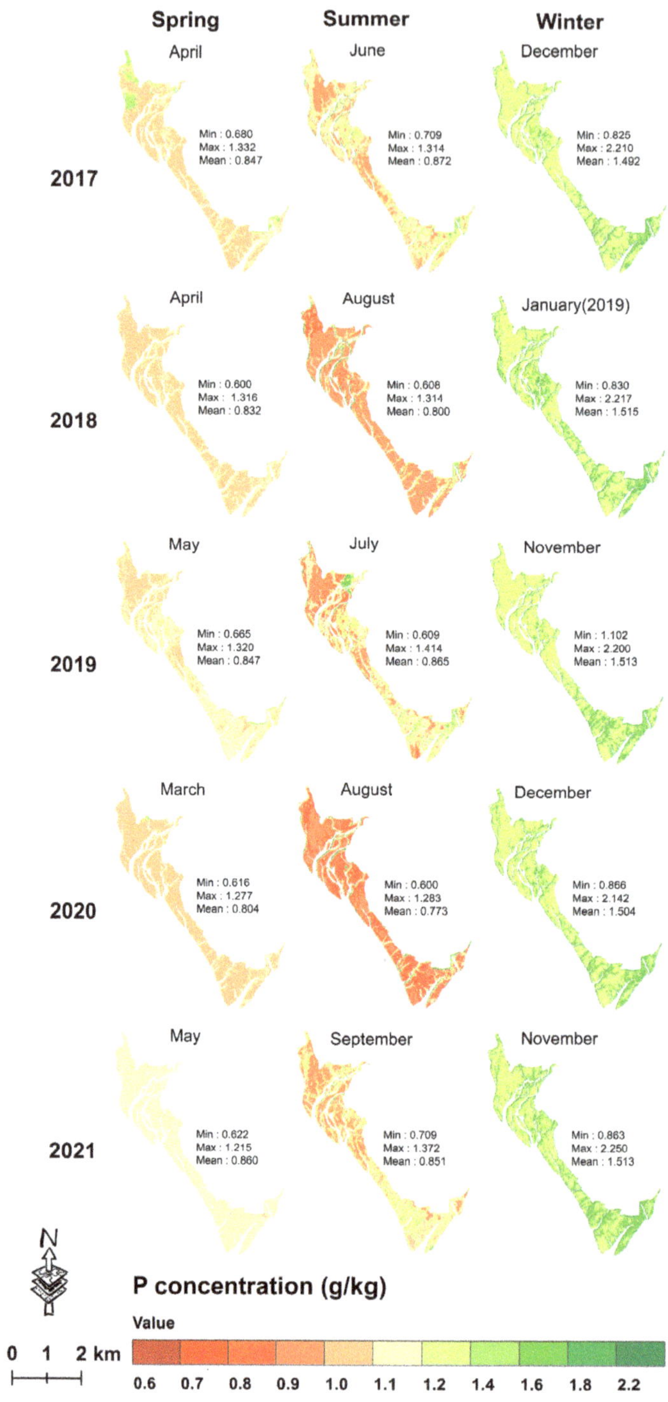

Figure 10. Seasonal leaf P mapping from 2017 to 2021 using the XGBoost model with the dataset of a single season.

4. Discussion

4.1. The Stoichiometry of Mangrove Leaf Nutrients across Different Seasons

The mean field-measured mangrove leaf C concentration of three seasons in Gaoqiao Mangrove Reserve was 466.33 g/kg, which is slightly higher than the globally averaged value of 492 terrestrial woody plants (464 g/kg) [51] and 11.53% higher than that of global coastal wetland plants (418.1 g/kg) [52]. The mean field-measured leaf N (12.13 g/kg) and P (1.13 g/kg) are 8.13–30.29% lower than the mean value of global coastal wetland vegetation (N = 16.1 g/kg, P = 1.6 g/kg) [52] and global terrestrial plants (N = 17.4 g/kg, P = 1.23 g/kg) [51]. These differences may be related to species types, phenology, and/or sampling strategies [53]. Such a comparison further confirms the fact that mangrove ecosystems have a strong C sequestration ability [54]. Moreover, the lower N/P ratios (<14) of the three seasons indicate that mangrove growth might be limited by N [55].

The seasonal trends of mapped leaf N and P concentration were similar (Figure 6) due to the strong correlation between them (Figure 3). Moreover, seasonality significantly affected the concentration of leaf N and P (Table 7), which agrees with the findings of Qin et al. [56]. However, Milla et al. [57] claimed that there was no significant correlation between the leaf N and P of woody plants, and Liu et al. [58] found that there was no strong seasonal variation in the leaf nutrients of *S. salsa* in the Yellow River Delta wetland. These results suggest that the nutrient utilization patterns could be affected by a variety of factors, such as climatic conditions, altitude, tidal levels, soil components, and species composition [59,60].

We found that leaf C was negatively correlated with N and P, and N was strongly correlated with P, which is in agreement with the findings of Michaels [61]. Such correlations also reflect the N and P utilization strategies in the C fixation process [62]. The fixation of C in the plant metabolism requires the participation of proteases (N storage), and the assembly of proteases requires the replication of nucleic acids (P storage) [61].

4.2. Sensitive Features Related to Mangrove Leaf Nutrients

The importance score ranking results (Figure 4) demonstrated that the red-edge bands (B5, B6, and B7, 705–783 nm) and near-infrared bands (B8 and B8a, 842–865 nm) performed better than other Sentinel-2 bands in estimating mangrove leaf nutrients, and the most sensitive VIs to leaf nutrients were mainly constructed by these bands. Moreover, B6 (740 nm) showed a higher correlation with the three nutrients than B5 (705 nm) and B7 (783 nm) (Figure 3), suggesting the superiority of B6 in correlating with mangrove leaf nutrients in different seasons, which agrees with the findings of Zhang et al. [63]. Many studies also demonstrated that red-edge bands are sensitive to leaf N and P across various plant species [13]. Moreover, the near-infrared bands are always used for developing NDVI, and the bands are less susceptible to saturation at high LAI and insensitive to unhealthy vegetation [12].

The simple ratio index (SRa, SRb, and SRchl) constructed by the ratio of two or three bands from B3, B4, B5, and B8 played an important role in estimating the three nutrients (Figure 4). Moreover, MCARI and its ratio with OSAVI were also sensitive to leaf C and N, and they have been widely used in the estimation of leaf chlorophyll due to the effective resistance to background interference and sensitivity to LAI saturation [34]. Several studies claimed that leaf chlorophyll is strongly correlated with leaf N [64], and leaf N and P had a close correlation, suggesting that chlorophyll content might be closely related to leaf C, N, and P estimation.

4.3. The Advantage of XGBoost in Estimating Mangrove Leaf Nutrients

Among the three machine learning models, in most cases, the XGBoost model was found to be optimal in estimating seasonal leaf nutrients using two modeling strategies (Tables 8 and 9). To our knowledge, this study was the first to estimate leaf C, N, and P in a mangrove forest using the XGBoost method and seasonal Sentinel-2 images. According to the interpretation of RPD [65], the XGBoost model using the dataset of a single season had approximate quantitative estimations (RPD = 2.0–2.5) of leaf C in summer and winter and reported the possibility of distinguishing between high and low values of leaf C in spring and leaf N and P in three seasons (RPD = 1.5–2.0).

In most cases, it is difficult to obtain many sampling plots (15 m × 15 m) in mangrove forests due to the rough field accessibility, leading to a limited data range of leaf nutrients and underestimation of the concentrations of leaf nutrients to some extent. Hence, inevitable sparse sampling and the relatively weak spectral information of leaf N and P with relatively low concentrations are the greatest challenges to the accurate mapping of leaf nutrients in mangrove forests. Compared to other machine learning algorithms (e.g., RF), XGBoost uses a sparse-aware split lookup method to process sparse data, which is practical when dealing with the sparse sampling in a mangrove forest. Our results agree with the findings of Tian et al. [20] and Mohammadi et al. [66], who also claimed that XGBoost outperformed RF and LightGBM in estimating grapevine leaf N and hydrogen solubility in hydrocarbons. The reason might be that XGBoost is more capable of solving the problems of feature selection, overfitting, hyperparameter tuning, and local optimality [67].

4.4. Limitation of Leaf Nutrients Estimation with Seasonal Sentinel-2 Images

We found that the model performance in estimating leaf nutrients was weaker in spring than in summer and winter, and the model performance was extremely poor in spring with the pooled dataset of three seasons (Table 9). One possible reason might be that the field sampling in April 2018 covered a smaller portion of the study area (Figure 1) with a narrower data range of leaf nutrients (Figures 5 and 6). Hence, it is necessary to improve the field sampling strategy with more sampling plots and larger sampling areas to increase the model performance and transferability.

Though Sentinel-2 images could provide a convenient way to monitor seasonal leaf nutrients, the 10 m pixels of Sentinel-2 images largely contain more than one species, and a pixel with low coverage or nearshore mangroves is also affected by sediment and seawater, which could influence the spectral features of mangroves and further lead to errors in the estimation of leaf nutrients. We mapped the leaf nutrients of 15 seasons from 2017 to 2021; however, the extended XGBoost model was developed by only one image of a single season, which might lead to a lack of field validation of the mapped results for other years.

5. Conclusions

We compared three machine learning models to estimate mangrove leaf C, N, and P with Sentinel-2 images in spring, summer, and winter, and the best-performing model was extended to map leaf nutrients of 15 seasons from 2017 to 2021. The main conclusions could be drawn as follows:

(1) The XGBoost method had great potential for accurate estimation of mangrove leaf nutrients with seasonal Sentinel-2 images.
(2) Among the three nutrients, leaf C concentrations were the most accurately estimated, followed by leaf N and P.
(3) Red-edge (especially B6) and near-infrared bands (B8 and B8a) of Sentinel-2 images were efficient estimators of mangrove leaf nutrients.

Long-time seasonal monitoring of leaf nutrients could facilitate an understanding of the dynamic variation of C fixation, nutrient utilization, and growth status of mangrove forests. To achieve efficient monitoring with time series Sentinel-2 images, it is necessary to establish an inversion model of leaf nutrients with high accuracy and strong transferability. In future work, the species composition, elevation, LAI, and canopy height with more sampling plots in wider areas will be incorporated into the present model to improve model accuracy and applicability.

Author Contributions: Conceptualization and methodology, J.W. and J.M.; validation, J.W., J.M. and J.Z.; formal analysis, J.W., J.M. and J.Z.; investigation, J.W.; resources, J.W., J.M., J.Z., D.Z., X.J., Z.S., C.G. and G.W.; writing—original draft preparation, J.M.; writing—review and editing, J.W.;

supervision, G.W.; project administration, J.W.; funding acquisition, J.W. All authors have read and agreed to the published version of the manuscript.

Funding: This research was funded by the Guangdong Basic and Applied Basic Research Foundation (No. 2019A1515010741) and the Shenzhen Science and Technology Program (No. JCYJ20210324093210029).

Data Availability Statement: Not applicable.

Conflicts of Interest: The authors declare no conflict of interest. The funders had no role in the design of the study; in the collection, analyses, or interpretation of data; in the writing of the manuscript; or in the decision to publish the results.

References

1. Xing, K.X.; Zhao, M.F.; Niinemets, U.; Niu, S.L.; Tian, J.; Jiang, Y.; Chen, H.Y.H.; White, P.J.; Guo, D.L.; Ma, Z.Q. Relationships between Leaf Carbon and Macronutrients across Woody Species and Forest Ecosystems Highlight How Carbon Is Allocated to Leaf Structural Function. *Front. Plant Sci.* **2021**, *12*, 674932. [CrossRef] [PubMed]
2. Leghari, S.J.; Wahocho, N.A.; Laghari, G.M.; HafeezLaghari, A.; MustafaBhabhan, G.; HussainTalpur, K.; Bhutto, T.A.; Wahocho, S.A.; Lashari, A.A. Role of nitrogen for plant growth and development: A review. *Adv. Environ. Biol.* **2016**, *10*, 209–219.
3. Friess, D.A.; Rogers, K.; Lovelock, C.E.; Krauss, K.W.; Hamilton, S.E.; Lee, S.Y.; Lucas, R.; Primavera, J.; Rajkaran, A.; Shi, S.H. The State of the World's Mangrove Forests: Past, Present, and Future. *Annu. Rev. Environ. Resour.* **2019**, *44*, 89–115. [CrossRef]
4. McLeod, E.; Chmura, G.L.; Bouillon, S.; Salm, R.; Bjork, M.; Duarte, C.M.; Lovelock, C.E.; Schlesinger, W.H.; Silliman, B.R. A blueprint for blue carbon: Toward an improved understanding of the role of vegetated coastal habitats in sequestering CO_2. *Front. Ecol. Environ.* **2011**, *9*, 552–560. [CrossRef]
5. Chen, H.; Xu, B.B.; Wei, S.D.; Zhang, L.H.; Zhou, H.C.; Lin, Y.M. Nutrient Resorption and Phenolics Concentration Associated with Leaf Senescence of the Subtropical Mangrove Aegiceras corniculatum: Implications for Nutrient Conservation. *Forests* **2016**, *7*, 290. [CrossRef]
6. Howarth, R.W.; Marino, R. Nitrogen as the limiting nutrient for eutrophication in coastal marine ecosystems: Evolving views over three decades. *Limnol. Oceanogr.* **2006**, *51 Pt 2*, 364–376. [CrossRef]
7. Kuenzer, C.; Bluemel, A.; Gebhardt, S.; Quoc, T.V.; Dech, S. Remote Sensing of Mangrove Ecosystems: A Review. *Remote Sens.* **2011**, *3*, 878–928. [CrossRef]
8. Guo, X.X.; Wang, M.; Jia, M.M.; Wang, W.Q. Estimating mangrove leaf area index based on red-edge vegetation indices: A comparison among UAV, WorldView-2 and Sentinel-2 imagery. *Int. J. Appl. Earth Obs. Geoinf.* **2021**, *103*, 102493. [CrossRef]
9. Jiang, X.; Zhen, J.; Miao, J.; Zhao, D.; Shen, Z.; Jiang, J.; Gao, C.; Wu, G.; Wang, J. Newly-developed three-band hyperspectral vegetation index for estimating leaf relative chlorophyll content of mangrove under different severities of pest and disease. *Ecol. Indic.* **2022**, *140*, 108978. [CrossRef]
10. Pham, L.T.H.; Brabyn, L. Monitoring mangrove biomass change in Vietnam using SPOT images and an object-based approach combined with machine learning algorithms. *ISPRS J. Photogramm. Remote Sens.* **2017**, *128*, 86–97. [CrossRef]
11. Curran, P.J. Remote sensing of foliar chemistry. *Remote Sens. Environ.* **1989**, *30*, 271–278. [CrossRef]
12. Daughtry, C.S.T.; Walthall, C.L.; Kim, M.S.; de Colstoun, E.B.; McMurtrey, J.E. Estimating Corn Leaf Chlorophyll Concentration from Leaf and Canopy Reflectance. *Remote Sens. Environ.* **2000**, *74*, 229–239. [CrossRef]
13. Clevers, J.; Gitelson, A.A. Remote estimation of crop and grass chlorophyll and nitrogen content using red-edge bands on Sentinel-2 and-3. *Int. J. Appl. Earth Obs. Geoinf.* **2013**, *23*, 344–351. [CrossRef]
14. Diaz, B.M.; Blackburn, G.A. Remote sensing of mangrove biophysical properties: Evidence from a laboratory simulation of the possible effects of background variation on spectral vegetation indices. *Int. J. Remote Sens.* **2003**, *24*, 53–73. [CrossRef]
15. Xiao, H.; Su, F.Z.; Fu, D.J.; Lyne, V.; Liu, G.H.; Pan, T.T.; Teng, J.K. Optimal and robust vegetation mapping in complex environments using multiple satellite imagery: Application to mangroves in Southeast Asia. *Int. J. Appl. Earth Obs. Geoinf.* **2021**, *99*, 102320. [CrossRef]
16. Pastor-Guzman, J.; Atkinson, P.M.; Dash, J.; Rioja-Nieto, R. Spatiotemporal Variation in Mangrove Chlorophyll Concentration Using Landsat 8. *Remote Sens.* **2015**, *7*, 14530–14558. [CrossRef]
17. Younes, N.; Joyce, K.E.; Northfield, T.D.; Maier, S.W. The effects of water depth on estimating Fractional Vegetation Cover in mangrove forests. *Int. J. Appl. Earth Obs. Geoinf.* **2019**, *83*, 101924. [CrossRef]
18. Annala, L.; Ayramo, S.; Polonen, I. Comparison of Machine Learning Methods in Stochastic Skin Optical Model Inversion. *Appl. Sci.* **2020**, *10*, 7097. [CrossRef]
19. Chlingaryan, A.; Sukkarieh, S.; Whelan, B. Machine learning approaches for crop yield prediction and nitrogen status estimation in precision agriculture: A review. *Comput. Electron. Agric.* **2018**, *151*, 61–69. [CrossRef]
20. Tian, Y.C.; Huang, H.; Zhou, G.Q.; Zhang, Q.; Tao, J.; Zhang, Y.L.; Lin, J.L. Aboveground mangrove biomass estimation in Beibu Gulf using machine learning and UAV remote sensing. *Sci. Total Environ.* **2021**, *781*, 146816. [CrossRef]
21. Yang, X.; Yang, R.; Ye, Y.; Yuan, Z.R.; Wang, D.Z.; Hua, K.K. Winter wheat SPAD estimation from UAV hyperspectral data using cluster-regression methods. *Int. J. Appl. Earth Obs. Geoinf.* **2021**, *105*, 102618. [CrossRef]

22. Blair, G.J.; Lefroy, R.D.B.; Lise, L. Soil carbon fractions based on their degree of oxidation, and the development of a carbon management index for agricultural systems. *Aust. J. Agric. Res.* **1995**, *46*, 1459–1466. [CrossRef]
23. Hiller, A.; Plazin, J.; Vanslyke, D.D. A study of conditions for Kjeldahl determination of nitrogen in proteins; description of methods with mercury as catalyst, and titrimetric and gasometric measurements of the ammonia formed. *J. Biol. Chem.* **1948**, *176*, 1401–1420. [CrossRef]
24. Harborne, J.B. (Ed.) Methods of Plant Analysis. In *Phytochemical Methods: A Guide to Modern Techniques of Plant Analysis*; Springer: Dordrecht, The Netherlands, 1984; pp. 1–36.
25. Brodu, N. Super-Resolving Multiresolution Images With Band-Independent Geometry of Multispectral Pixels. *IEEE Trans. Geosci. Electron.* **2017**, *55*, 4610–4617. [CrossRef]
26. El-Mejjaouy, Y.; Lahrir, M.; Naciri, R.; Zeroual, Y.; Mercatoris, B.; Dumont, B.; Oukarroum, A. How far can chlorophyll a fluorescence detect phosphorus status in wheat leaves (*Triticum durum* L.). *Environ. Exp. Bot.* **2022**, *194*, 104762. [CrossRef]
27. Datt, B. Remote Sensing of Chlorophyll *a*, Chlorophyll *b*, Chlorophyll *a+b*, and Total Carotenoid Content in Eucalyptus Leaves. *Remote Sens. Environ.* **1998**, *66*, 111–121. [CrossRef]
28. Rondeaux, G.; Steven, M.; Baret, F. Optimization of soil-adjusted vegetation indices. *Remote Sens. Environ.* **1996**, *55*, 95–107. [CrossRef]
29. Horler, D.N.H.; Dockray, M.; Barber, J. The red edge of plant leaf reflectance. *Int. J. Remote Sens.* **2007**, *4*, 273–288. [CrossRef]
30. Gitelson, A.A.; Gritz, Y.; Merzlyak, M.N. Relationships between leaf chlorophyll content and spectral reflectance and algorithms for non-destructive chlorophyll assessment in higher plant leaves. *J. Plant Physiol.* **2003**, *160*, 271–282. [CrossRef]
31. Huete, A.R.; Liu, H.Q.; Batchily, K.; van Leeuwen, W. A comparison of vegetation indices global set of TM images for EOS-MODIS. *Remote Sens. Environ.* **1997**, *59*, 440–451. [CrossRef]
32. Haboudane, D. Hyperspectral vegetation indices and novel algorithms for predicting green LAI of crop canopies: Modeling and validation in the context of precision agriculture. *Remote Sens. Environ.* **2004**, *90*, 337–352. [CrossRef]
33. Haboudane, D.; Miller, J.R.; Tremblay, N.; Zarco-Tejada, P.J.; Dextraze, L. Integrated narrow-band vegetation indices for prediction of crop chlorophyll content for application to precision agriculture. *Remote Sens. Environ.* **2002**, *81*, 416–426. [CrossRef]
34. Wu, C.; Niu, Z.; Tang, Q.; Huang, W. Estimating chlorophyll content from hyperspectral vegetation indices: Modeling and validation. *Agric. For. Meteorol.* **2008**, *148*, 1230–1241. [CrossRef]
35. Baret, F.; Guyot, G.; Begue, A.; Maurel, P.; Podaire, A. Complementarity of middle-infrared with visible and near-infrared reflectance for monitoring wheat canopies. *Remote Sens. Environ.* **1988**, *26*, 213–225. [CrossRef]
36. Blackburn, G.A. Spectral indices for estimating photosynthetic pigment concentrations: A test using senescent tree leaves. *Int. J. Remote Sens.* **2010**, *19*, 657–675. [CrossRef]
37. Gitelson, A.A.; Vina, A.; Ciganda, V.; Rundquist, D.C.; Arkebauer, T.J. Remote estimation of canopy chlorophyll content in crops. *Geophys. Res. Lett.* **2005**, *32*, L08403. [CrossRef]
38. Galvão, L.S.; Formaggio, A.R.; Tisot, D.A. Discrimination of sugarcane varieties in Southeastern Brazil with EO-1 Hyperion data. *Remote Sens. Environ.* **2005**, *94*, 523–534. [CrossRef]
39. Hunt, E.R.; Rock, B.N. Detection of changes in leaf water content using near-and middle-infrared reflectances. *Remote Sens. Environ.* **1989**, *30*, 43–54.
40. Zarcotejada, P.; Berjon, A.; Lopezlozano, R.; Miller, J.; Martin, P.; Cachorro, V.; Gonzalez, M.; Defrutos, A. Assessing vineyard condition with hyperspectral indices: Leaf and canopy reflectance simulation in a row-structured discontinuous canopy. *Remote Sens. Environ.* **2005**, *99*, 271–287. [CrossRef]
41. Carter, G.A. Ratios of leaf reflectances in narrow wavebands as indicators of plant stress. *Int. J. Remote Sens.* **2007**, *15*, 697–703. [CrossRef]
42. Fang, H.L.; Liu, W.W.; Li, W.J.; Wei, S.S. Estimation of the directional and whole apparent clumping index (ACI) from indirect optical measurements. *ISPRS J. Photogramm. Remote Sens.* **2018**, *144*, 1–13. [CrossRef]
43. Imran, A.B.; Khan, K.; Ali, N.; Ahmad, N.; Ali, A.; Shah, K. Narrow band based and broadband derived vegetation indices using Sentinel-2 Imagery to estimate vegetation biomass. *Glob. J. Environ. Sci. Manag.* **2020**, *6*, 97–108.
44. Darmawan, S.; Takeuchi, W.; Vetrita, Y.; Winarso, G.; Wikantika, K.; Sari, D.K. Iop Characterization. In Proceedings of the 7th IGRSM International Remote Sensing and GIS Conference and Exhibition, Kuala Lumpur, Malaysia, 22–23 April 2014.
45. Jia, M.; Wang, Z.; Wang, C.; Mao, D.; Zhang, Y. A New Vegetation Index to Detect Periodically Submerged Mangrove Forest Using Single-Tide Sentinel-2 Imagery. *Remote Sens.* **2019**, *11*, 2043. [CrossRef]
46. Chen, T.; Guestrin, C. XGBoost: A Scalable Tree Boosting System. In Proceedings of the 22nd ACM SIGKDD International Conference on Knowledge Discovery and Data Mining, San Francisco, CA, USA, 13–17 August 2016; pp. 785–794.
47. Sheridan, R.P.; Wang, W.M.; Liaw, A.; Ma, J.S.; Gifford, E.M. Extreme Gradient Boosting as a Method for Quantitative Structure-Activity Relationships. *J. Chem. Inf. Model.* **2016**, *56*, 2353–2360. [CrossRef] [PubMed]
48. Cui, Y.; Meng, F.; Fu, P.J.; Yang, X.Y.; Zhang, Y.X.; Liu, P.D. Application of hyperspectral analysis of chlorophyll a concentration inversion in Nansi Lake. *Ecol. Inform.* **2021**, *64*, 101360. [CrossRef]
49. Yuan, Z.R.; Ye, Y.; Wei, L.F.; Yang, X.; Huang, C. Study on the Optimization of Hyperspectral Characteristic Bands Combined with Monitoring and Visualization of Pepper Leaf SPAD Value. *Sensors* **2022**, *22*, 183. [CrossRef] [PubMed]

50. Ke, G.L.; Meng, Q.; Finley, T.; Wang, T.F.; Chen, W.; Ma, W.D.; Ye, Q.W.; Liu, T.Y. LightGBM: A Highly Efficient Gradient Boosting Decision Tree. In Proceedings of the 31st Annual Conference on Neural Information Processing Systems (NIPS), Long Beach, CA, USA, 4–9 December 2017.
51. Elser, J.J.; Fagan, W.F.; Denno, R.F.; Dobberfuhl, D.R.; Folarin, A.; Huberty, A.; Interlandi, S.; Kilham, S.S.; McCauley, E.; Schulz, K.L.; et al. Nutritional constraints in terrestrial and freshwater food webs. *Nature* **2000**, *408*, 578–580. [CrossRef]
52. Hu, Y.K.; Liu, X.Y.; He, N.P.; Pan, X.; Long, S.Y.; Li, W.; Zhang, M.Y.; Cui, L.J. Global patterns in leaf stoichiometry across coastal wetlands. *Glob. Ecol. Biogeogr.* **2021**, *30*, 852–869. [CrossRef]
53. Abbas, M.; Ebeling, A.; Oelmann, Y.; Ptacnik, R.; Roscher, C.; Weigelt, A.; Weisser, W.W.; Wilcke, W.; Hillebrand, H. Biodiversity Effects on Plant Stoichiometry. *PLoS ONE* **2013**, *8*, e58179. [CrossRef]
54. Duarte, C.M.; Cebrian, J. The fate of marine autotrophic production. *Limnol. Oceanogr.* **1996**, *41*, 1758–1766. [CrossRef]
55. Koerselman, W.; Meuleman, A.F.M. The vegetation N:P ratio: A new tool to detect the nature of nutrient limitation. *J. Appl. Ecol.* **1996**, *33*, 1441–1450. [CrossRef]
56. Qin, J.; Shangguan, Z.; Xi, W. Seasonal variations of leaf traits and drought adaptation strategies of four common woody species in South Texas, USA. *J. For. Res.* **2019**, *30*, 1715–1725. [CrossRef]
57. Milla, R.; Maestro-Martinez, M.; Montserrat-Marti, G. Seasonal branch nutrient dynamics in two Mediterranean woody shrubs with contrasted phenology. *Ann. Bot.* **2004**, *93*, 671–680. [CrossRef] [PubMed]
58. Liu, F.D.; Liu, Y.H.; Wang, G.M.; Song, Y.; Liu, Q.; Li, D.S.; Mao, P.L.; Zhang, H. Seasonal Variations of C: N: P Stoichiometry and Their Trade-offs in Different Organs of Suaeda salsa in Coastal Wetland of Yellow River Delta, China. *PLoS ONE* **2015**, *10*, e0138169. [CrossRef] [PubMed]
59. Almahasheer, H.; Duarte, C.M.; Irigoien, X. Nutrient Limitation in Central Red Sea Mangroves. *Front. Mar. Sci.* **2016**, *3*, 271. [CrossRef]
60. Ma, S.H.; He, F.; Tian, D.; Zou, D.T.; Yan, Z.B.; Yang, Y.L.; Zhou, T.C.; Huang, K.Y.; Shen, H.H.; Fang, J.Y. Variations and determinants of carbon content in plants: A global synthesis. *Biogeosciences* **2018**, *15*, 693–702. [CrossRef]
61. Michaels, A.F. Ecological stoichiometry—The biology of elements from molecules to the biosphere. *Science* **2003**, *300*, 906–907. [CrossRef]
62. Wright, I.J.; Reich, P.B.; Westoby, M.; Ackerly, D.D.; Baruch, Z.; Bongers, F.; Cavender-Bares, J.; Chapin, T.; Cornelissen, J.H.C.; Diemer, M.; et al. The worldwide leaf economics spectrum. *Nature* **2004**, *428*, 821–827. [CrossRef]
63. Zhang, C.H.; Kovacs, J.M.; Wachowiak, M.P.; Flores-Verdugo, F. Relationship between Hyperspectral Measurements and Mangrove Leaf Nitrogen Concentrations. *Remote Sens.* **2013**, *5*, 891–908. [CrossRef]
64. Zhen, J.; Jiang, X.; Xu, Y.; Miao, J.; Zhao, D.; Wang, J.; Wang, J.; Wu, G. Mapping leaf chlorophyll content of mangrove forests with Sentinel-2 images of four periods. *Int. J. Appl. Earth Obs. Geoinf.* **2021**, *102*, 102387. [CrossRef]
65. Saeys, W.; Mouazen, A.M.; Ramon, H. Potential for Onsite and Online Analysis of Pig Manure using Visible and Near Infrared Reflectance Spectroscopy. *Biosyst. Eng.* **2005**, *91*, 393–402. [CrossRef]
66. Mohammadi, M.R.; Hadavimoghaddam, F.; Pourmahdi, M.; Atashrouz, S.; Munir, M.T.; Hemmati-Sarapardeh, A.; Mosavi, A.H.; Mohaddespour, A. Modeling hydrogen solubility in hydrocarbons using extreme gradient boosting and equations of state. *Sci. Rep.* **2021**, *11*, 17911. [CrossRef] [PubMed]
67. Jia, Y.; Jin, S.G.; Savi, P.; Gao, Y.; Tang, J.; Chen, Y.X.; Li, W.M. GNSS-R Soil Moisture Retrieval Based on a XGboost Machine Learning Aided Method: Performance and Validation. *Remote Sens.* **2019**, *11*, 1655. [CrossRef]

remote sensing

MDPI

Article

Carbon Stock Prediction in Managed Forest Ecosystems Using Bayesian and Frequentist Geostatistical Techniques and New Generation Remote Sensing Metrics

Tsikai Solomon Chinembiri [1,*], Onisimo Mutanga [1] and Timothy Dube [2]

[1] College of Agricultural, School of Agricultural Earth and Environmental Sciences, University of KwaZulu-Natal, Private Bag X01, Pietermaritzburg 3209, South Africa
[2] Institute of Water Studies, Department of Earth Sciences, University of the Western Cape, Private Bag X17, Bellville 7535, South Africa
* Correspondence: chinembiri24500@alumni.itc.nl

Abstract: The study compares the performance of a hierarchical Bayesian geostatistical methodology with a frequentist geostatistical approach, specifically, Kriging with External Drift (KED), for predicting C stock using prediction aides from the Landsat-8 and Sentinel-2 multispectral remote sensing platforms. The frequentist geostatistical approach's reliance on the long-run frequency of repeated experiments for constructing confidence intervals is not always practical or feasible, as practitioners typically have access to a single dataset due to cost constraints on surveys and sampling. We evaluated two approaches for C stock prediction using two new generation multispectral remote sensing datasets because of the inherent uncertainty characterizing spatial prediction problems in the unsampled locations, as well as differences in how the Bayesian and frequentist geostatistical paradigms handle uncertainty. Information on C stock spectral prediction in the form of NDVI, SAVI, and EVI derived from multispectral remote sensing platforms, Landsat-8 and Sentinel-2, was used to build Bayesian and frequentist-based C stock predictive models in the sampled plantation forest ecosystem. Sentinel-2-based C stock predictive models outperform their Landsat-8 counterparts using both the Bayesian and frequentist inference approaches. However, the Bayesian-based Sentinel-2 C stock predictive model ($RMSE = 0.17$ MgCha^{-1}) is more accurate than its frequentist-based Sentinel-2 ($RMSE = 1.19$ MgCha^{-1}) C stock equivalent. The Sentinel-2 frequentist-based C stock predictive model gave the C stock prediction range of $1 \leq$ MgCha^{-1} ≤ 290, whilst the Sentinel-2 Bayesian-based C stock predictive model resulted in the prediction range of $1 \leq$ MgCha^{-1} ≤ 285. However, both the Bayesian and frequentist C stock predictive models built with the Landsat-8 sensor overpredicted the sampled C stock because the range of predicted values fell outside the range of the observed C stock values. As a result, we recommend and conclude that the Bayesian-based C stock prediction method, when it is combined with high-quality remote sensing data such as that of Sentinel-2, is an effective inferential statistical methodology for reporting C stock in managed plantation forest ecosystems.

Keywords: Bayesian methodology; classical geostatistics; multispectral remote sensing; carbon stock; plantation forest; managed ecosystem

1. Introduction

Plantation and natural forests serve as one of the main economic pillars in sub-Saharan Africa because they support both economic growth and human livelihoods. Due to the vulnerabilities caused by climate change, economies in Sub-Saharan Africa and other regions of the world must balance the need to protect the environment against the economic pressures brought on by population growth and poverty [1]. Nearly 50% of Africans live without access to electricity, and at least 60% of them still rely on wood for cooking and heating [2]. This disproportionate reliance on climate-vulnerable industries such as energy and agriculture for economic survival and expansion leads to significant increases in the supply of primary energy and greenhouse gas emissions [3]. Forest biomass is regarded as a crucial part of monitoring forest resources according to the Food and Agriculture Organization's (FAO) 1994 International Forest Resources Monitoring Program [4]. Therefore, accurate monitoring and estimation of the forest's aboveground biomass (AGB) at the local and regional scales are essential for understanding how the forest's AGB contributes to the regional and global carbon cycles [5].

Zimbabwe is one of several African countries that, in their Nationally Determined Contributions (NDCs) to the Paris Agreement, have made bold proposals for establishing low-carbon and climate-resilient economies [3]. Yet Agriculture, Forestry and Other Land Use (AFOLU) still remain the biggest contributors of Greenhouse Gas Emissions (GHG), accounting for 54% of GHG in 2017 in Zimbabwe [6]. Deforestation resulting from agriculture expansion, increased stocking levels, the fetching of fuelwood, veld fires, the harvesting of timber for construction, mining, illegal settlements, tobacco curing, charcoal making, and commercial logging are some of the major drivers of GHG emissions in the AFOLU sector [7]. The Zimbabwean government also introduced the Greenhouse Gas Abatement Cost Model (GACMO) for establishing a GHG database [8]. The model can also be used as a tool for reporting, monitoring, and verifying transparency in mitigation actions for climate change.

The commonly used methods for estimating aboveground biomass (AGB) include mean biomass density, allometric equations, remote sensing, forest identity, geostatistics, and biomass expansion factors [9]. The method of statistical inquiry for any of these AGB estimation and prediction methods can be comprehensively categorised as either Bayesian or frequentist ones, depending on the circumstances of the investigation and assumptions underlying the inference. Notable differences between the Bayesian and the frequentist statistical methodologies regard the nature of the unknown parameters under investigation [10]. The frequentist paradigm treats parameters of interest as unknown and fixed ones, whilst the Bayesian framework regards all unknown parameters as uncertain ones, and therefore, should be characterised by a probability distribution [11].

The performance of Maximum Likelihood (ML), Least Squares and Bayesian approaches in Bogota, Columbia, was tested by Ghosh and Carriazo [12] in a hedonic estimation context and concluded that none of the aforementioned approaches are better than the other ones. However, because of the philosophical differences governing Bayesian and frequentist statistical techniques, the authors recommend the choice of the estimation technique to be grounded on the peculiarities of the policy problem at hand [12]. Some studies in the literature dwell at one of the two approaches as the principal methodology of inference. Notable work in the realm of biomass estimation using the Bayesian techniques include [13–16]. Recent studies assessing AGB distribution established accuracies of 17.52 Mg/ha [14] and 1.16 $MgCha^{-1}$ and 2.69 $MgCha^{-1}$ for Sentinel-2-based and Landsat-8-based Carbon (C) stock predictive models, respectively [16]. Other remote sensing- and machine learning-based efforts towards the estimation and prediction of AGB in recent times include those by Do et al. [17], who established mangrove AGB predictions in Vietnam at ranges from 6.51 to 368 $Mgha^{-1}$ and from 13.70 to 320.1 $Mgha^{-1}$ for remote sensing and Artificial Neural Networks, respectively.

Addressing a geostatistical research question from the Bayesian view point makes it possible to provide definitions of spatial predictors contributing to uncertainty in the unknown spatial covariance structure [18]. Kriging provided an optimal geostatistical technique under the frequentist paradigm that is employed in the description of spatial patterns and predicting values of a variable at unsampled locations and, consequently, evaluated the uncertainty associated with the predicted values [19].

Research making use of the frequentist geostatistical approach independent of the Bayesian technique for the estimation of C stock are also well documented in the literature. The authors of [20] mapped AGB in the Brazilian Amazon using Kriging with External Drift and established prediction accuracies for different sample sizes ranging from 0 to

110 for distances within 300 km radii from the prediction locations. The lowest RMSE for the estimated AGB for a sample size of 110 was 32.8 Mgha^{-1}, whilst the lowest accuracy for the lowest sample size of $n > 0$ was 48.06 Mgha^{-1}. To add to this, the authors of [21] predicted the AGB in a Wangyedia forest farm in China using Landsat-8 and the newly launched Landsat-9 datasets and arrived at RMSEs of 16.83 tha^{-1} and 17.91 tha^{-1}, respectively. The authors of [22] coupled remote sensing-derived Sentinel-2 explanatory variables with geostatistics and machine learning algorithms in Mayanmar for predicting the aboveground biomass and established accuracies of 24.91 Mgha^{-1} and 34.72 Mgha^{-1} for Random Forest-based ordinary kriging and Random Forest-based co-kriging, respectively. As demonstrated by the results of Jiang et al. [21], Landsat-8 built AGB estimation models can still be superior to AGB models built from the successor Landsat-9 sensor, despite the relative spectral and radiometric improvements in the latest Landsat-9 sensor.

It is worthwhile to explore avenues for improving AGB prediction and estimation accuracy through the adoption of befitting statistical methods of inference for the production of high-quality reports applicable to climate change mitigation and climate change action. A significantly higher proportion of studies assessing the predictive performance of Landsat-8 and Sentinel-2 in aboveground biomass estimation favour Sentinel-2 over Landsat-8, though the differences in prediction performances are insignificant [21–23]. On the other hand, studies that have employed the Bayesian spatial hierarchical technique coupled with remote sensing-derived ancillary data have always outperformed similar studies conducted using the frequentist approach for predicting the AGB [14,24–27].

From the frequentist viewpoint, the interpretation of a confidence interval (CI) is hypothesised based on Neyman understanding, where the CI gives a measure of uncertainty by taking into account the long-run frequency of replicated experiments [28,29]. This suggests that if a practitioner/forester gathers 500 datasets on tree *dbh* from independent trials for estimating the parameter of a C stock prediction model and constructs a, with a 97% confidence interval for the parameter estimate for each dataset, at least 97 of the CIs would be expected to contain the true (but fixed) unknown model parameter [28,29]. However, in most practical settings, foresters or practitioners would not have access to multiple datasets and usually have a single dataset, as it is rather costly and unfeasible to undertake multiple experiments. The constructed CI may or may not contain the true (but fixed) unknown model parameter. On the other hand, the Bayesian credibility interval is a more pragmatic proposition as the credible interval is built in a manner that guarantees that there is a certain probability associated with getting the true (but random) unknown model parameter [30,31]. By the same reasoning, if a practitioner estimates the C stock predictive model using a single tree *dbh* dataset and constructs a, with a 97% credible interval, there would be a 97% probability that the true (but random) unknown model parameter is contained within that credible interval [32].

Since uncertainty is inherent in spatial prediction problems at unsampled locations, the Bayesian approach handles it better than the frequentist approach does because it benefits from having access to the full posterior predictive distribution of the modelled variable [31,33,34]. The principled way in which Bayesian inference incorporates pre-experimental information in the form of priors and experimental data, combined with the real-world benefits of parameters of carbon accounting and climate change action, makes it worthwhile to compare the two inferential paradigms when one is using freely available and new generation remote sensing data. As a result, the current study is an extension of the current earth observation-based inferential techniques for C stock accounting in climate change adaptation and mitigation within the United Nations Framework Convention on Climate Change (UNFCCC). As previously hypothesised, we set out to determine whether the Bayesian inferential approach can handle the uncertainty inherent in spatial prediction phenomena better than its frequentist counterpart can.

2. Methods

2.1. Study Area

We undertook the study at Lot 75A of Nyanga Downs in the Manicaland province of the eastern highlands of Zimbabwe. The area of interest is in the Nyanga district of the aforementioned province and has dominant tree species comprised of *Pinus patula*, *Eucalyptus grandis*, and *Eucalyptus camaldulensis*. Dotted patches of the area lying between latitude 32°40′ E and 32°54′ E and longitude 18°10′12″ S and 18°25′4″ S as illustrated in Figure 1 have undergone changes in land use and are currently used for gold panning, agriculture, and grazing [35].

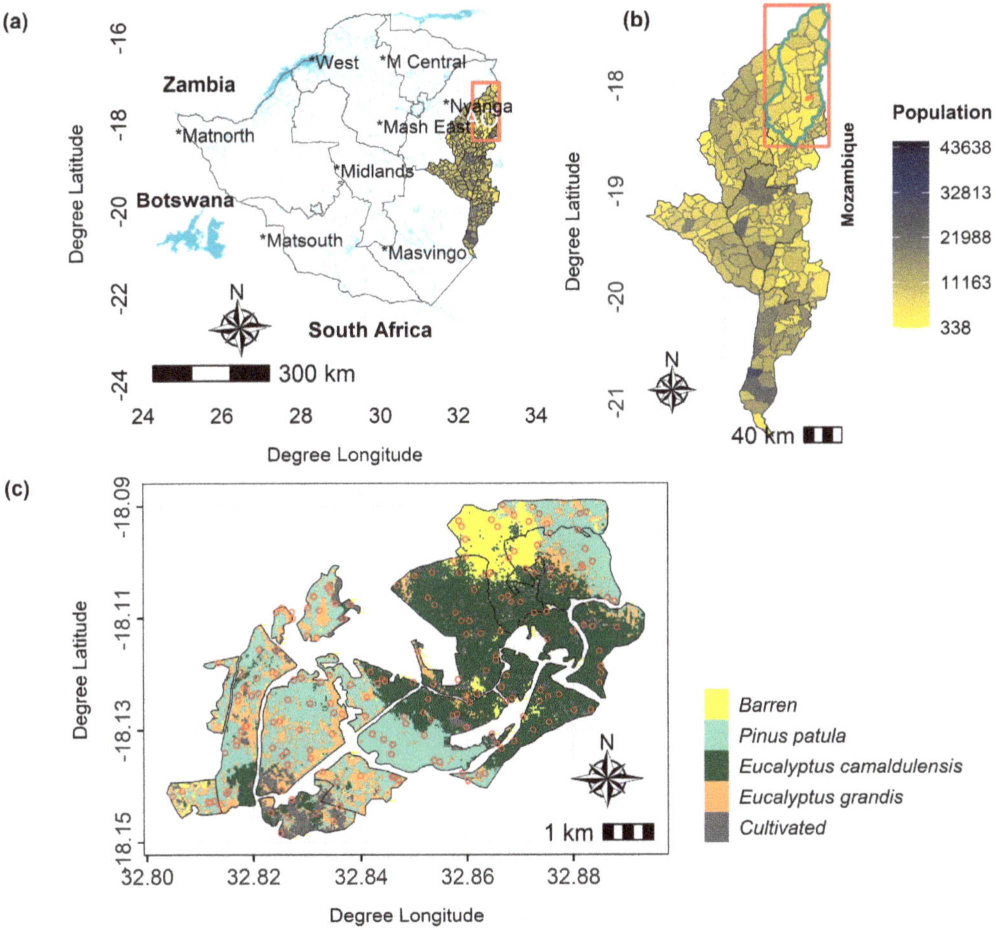

Figure 1. Study map showing (**a**) location of the province where samples for the study were collected, (**b**) study area location within the sampled province, and (**c**) plantation forest species distribution in the area of interest (AOI). * Shows provincial names.

Changes in land use to grazing and agriculture came after some parts of the commercially owned plantation forests were redistributed by the government of Zimbabwe to communal farmers in 2000. This development removed access barriers that were in place within the commercial plantation forests before the land redistribution, giving settlers more access to forest resources [36]. The area of study which covers an approximate area of

2767 ha, receives variable amounts of rainfall ranging from 741 mm to 2997 mm, and a mean annual precipitation of 1200 mm [37]. Annual mean temperatures are also varied, with a minimum annual range from 9 °C to 12 °C and a maximum range from 25 °C to 28 °C. Extensive wild fires occur at high altitudes due to the hot weather experienced during the summer season of the year between the months of August and November [38]. Rapid population growth (Figure 1b) within the jurisdiction of the studied region is also a possible factor driving rapid land use change, leading to veld fires and logging for opening up more land for agriculture.

2.2. *Remote Sensing Covariates*

The relatively low cost of acquiring Landsat and other freely available earth observation sensors, in addition to their spatial coverage, makes them indispensable for natural resource modelling. The recently launched Sentinel-2 satellite sensor is made up of six bands, which can be compared to the Landsat-8 bands, and also carries three additional bands comprising the red edge (RE) spectrum [39]. The red edge bands are positioned at 704, 740, and 782 nm, whose band widths are 15, 15, and 20 nm, respectively. The red edge forms the major spectral feature of vegetation located between the high reflectance in the NIR (750 nm) and the red absorption maximum (680 nm) [40]. Sentinel-2′s surface bands have 10 m and 20 m spatial resolutions as compared to Landsat-8′s 30 m bands. These differences in sensor configuration and properties form the basis for assessing their mapping accuracies.

Landsat OLI and Sentinel-2 MSI Imagery

We obtained Landsat-8 images from the United States Geological Survey Earth Explorer site as georeferenced and analysis ready data (ARD) (http://earthexplorer.usgs.gov, accessed on 13 February 2023). We filtered the datasets for cloud cover and set cloud shadow thresholds to below 10%. We downloaded Sentinel-2 cloud free images on the 20 September 2020, coinciding with the time that we acquired the Landsat-8 data, which covered the whole area of interest, including Lot 75A in the Nyanga Downs in the eastern highlands of Zimbabwe. Sentinel-2 imagery with 13 spectral bands was acquired as level-1C 12-bit fixed Top of the Atmosphere (TOA) reflectance values. We carried out the orthorectification and pre-processing of the Sentinel-2 level 1-C data using the *sen2r* package in the R Statistical and Computing Environment [41].

We derived the Enhanced Vegetation Index (EVI), Soil-Adjusted Vegetation Index ($SAVI$), and Normalised Difference Vegetation Index ($NDVI$) from each of the two sensors as independent variables for use in C stock prediction in a managed plantation forest in Zimbabwe. The authors of [40,41] employed the aforementioned vegetation indices as independent variables in AGB estimation and assessment. We therefore employ these variables using philosophically different statistical research methodologies in order to assess the best framework that can be applied in C accounting and reporting for climate change studies.

2.3. *Sampling Design*

Spatial Coverage Sampling and Mapping of Regionalised Variables

We carried out sampling in an area that had not been sampled before, and hence, the scales of spatial variability were not known beforehand. Under such circumstances, the Mean Squared Shortest Distance (MSSD) is a befitting objective function, which we utilised in order to optimize the sample locations. We utilised the *k-means* clustering algorithm for uniform area coverage sampling. According to Walvoort, Brus, and de Gruijter [42], the even distribution of sampling locations within a study domain can enhance the mapping and estimation of regionalised variables. This assertion is further confirmed by Brus, de Gruijter, and van Groenigen [43], who demonstrated how even coverage of the study area with sampling observations can be utilised for the dual role of estimating the spatial means of regionalised variables and resolving mapping in forestry, soil, and environmental research. The MSSD remains a dominant methodology for optimizing the sampling pattern over other methods such as Spatial Simulated Annealing (SSA) for both prediction and estimation designs for regionalised variables. The suitability of this design for locations where sampling schemes cannot be extended beyond a single phase is well documented.

We subdivided the study domain (D) into compact subunits through the clustering of building blocks making up the sampling domain using the *k-means* optimization procedure [44,45]. x and y coordinates of the central points of the building blocks are the classification variables for the *k-means* optimization function. We utilised the centroids of clusters as sample locations, where the sampling plots for C stock were set up.

2.4. Carbon Stock Data

2.4.1. Aboveground Tree Biomass (AGTB) Field Measurement

Measurements of all trees with at least 10 cm diameter at breast height (*DBH*) (at 1.3 m) were taken using 500 m^2 circular plots from the 19 September to the 24 October 2021. Diameter and linear tapes were used for the tree measurements, and trees with dbh less than 10 cm were excluded as they are generally regarded to have an insignificant C stock [3]. As the average slope within the study area was generally less than 30%, we did not consider slope correction for the measured outcome variable [46]. We carried out optimization of the resulting 200 sampling points using the ***spcosa*** package implemented using the R Statistical and Computing Environment [44,45]. We pre-uploaded the 200 probable sampling observations into a 72 H handheld Garmin GPS before setting out for the field work programme. The number of actual sampling points of forest biomass obtained during the field exercise was one hundred and nine, as nine of the pre-loaded sampling locations fell outside the boundaries of the defined study domain (Figure 1).

2.4.2. Biomass Calculation and Derivation of C Stock

Allometric equations used by Brown [7] were applied in the calculation of AGB *Pinus* species, whilst the AGB of *Eucalyptus* species was calculated using allometric equations developed by the authors of [47]. The same allometric equations used for *Pinus* and *Eucalyptus* species were also applied to *Eucalyptus* and *Pinus* species of the Manica province in Mozambique, whose climatic and weather conditions largely resemble those of the studied region in the eastern highlands of Zimbabwe. We then converted the AGB of every individual tree to C stocks per species through the conversion factor in [4]. Estimated per plot AGB values were then expanded to a standardised unit area of a hectare measured in $MgCha^{-1}$.

2.5. The Bayesian Geostatistical Modelling Framework

The modelling framework for the Bayesian and frequentist geostatistical approaches using new generation remote sensing Landsat-8 and Sentinel-2 as data sources is illustrated in Figure 2. Both approaches culminate in C stock predictions whose qualities were evaluated using cross-validation statistics (Figure 2). We assumed the Bayesian hierarchical methodology in order to have a full account of the parameter uncertainty for the measured C stock as given in Equation (1) [48].

$$Y(s) = X^T(s)\beta + w(s) + \varepsilon(s) \tag{1}$$

where:

$w(s)$ represents the spatial random effects term;

β denotes a vector of covariate coefficients;

$X^T(s)$ denotes a vector of predictors measured at the same location as $Y(s)$;

$Y(s)$ denotes the sampled C stock variable;

$\varepsilon(s)$ denotes the white-noise-assumed independent and identically distributed ($i.i.d.N(0, \sigma_\varepsilon^2)$).

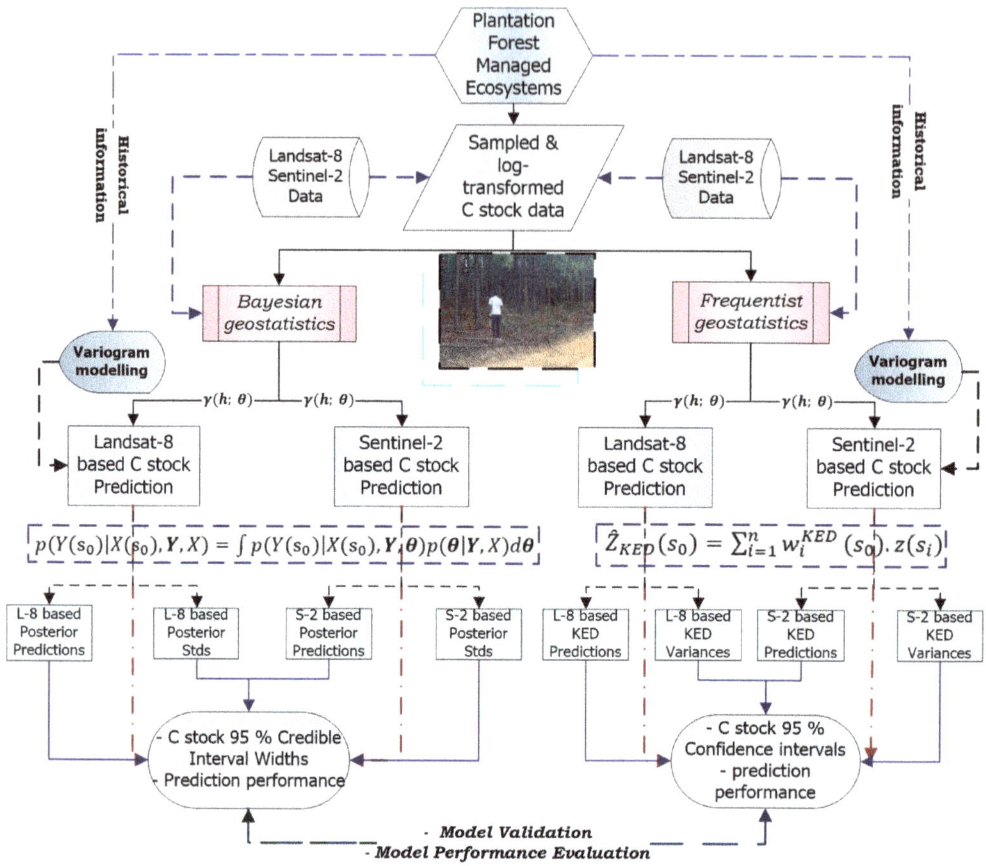

Figure 2. Bayesian and frequentist geostatistical modelling framework.

We performed the simultaneous estimation and prediction of C stock parameters by making use of the Markov Chain Monte Carlo (MCMC) technique to derive and calculate C stock predictions at unvisited locations, as in Equation (2).

$$\hat{Y}(s)' = X^T(s)'\hat{\beta} + \hat{w}(s)' \quad (2)$$

All hierarchical models were made using the *spBayes* package (Finley, Sudipto, and Carlin [49]) in the R Statistical and Computing Environment [50]. According to Gelfand [51], the vector of model parameters, $\boldsymbol{\theta} = \beta, \sigma^2, \phi, \tau^2$, is treated as random and mutually independent variables and is assigned prior distributions. We therefore sampled the posterior distribution of the parameters of interest, $\boldsymbol{\theta}$, as in Equation (3).

$$p(\boldsymbol{\theta}|\boldsymbol{y}, X) \propto p(\boldsymbol{\theta}) \times N(\boldsymbol{w}|0, \Sigma_w) \times N(\boldsymbol{y}|X^t\boldsymbol{\beta} + \boldsymbol{w}, \Sigma_\varepsilon) \quad (3)$$

Equation (3) was employed in the quantification of uncertainties in model parameters and C stock predictions at unsampled locations derived using Equation (4).

$$p(y_0|\boldsymbol{y}, \mathbf{X}, x_0) \propto \int p(y_0|\boldsymbol{y}, \boldsymbol{\theta}, x_0)p(\boldsymbol{\theta}|\boldsymbol{y}, \mathbf{X})d\boldsymbol{\theta} \quad (4)$$

where:

y_0 represents the predicted C stock at a site s_0 and x_0 are the predictor values at site s_0.

The overall mean of the sampled C stock was assigned a normal prior, whilst the regression coefficients were assigned a multivariate normal prior. Since the study made use of ancillary data from Landsat-8 and Sentinel-2 sensors, we specified two classes of priors for hierarchical modelling of the outcome variable. We assigned an inverse gamma distribution for the C stock data and measurement error variance, whilst the spatial decay parameter, ϕ, was assigned a uniform prior as indicated by $p(\boldsymbol{\theta}_1)$ and $p(\boldsymbol{\theta}_2)$ for the Landsat-8 and Sentinel-2 sets of priors, respectively. The assignment of the prior distribution on the spatial decay parameter was guided by the maximum distance between the sampling locations (2413 m) within the geographic domain ($\boldsymbol{\theta}_1, \boldsymbol{\theta}_2 \in D$) of the studied region.

As we expected the white noise error variance (nugget, σ_ε^2) to be smaller than the structured variance was, σ_w^2, scale parameter values were adopted in order to express the preference that $\sigma_\varepsilon^2 < \sigma_w^2$ [52]. A uniform prior with support covering the geographic domain of the study area was assigned to the spatial decay parameter, ϕ. Prior distributions on the modelled parameters were derived from covariance parameters from the exploratory variograms of the two multispectral remote sensing sensors. A Metropolis–Hastings algorithm for MCMC was utilised [48]. We then specified an algorithm of one chain comprised of 20,000 MCMC iterations for the posterior densities of the model parameters. Fifteen thousand chains were discarded as burn-in.

Bayesian Model Validation and Diagnostic Evaluation

We compared the spatial model, the spatial-intercept-only model, and the independent error model (simple multiple linear regression) for assessing the performance of predictions from the Bayesian hierarchical modelling approach with the Uniform (***Unif***) and Inverse Gamma (***IG***) priors on the spatial decay and the spatial random effects as illustrated for θ_1 and θ_2, respectively.

$$p(\boldsymbol{\theta}_1) = \boldsymbol{Unif}(\phi|0.38, 0.0012) \times \boldsymbol{IG}\left(\sigma^2 \middle| 0.52, 1.58\right) \times \boldsymbol{IG}\left(\tau^2 \middle| 0.1, 1.58\right) \times \boldsymbol{MVN}\left(\beta \middle| 0, \Sigma_\beta\right)$$

$$p(\boldsymbol{\theta}_2) = \boldsymbol{Unif}(\phi|0.38, 0.0012) \times \boldsymbol{IG}\left(\sigma^2 \middle| 0.052, 0.0028\right) \times \boldsymbol{IG}\left(\tau^2 \middle| 0.1, 1.52\right) \times \boldsymbol{MVN}\left(\beta \middle| 0, \Sigma_\beta\right)$$

The predictive performance of each of the three models was tested using a *k-fold* cross-validation algorithm, which performed the cross-validation through random splitting of the measured 191 C stock observations into approximately ten equally seized segments [53]. We calculated validation metrics of the Mean Absolute Error (MAE), Root Mean Square Error (RMSE) and other goodness-of-fit statistics such as the Deviance Information Criterion (DIC) for ranking candidate models on their ability to fit data [54]. Desirable and better fitting models display lower values of the DIC, whilst better performing models would have the lowest $k(k = 10)$-fold MAE and RMSE. We also assessed the conformance of the models to the modelling assumptions using graphical diagnostic plots of model residuals [15,53].

2.6. *The Frequentist Geostatistical Modelling Framework*

2.6.1. Carbon Stock Spatial Interpolation

We utilised ordinary kriging (OK) as a baseline for assessing how covariates derived from Landsat-8 and Sentinel-2 impact the C stock model using a frequentist geostatistical methodology. Covariates are a means of enhancing the predictive properties of spatial models [55]. The manner in which auxiliary variables lead to better kriging estimates than the ordinary kriging algorithm can produce is shown by the work in [54,56]. We therefore employed kriging with External Drift (KED) as a geostatistical methodology for modelling the spatial distribution of C stock in managed plantation forest ecosystems using vegetation indices derived from Landsat-8 and Sentinel-2. As Equation (5) shows, the KED algorithm restricts stationarity within a search neighbourhood and provides more detailed information compared to that achieved by ordinary kriging [57]. The standard measurement unit for forest C stock accounting is a hectare [3], and hence, we utilised a 100 m $\times$ 100 m resolution grid for kriging, with the normality assumptions tested with residuals of the selected liner model [58].

$$Z_{KED}^*(\mu) = \sum_{\alpha=1}^{n(\mu)} \lambda^{KED}(\mu) Z(\mu_\alpha) \tag{5}$$

where:

$Z^*_{KED}(\mu)$ denotes the KED estimated value at site μ;

$\lambda^{KED}(\mu)$ denotes the KED weights pertaining to n samples at site μ;

$Z(\mu_\alpha)$ denotes the sample values inside the search neighbourhood at alpha.

2.6.2. Frequentist Model Validation and Diagnostics

We followed the method of cross-validation as outlined in [59] for assessing the quality of predictions of each of the C stock predictive models constructed from Landsat-8 and Sentinel-2 satellite sensors. We therefore presented the validation statistical metrics in the form of RMSE, MSE, ME, and the Predicted Residual Sum of Squares (PRESS).

2.7. Variogram Modelling of the Regionalised Variable

The Bayesian and frequentist geostatistical modelling philosophies both make use of the variogram as the basis for establishing spatial covariance parameters. We presented the variogram of residuals from the linear modelling of C stock with predictors from both Landsat-8- and Sentinel-2-derived vegetation indices. Variogram modelling was, therefore, utilised as a basis for assessing the robustness of the spatial correlation structure of the modelled C stock response variable [60–62]. The regionalised variable was also transformed into a logarithmic scale to ensure conformance to the normality of residuals modelling assumptions using the Box–Cox transformation technique [63].

3. Results

3.1. C Stock Descriptive Statistics

Descriptive statistics of the sampled C stock data relating to the measured forest parameters are illustrated in Table 1. The mean C stock for *Eucalyptus camaldulensis*, *Eucalyptus grandis*, and *Pinus patula* species were 2485.3 $MgCha^{-1}$, 405.7 $MgCha^{-1}$, and 377.9 $MgCha^{-1}$, respectively. *Eucalyptus camaldulensis* had the highest C stock density in the sampled plantation forest, as illustrated in Table 1.

Table 1. Summary statistics of the measured C stock plantation forest parameters.

Statistic ($MgCha^{-1}$)	*Eucalyptus camaldulensis*			*Eucalyptus grandis*			*Pinus patula*		
	DBH	Height	C Stock	DBH	Height	C Stock	DBH	Height	C Stock
Mean	81.4	60.6	2485.3	67.4	70.6	405.7	56.8	58.6	377.9
Median	77.4	52.7	1470.3	51.4	49.7	327.8	43.5	38.7	295.4
Max	231.9	88.9	8998.2	97.9	90.1	429.8	64.3	66.6	600.3
Min	11.4	23.8	13.7	14.7	27.8	111.3	10.6	19.4	9.7
n	97	-	-	60	-	-	34	-	-
s.td	57.6	-	-	51.7	-	-	48.9	-	-

3.2. Hierarchical Bayesian Geostatistical Approach

The variogram of residuals of predictors obtained from the Landsat-8 and the Sentinel-2 satellite sensors provided priors of the modelled parameter specifications of σ^2_ε and σ^2_w for the geostatistical approach, Figure 3a,b, respectively. Sentinel-2-derived predictors have more influence on the spatial distribution of the modelled regionalised variable than what we see with its counterpart from the Landsat-8-derived vegetation indices. The spatial dependence is greatly reduced in the Sentinel-2-derived modelled variogram (Figure 3a,b), as the variable displays a strong spatial structure attributed to the finer spatial and spectral resolutions of Sentinel-2 than that of the Landsat-8 data. The modelled C stock using the Bayesian hierarchical approach, therefore, adopted scale parameters derived from the variogram exploratory analysis of the outcome variable using ancillary data from Landsat-8 and Sentinel-2 [53].

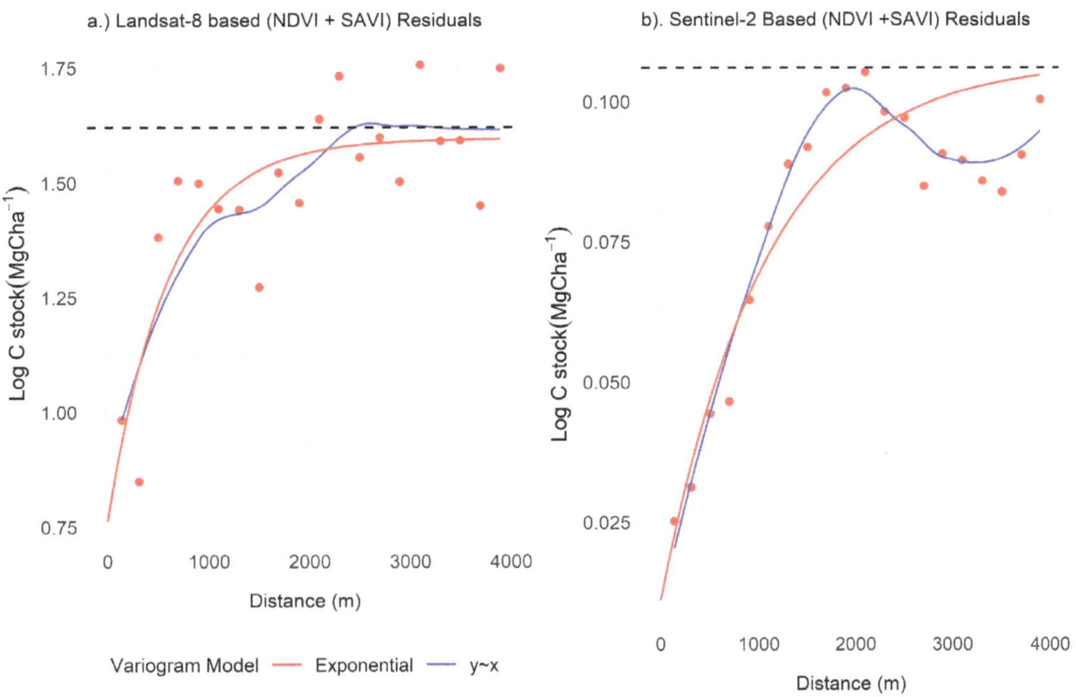

Figure 3. Variogram modelling for Landsat-2- and Sentinel-2-derived vegetation indices. The asymptote for the theoretical variogram model is shown by the black dotted line.

In accordance with the foregoing logic, the scale parameter values derived from Figure 3 put on a constraint on the parameter space of the probability distribution function [34]. We fitted the variogram of residuals for both sensors using the exponential covariance model, as illustrated in Figure 3.

3.2.1. C Stock and Medium Resolution Sensor-Derived Vegetation Indices

Medium resolution-derived vegetation indices utilised as predictors for C stock modelling in the form of $NDVI$, $SAVI$, and EVI showed different results for the two sensors. The Landsat-8-based C stock model demonstrated $NDVI$ as the only predictor for C stock, whilst the Sentinel-based C stock employing the same covariates illustrated both $NDVI$ and $SAVI$ as significant predictors of the C stock. Amongst the tested Landsat-8- and Sentinel-2-derived vegetation indices, only $NDVI$ was significantly different from zero, as 95% Credible Intervals (CI) for both sensors exclude zero, as illustrated in Table 2.

The 95% CI of $2.63 \leq NDVI \leq 6.30$ and $6.06 \leq NDVI \leq 6.51$ for the Landsat-8- and Sentinel-2-based C stock prediction models, respectively, puts the Sentinel-2-based predictive model in a stronger position due to the strength of the predictor coefficient. It is evident from Table 2 that the posterior distribution of independent variable coefficients for Landsat-8 and Sentinel-2 displayed marked differences. The larger $NDVI$ coefficients in the Sentinel-2-derived C stock model signify the relative preference of using Sentinel-2 for C stock prediction over Landsat-8-derived vegetation indices.

The Landsat-8-derived C stock model has a slightly stronger spatial correlation compared to that of its Sentinel-2 C stock counterpart, as evidenced by the estimates of the effective ranges of the models. We observed an effective range ($\approx \frac{3}{\phi}$) of 2500 m with a ($2142 \leq \phi \leq 2307$) 95% CI for the Landsat-8 C stock, whilst the Sentinel-2-based C stock model gave an effective range of 1667 m with a ($1304 \leq \phi \leq 2142$) 95% CI, as illustrated in Table 2. The Sentinel-2-based $NDVI$ takes away most of the spatial correlation structure in C stock than the same predictor derived from the Landsat-8 OLI does. The differences in the spectral properties between Landsat-8 (11 spectral bands) and Sentinel-2

(13 spectral bands) vindicates this observation [64,65]. In both models of Landsat-8 and Sentinel-2, the spatially structured variance, σ^2_w, is higher than the white noise variance is [66,67].

Table 2. Landsat-8- and Sentinel-2-derived predictors of C stock. NDVI =Normalised Difference Vegetation Index; SAVI = Soil-Adjusted Vegetation Index; EVI = Enhanced Vegetation Index; σ^2_w = spatially structured variance; σ^2_ε = White noise; ϕ = spatial decay parameter.

Parameter	Landsat-8 OLI C Stock Model				Sentinel-2 MSI C Stock Model			
	Mean	s.d	2.5%	97.5%	Mean	s.d	2.5%	97.5%
Intercept	1.34	0.49	0.37	2.27	0.93	0.24	1.42	−0.49
NDVI	4.49	0.94	2.63	6.30	6.30	0.11	6.06	6.51
SAVI	−0.50	0.72	−1.55	1.26	0.02	0.38	−0.72	0.77
EVI	−0.50	0.55	−1.65	0.53	0.01	0.11	−0.19	0.22
σ^2_w	1.47	0.39	0.76	2.22	0.07	0.01	0.053	0.10
σ^2_ε	0.39	0.15	0.13	0.68	0.005	0.004	0.0005	0.01
ϕ	0.0013	0.000	0.0013	0.0014	0.0012	0.0003	0.0014	0.0023

3.2.2. Bayesian-Based C Stock Predictions

We fitted models with Landsat-8- and Sentinel-2-derived spectral auxiliary variables for predicting the C stock at unsampled sites within the studied region. Covariates in the form of $NDVI$, $SAVI$, and EVI were derived from a 10,000 m^2 gridded raster, thereby making the predicted C stock represent the average values in every raster pixel. $NDVI$ is the only significant predictor for C stock prediction in managed plantation forest ecosystems, as the 95% CIs of the other covariates contain zero (Table 2). The significance of $NDVI$ as a vegetation index correlated with the biophysical properties of vegetation, leaf area index being one of them, is well established [68]. Sentinel-2-derived C stock predictions are more credible than their Landsat-8 derived C stock predictions are. Landsat-8-based C stock predictions have higher uncertainty compared to that of Sentinel-2-based C stock predictions.

This is demonstrated in the 95% posterior predictions illustrated in Figure 4a, indicating the C stock 95% CI to be greater than the Sentinel-2 predicted values are. This makes Landsat-8-based predictions highly uncertain and less precise than the Sentinel-2-based C stock predictions are. Landsat-8 and Sentinel-2-based C stock predictions alongside their 95% CIs are illustrated in Figure 4a,b and Figure 5a,b, respectively. The C stock predictions range between 1 MgCha^{-1} and 497 MgCha^{-1} and between 1 MgCha^{-1} and 285 MgCha^{-1} for the Landsat-8 and Sentinel-2 sensors, respectively. Higher C stock values are predicted in the southern part of the studied region in the Landsat-8-based predictive models than those in the Sentinel-2 models. In spite of this trend, the Sentinel-2-based C stock predictive model displays higher C stock values uniformly across the study area with a smaller magnitude than the Landsat-8-based models do (Figure 4a,b and Figure 5a,b).

As such, the Sentinel-2 model seems to underpredict more of the C stock at unsampled locations compared to that of its Landsat-8-based C stock predictive model counterpart. The slight underprediction by Sentinel-2 can partly be attributed to the finer spatial and spectral resolutions of the sensor within the visible and near-infrared segments of the electromagnetic spectrum (EMS) [16,69,70]. Enhancements in the spectral and spatial resoultion of Sentinel-2 confirms the much shorter 95% Credible Interval Widths (CIWs) displayed by the Sentinel-2-based C stock predictive model in Figure 4d (0.40–1.78 MgCha^{-1}) than those of the Landsat-8-based predictive model in Figure 4b (2.0–4.7 MgCha^{-1}). Predicted values in both new generation remote sensing-based models look better compared to the ones reported in previous research including Jiang et al. The authors of [71] established a mean AGB RMSE of 40.9 MgCha^{-1} in north-east China, and the authors of [72] determined an AGB RMSE of 36.67 MgCha^{-1} in Vietnam using the Random Forest (RF) algorithm.

Furthermore, Takagi et al. [73] employed LiDAR for the prediction of forest biomass in Hokkaido, Japan, and determined an RMSE biomass prediction of 19.1 MgCha^{-1}. The differences between the prediction accuracy results reported in the literature and our study can also be justified by the differences in forest density, since the erstwhile studies were carried out in subtropical rainforest biomes.

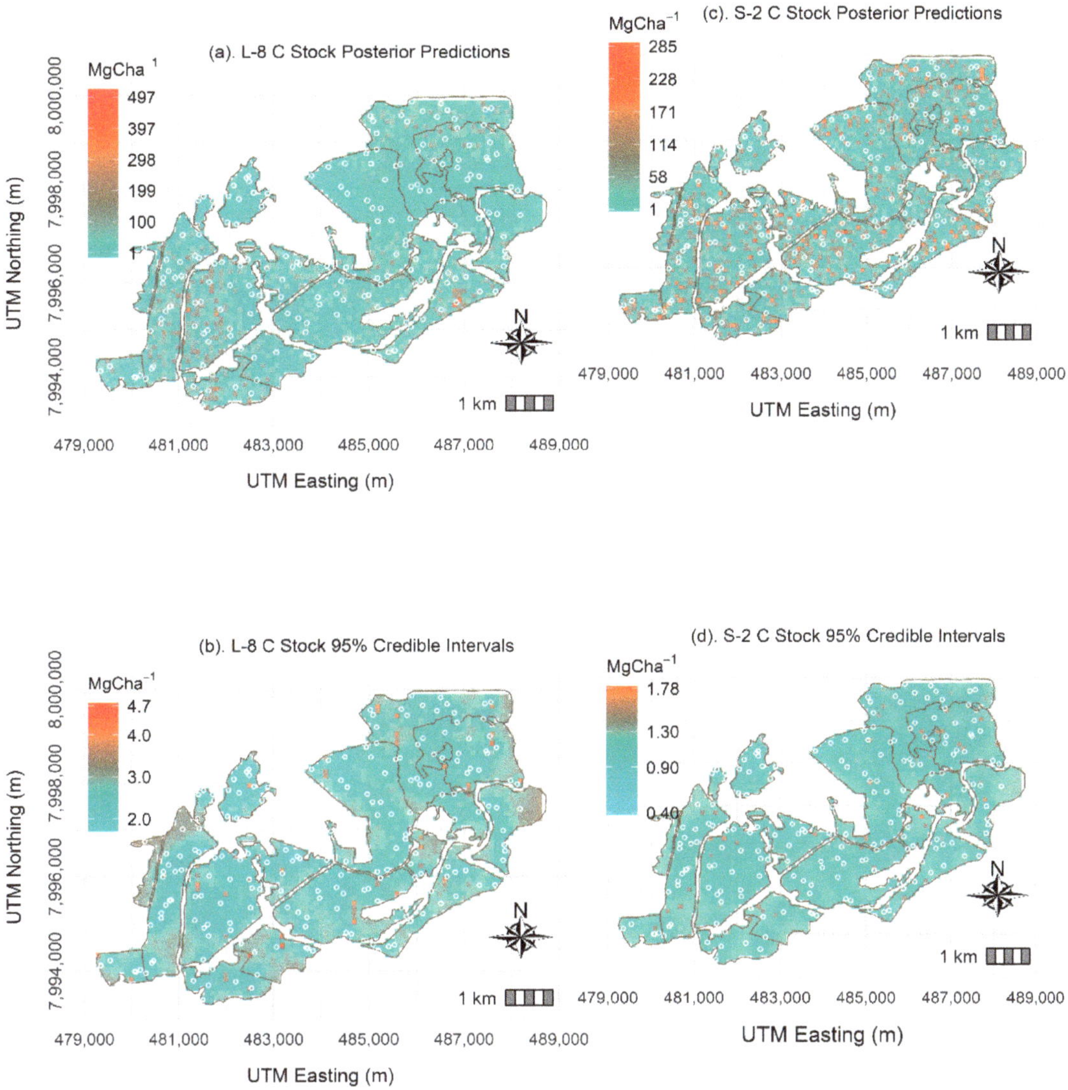

Figure 4. Bayesian Landsat-8 (L-8)- and Sentinel-2 (S-2)-based C stock posterior predictions alongside 95% credible intervals.

3.2.3. Model Validation and Diagnostics

The *k-fold* cross-validation metrics employed in the assessment of the Bayesian-based predictive models are presented in Table 3. The Sentinel-2-based C stock predictive model is the top performing model in terms of the RMSE (0.17 MgCha^{-1}) and Mean Absolute Error (MAE) (0.13 MgCha^{-1}). The Sentinel-2 model presents a predictive ability that is almost at the benchmark nominal coverage of 95% [60,73]. The Landsat-8-based C stock predictive model has an 85.4% coverage for the 95% prediction intervals coupled with higher RMSE and MAE values (Table 3).

Figure 5a,b illustrates the scatterplots of observed C stock against the predicted C stock, alongside the 95% intervals for both Landsat-8 and Sentinel-2 C stock-based predictive models. Evidence of the Sentinel-2-based C stock predictive model performing better than its Landsat-8 C stock-based counterpart did is clear from the scatter plot of the model in Figure 5b. It is evident from the model diagnostics illustrated in Figure 5a that the Landsat-8-based C stock predictive model tends to over-predict more C stock values compared to those of the Sentinel-2 C stock-based predictive model (Figure 5b). This makes the Sentinel-2 C stock-based model favourable compared to the Landsat-8 predictive model.

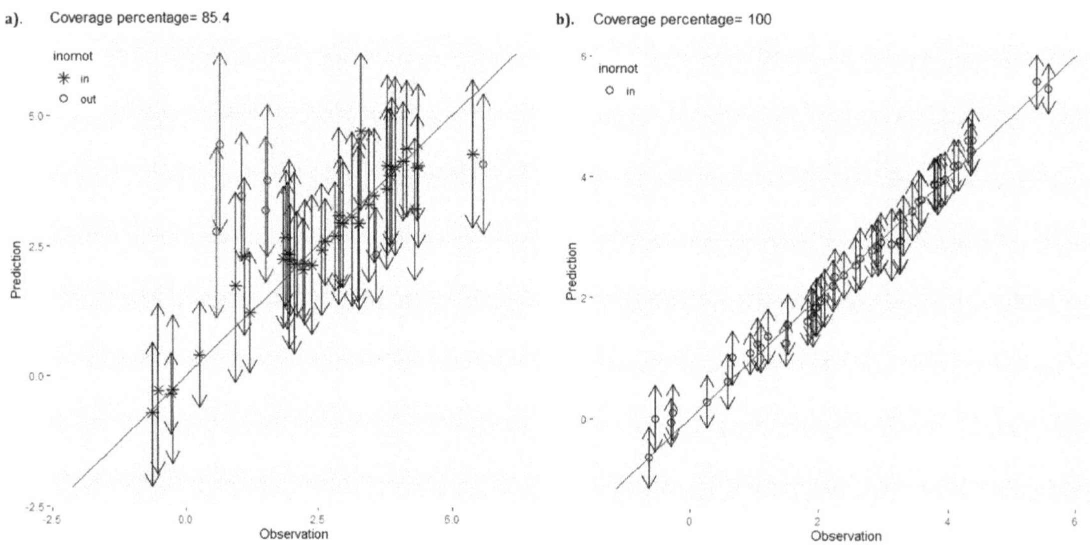

Figure 5. (**a**) Predictions (MgC ha^{-1}) against observed C stock (MgCha^{-1}) for the Landsat-8-based spatial model. (**b**) Predictions against observed C stock for the Sentinel-2-based spatial model alongside 95% intervals.

Table 3. Validation statistics for C stock Bayesian-based C stock predictive models.

Model Evaluation Criterion	Landsat-8-Derived Predictors			Sentinel-2-Derived Predictors		
	Independent Error Model	Spatial Intercept Only Model	Spatial Model	Independent Error Model	Spatial Intercept Only Model	Spatial Model
RMSE (Mgha^{-1})	1.23	0.97	0.97	0.31	1.18	0.17
MAE (Mgha^{-1})	0.93	0.53	0.57	0.26	0.77	0.13
CRPS (Mgha^{-1})	0.72	0.38	0.38	0.20	0.56	0.14
CVG (%)	91.67	85.42	85.42	95.83	89.58	100.00
DIC	220.8	48.40	71.0	−201.3	283.5	−564.5

3.3. Frequentist Geostatistical Modelling

3.3.1. C Stock Density

We assessed the C stock density of the studied region using species type as a possible source of C stock density variability and established the mean C stock of *Eucalyptus camaldulensis*, *Eucalyptus grandis*, and *Pinus patula* species to be 35.10 MgCha^{-1}, 37.38 MgCha^{-1}, and 29.45 MgCha^{-1}, respectively. Despite *Pinus patula* being the most dominant species in the sampled region (Figure 1), Eucalyptus *grandis* has the highest concentration of C stock. An evaluation of the C stock density of the various plantation forest species making up the study area using Analysis of Variance (ANOVA) shows that the C stock densities of the different tree species are not significantly different from each other ($F_{2188} = 0.21,\ p = 0.811$). Tukey–Kramer [74] multiple comparison test was conducted in order to avoid the risk of accumulating false positives as a result of multiple tests being carried out at the same time, and the results, also not significantly different. An evaluation of C stock density categorised by management style conducted in Nepal using frequentist ordinary kriging and KED geostatistical methodologies demonstrated significant differences between the C stock densities stored by different community forests [75].

However, the present study evaluated the prediction performance of geostatistical models using probabilistic- and likelihood-based frameworks, without stratifying the prediction performances by forest species as was performed by the authors of [75]. This would be an interesting research question for the future, as it is evident that different forest species of tropical and subtropical biomes store significantly different biomass and C stocks, an issue that forest practitioners would be interested in knowing for their investment options and decisions [9].

3.3.2. Landsat-8- and Sentinel-2-Based C Stock Linear Modelling

The linear modelling of C stock using sampled C stock data and multispectral remotely sensed data showed $NDVI$ to be a significant predictor for the Landsat-8-derived vegetation indices, whilst $NDVI$ and $SAVI$ were significant predictors for the Sentinel-2-derived vegetation indices (Table 4).

Table 4. Landsat-8 and Sentinel-2 best linear models of feature space.

Predictors	Landsat-8-Based Linear Model			Sentinel-2-Based Linear Model		
	Coefficient	p-Value	$\alpha = 0.05$	Coefficient	p-Value	$\alpha = 0.05$
Intercept	0.03	0.93	*Insignificant*	−0.30	0.31	*Insignificant*
$NDVI$	7.67	0.00	*Significant*	6.69	0.00	*Significant*
$SAVI$	1.04	0.11	*Insignificant*	1.24	0.01	*Significant*
EVI	0.33	0.57	*Insignificant*	0.20	0.15	*Insignificant*

3.3.3. Landsat-8 and Sentinel-2-Based KED Predictions

Figure 6a illustrates the Landsat-8-based C stock KED predictions alongside their prediction variances. The results of KED using the spatial dependencies in the outcome variable gave C stock predictions with a range of $1 \leq \text{MgCha}^{-1} \leq 327$ and a corresponding standard error range of $1.1 \leq \text{MgCha}^{-1} \leq 5.1$. As shown in Figure 6b, C stock predictions similar to the sampled data displayed less uncertainty compared to that of the predictions made at remote locations within the study domain. Thus, the Landsat-8-based KED C stock variances illustrated in Figure 6b are not better than 26.01 ($=5.1^2$) MgCha^{-1} for the studied managed plantation forest ecosystem. A high C stock density dominates the predictions at the original support (500 m^2), where primary data were derived using sampled field data. This is because kriging is a geostatistical method that characterises the values of an outcome variable (C stock) similar to the original data locations, which tend to have more similar statistical properties to the sampled value at that point than those of the values obtained in remote locations [75,76].

As illustrated in Table 4, we made Sentinel-2-based C stock KED predictions using $NDVI$ and $SAVI$ as aides of the predictive model, and these were made after ordinary variogram modelling of the primary variable. Incorporation of $NDVI$ and $SAVI$ as independent variables gave a reduced total sill of the modelled variogram (Figure 3b) and the subsequent shortening of the range of spatial dependence. Hence, Sentinel-2-derived vegetation indices utilised as predictors in the C stock model predicted C stock with a $1 \leq \text{MgCha}^{-1} \leq 290$ range and an accompanying standard error ranging from $2 \leq \text{MgCha}^{-1} \leq 11$ (Figure 6c,d). Finer spatial and spectral characteristics within the visible and NIR of the Sentinel-2 satellite sensor resulted in $SAVI$ being incorporated in the C stock predictive model, in addition to $NDVI$ [77,78].

Locations closer to the margins of the sampled domain show an increasing trend in the prediction standard error, as these locations are farther away from the sampled C stock observations. As Figures 3b and 6d show, the significant reduction of the range of spatial dependence in the modelled C stock data demonstrate how the Sentinel-2-driven model of feature space carry both $NDVI$ and $SAVI$ in the spatial correlation structure of the outcome variable. In line with theory, the KED calculated error variance appears to rely on the sampled data configuration, in which uncertainty decays towards the sampling sites [66,79].

As the results in Figure 6a–d show, the Landsat-8-based C stock model predicts more C stock in the southern and northern parts of the sampled region of the sampled domain. On the contrary, the Sentinel-2-based C stock predictive model predicts more C stock uniformly across the sampling domain. This can be explained by the precision with which the much-improved spatial resolution of

Sentinel-2 allows the sensor's sensitivity to the forest parameter spectral signal within the NIR and visible regions of the electromagnetic spectrum.

3.3.4. Frequentist Geostatistical Predictive Model Evaluation

We evaluated the frequentist-based C stock predictive models using leave-one-out cross-validation statistics and model residual diagnostics. As illustrated in Table 5, the Sentinel-2-based C stock predictive model has the lowest RMSE compared to that of the Landsat-8-based C stock model. The validation statistics shown in Table 5, therefore, suggest the Sentinel-2-based KED C stock model to be the most ideal under the frequentist geostatistical approach, as it has the best characteristics. This is evident from the predictive model's RMSE (1.19 $MgCha^{-1}$) and PRESS (6.06).

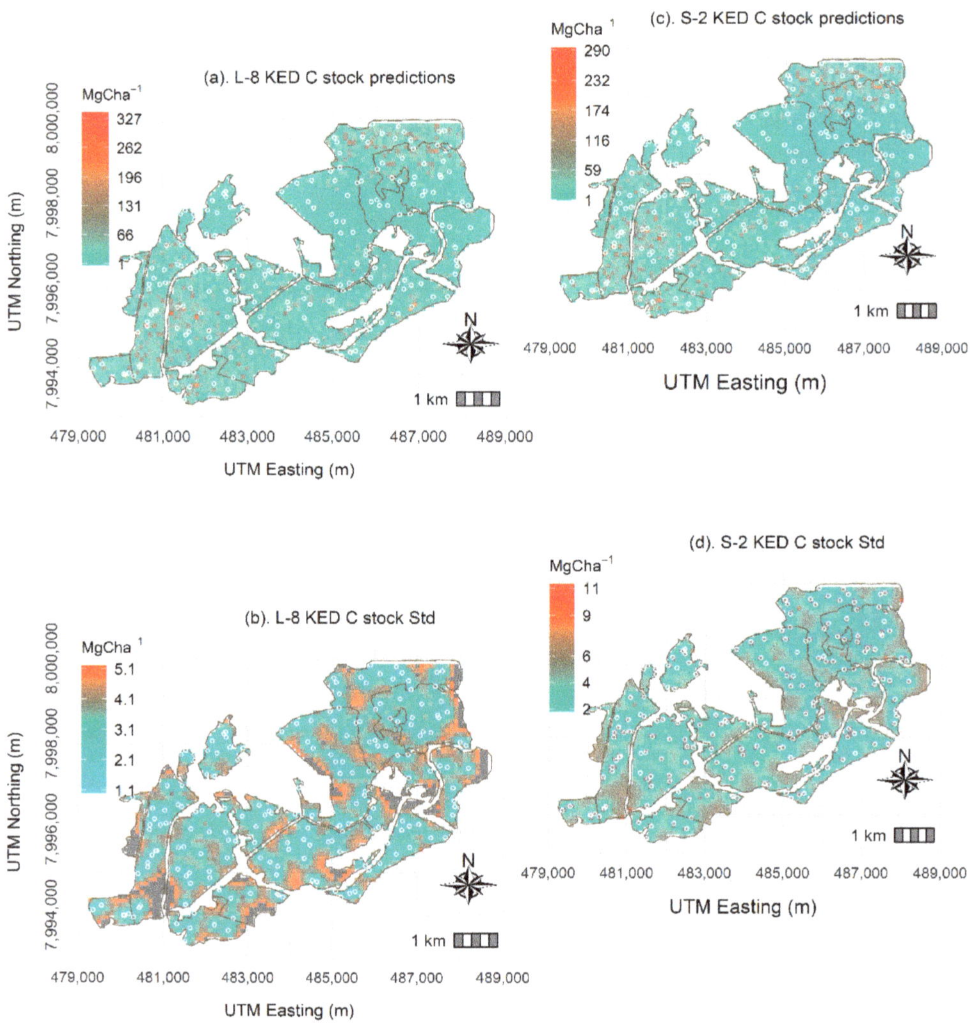

Figure 6. Frequentist Landsat-8- (L-8) and Sentinel-2 (S-2)-based C stock predictions and 95% confidence intervals.

The supremacy of the Sentinel-2-based C stock predictive model is also corroborated by the model diagnostics illustrated in Figure 7, showing robust model residuals that are evidently insensitive to outlying observations. Furthermore, the test statistics regarding the normality of the model residuals for the two kriging variants (Table 6) show the p-values of more than the 0.05 test statistic, leading to the rejection of the null hypothesis of non-normal residual errors. The rejection of the Shapiro normality null hypothesis is necessary for both Landsat-8 and Sentinel-2-based C stock predictive models (Table 6). The best linear model of feature space making use of Landsat-8- and Sentinel-2-derived vegetation indices gives the Sentinel-2-based C stock predictive model a more symmetrical error distribution than that of its Landsat-8-based counterpart (Figure 7). This fact is confirmed and reinforced by the residuals of the Sentinel-2 C stock predictive model.

Table 5. Frequentist geostatistical C stock prediction validation statistics.

Predictors	Landsat-8-Based C Stock Predictions		Sentinel-2-Based C Stock Predictions	
	Ordinary Kriging (OK)	Kriging with External Drift (KED)	Ordinary Kriging (OK)	Kriging with External Drift (KED)
ME	1.01	1.00	1.01	1.00
MSE	1.01	1.00	1.01	1.00
RMSE	2.94	2.84	2.91	1.19
PRESS	222.34	208.42	222.34	6.06

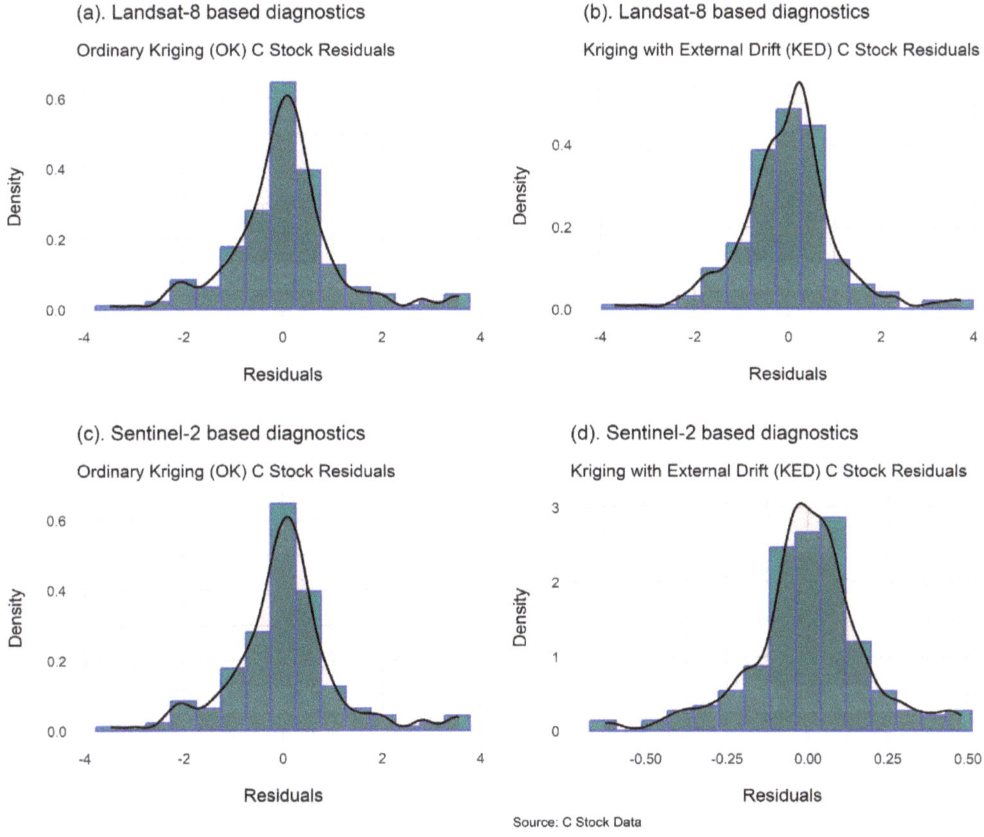

Figure 7. Frequentist-based Landsat-8 and Sentinel-2 C stock KED residual diagnostics.

Table 6. Frequentist geostatistical model diagnostics test statistics.

Modelling Approach	Test Statistic	p-Value	Modelling Technique
Frequentist approach	0.097	0.264	Landsat-8
Frequentist approach	0.132	0.136	Landsat-8
Hierarchical Bayesian approach	0.975	0.367	Sentinel-2
Hierarchical Bayesian approach	0.773	0.278	Sentinel-2

3.4. Bayesian- and Frequentist-Based C Stock Predictive Model Summaries

A summarised overview of the performances of the tested inferential approaches, one using the Bayesian framework and the other one using the frequentist approach, are illustrated in Table 7. We deduced from the summary statistics depicted in Table 7 that all the Landsat-8-based C stock predictive models from either the Bayesian or the frequentist paradigms tend to overpredict the modelled C stock values, whilst the Sentinel-2-based C stock-based predictive models gave C stock values that are just within the range. The Sentinel-2-based C stock predictive model from the Bayesian statistical paradigm offers the best qualities of C stock prediction model in terms of the quality of predictions and ability to predict true values outside the sampled data (Table 7).

Table 7. Summaries of the Bayesian and the frequentist geostatistical approaches. *CIWs* = Confidence Interval Widths; ME = Mean Error; RMSE = Root Mean Square Error.

Validation Criterion	Bayesian Geostatistical Approach		Frequentist Geostatistical Approach	
	Landsat-8-Based C Stock Model	Sentinel-2-Based C Stock Model	Landsat-8-Based C Stock Model	Sentinel-2-Based C Stock Model
RMSE	0.97	0.17	2.84	1.19
ME	0.57	0.13	1.01	1.00
Error/CIWs	$2 \leq MgC \leq 4.7$	$0.4 \leq MgC \leq 1.8$	$1.1 \leq MgC \leq 5.1$	$2 \leq MgC \leq 11$
Prediction range	$1 \leq MgC \leq 497$	$1 \leq MgC \leq 285$	$1 \leq MgC \leq 327$	$1 \leq MgC \leq 290$
Conclusion	Overprediction	Perfect	Overprediction	Perfect

4. Discussion

4.1. Bayesian Geostatistical Approach and C Stock Predictions

The hierarchical design of stochastic models is intrinsically related to Bayesian inference, where the probability distribution of observed C stock (in the present case) is specified in a hierarchical fashion [61]. From the three tested predictors of $NDVI$, $SAVI$, and EVI for C stock prediction, we established $NDVI$ as the only significant covariate of the modelled C stock variable for both the Landsat-8- and Sentinel-2-derived C stock predictive models. Because of the higher spatial resolution of Sentinel-2, $NDVI$ emerged as a stronger predictor of C stock for the Sentinel-2-based model than it was for the Landsat-8-based C stock predictive model [78].

The importance of $NDVI$ as a vegetation index correlated with the biophysical properties of vegetation such as the Leaf Area Index (LAI), is well known [68]. This shows the relative preference of employing Sentinel-2 for C stock prediction over Landsat-8 derived vegetation indices. Consequently, Sentinel-2 derived C stock predictions have better credibility than the Landsat-8-derived equivalent does. Landsat-8-based C stock predictions have more uncertainty compared to that of the Sentinel-2-based C stock predictions. The utilisation of new generation remote sensing-derived vegetation indices as predictors of C stock for a managed plantation forest ecosystem under a Bayesian framework is unique [80]. Previous studies, including Babcock et al. [14] and Babcock et al. [81], have used LiDAR on its own to obtain predictors of forest biomass rather than a comparative approach such as that we employed in this research.

The tendency of underprediction in Sentinel-2-derived C stock models can partly be attributed to the finer spatial resolutions of the sensor within the visible and near-infrared segments of the electromagnetic spectrum (EMS) [16,69,70]. As highlighted in [78], there are notable improvements in the radiometric resolution of the Landsat-8 OLI sensor from 8 bits to 12 bits.

Enhancements in the spatial resolution of Sentinel-2 confirms the much smaller 95% Credible Interval Widths (*CIWs*) displayed by the Sentinel-2-based C stock predictive model. The predicted values in both new generation remote sensing-based models look more attractive compared to the ones reported in previous research, including that of Jiang et al. [71], who established a mean AGB RMSE of 40.9 $MgCha^{-1}$ in north-east China, and that of Dang et al. [72], who determined an AGB RMSE of 36.67 $MgCha^{-1}$ in Vietnam using the Random Forest (RF) algorithm. Furthermore, Takagi et al. [73] employed LiDAR for the prediction of forest biomass in Hokkaido, Japan, and determined an RMSE biomass prediction of 19.1 $MgCha^{-1}$. Differences between the prediction accuracy results reported in the literature and our study can also be justified by the differences in forest densities, since the erstwhile studies were carried out in subtropical rainforest biomes [70–72]. For instance, Japan is regarded as the most forested country in the world, with approximately 70% of its land being forested, which is more than that of the location of the present study, Zimbabwe, where about 40% or 15,624,000 ha of the land is forested [82].

Machine learning methods for the mapping of AGB premised on Landsat-8 imagery were compared by Wu et al. [83] and Xiong and Wang [84], and a Random forest method with an RMSE of 26.43 tons/ha was found to be superior to the other methods such as stochastic gradient boosting, k-nearest neighbour, and support vector regression. The current study establishes AGB accuracies using the Bayesian and the frequentist geostatistical approach with the prediction accuracies of the former one outweighing those of the latter one. The Bayesian methods utilised in this study are both superior to both the frequentist approach and methods utilised in the literature for the mapping of AGB. However, Bayesian methods are not easy to implement and adopt by ordinary forest practitioner due to their complexity [51]. It is this complexity and the lack of simple software packages that greatly hinder their adoption and operationalization for forestry monitoring.

4.2. Frequentist Geostatistical Approach and C Stock Predictions

The geostatistical kriging variant, KED, provides an optimal geostatistical technique under the frequentist paradigm that is employed in the description of spatial patterns and predicting values of a variable at unsampled locations and, consequently, it evaluates the uncertainty associated with the predicted values [19]. We employed and subjected the C stock model to the same new generation remote sensing-derived predictors of vegetation indices as we did under the Bayesian approach and established some notable differences and similarities. Similar to the Bayesian C stock model for Landsat-8, $NDVI$ also presented as the only significant predictor in the frequentist Landsat-8-based C stock predictive data. However, both $SAVI$ and $NDVI$ were significant predictors for the frequentist Sentinel-2-based C stock model. We employed KED for both the Landsat-8- and the Sentinel-2-based C stock models. The finer spatial scale of the Sentinel-2 sensor, coupled with the wider spectrum of its NIR band (760–900 nm) compared to the narrow spectrum of Landsat-8 NIR band (850–880 nm), resulted in $SAVI$ and $NDVI$ being incorporated in the C stock predictive model [77,78]. The NIR band is widely known to be vital to the biophysical factors of vegetation assessment and monitoring. Consequently, the spectral response profile of Landsat-8 and Sentinel-2 display some minor differences in the NIR and visible regions of the electromagnetic spectrum, which could explain differences in the prediction performances, as established in the current study [62,81].

KED has been widely applied for the estimation of AGB using other factors as predictors of the forest biomass [18,76]. For instance, Korhonen et al. [20] made an average AGB prediction of 32 $Mgha^{-1}$ in the Brazilian Amazon. KED performed better in this study than the pure-based approach of the ordinary kriging algorithm did, as $NDVI$ and $SAVI$ formed the best linear model of feature space in the Sentinel-2 C stock predictive model. Most studies comparing the prediction performances of Landsat-8 and Sentinel-2, including those by Korhonen et al. [23] and Meyer et al. [85], have not established significant or systematic differences in the predictive performance of the aforementioned sensors. In the current study, both Bayesian and frequentist Landsat-8-based C models tend to overpredict the modelled and estimated C stock values across the sampled domain, as their range falls outside the range of observed data. This implies that forest practitioners can feasibly exploit the scale and band spectrum characteristics of Sentinel-2 for representative and accurate C reporting necessary for monitoring and verifying climate change mitigation actions.

4.3. Comparative Bayesian and Frequentist C Stock Predictive Model Evaluation

Sentinel-2-based prediction of C stock using the frequentist-based KED illustrates predictions within the range of measured C stock values, but the variances are high compared to those of the same predictions made using Sentinel-2 within a Bayesian inferential framework. On the contrary, both Bayesian and frequentist C stock predictive models constructed using Landsat-8 overpredicted the sampled C stock, as the range of predicted values fell outside the observed C stock values. This observation is further bolstered by the results of the diagnostic residuals of the models from the frequentist and the Bayesian techniques, where the plots of the observed versus the predicted values of C stock are perfectly predicted in the Bayesian-based Sentinel-2 models. As reported in the study, Sentinel-2-based C stock predictive models are more accurate than their Landsat-8 equivalents are for both the Bayesian and the frequentist inferential approaches. Previous studies mapping AGB accuracies using either the frequentist or the Bayesian approaches, coupled with satellite imagery data of Landsat-8 and Sentinel-2, report lower prediction accuracies [22,71]. As hypothesised, this confirms the superiority of the Bayesian geostatistical approach in handling the uncertainty and improving the accuracy for predicting C stock over the frequentist geostatistical approach.

Important research in the field of biomass estimation using the Bayesian techniques include [13–16]. Recent studies assessing AGB distribution in temperate European ecosystems established accuracies of 17.52 Mg/ha by Babcock et al. [14], and 1.16 $MgCha^{-1}$ and 2.69 $MgCha^{-1}$ accuracies were found for Sentinel-2-based and Landsat-8-based C stock predictive models, respectively [16]. Other remote sensing- and machine learning-based efforts for the estimation and prediction of AGB in recent times include those by Do et al. [17], who found that mangrove AGB predictions in Vietnam range from 6.51 to 368 $Mgha^{-1}$ and from 13.70 to 320.1 $Mgha^{-1}$ for remote sensing and Artificial Neural Networks, respectively. Because the present study utilised data from improved remote sensing platforms for predicting C stock, our reported accuracies surpass those reported in the literature utilising different methodologies with the same satellite sensors with lower quality.

Past studies utilising the frequentist geostatistical approach separate from the Bayesian technique for C stock estimation are also well documented in the literature. The authors of [20] mapped AGB in the Brazilian Amazon using Kriging with External Drift and established prediction accuracies for different sample sizes ranging from 0 to 110 km for distances within 300 km radii from the prediction locations. The lowest RMSE for the estimated AGB for a sample size of 110 was 32.8 $Mgha^{-1}$, whilst the lowest accuracy for the lowest sample size of $n > 0$ was 48.06 $Mgha^{-1}$. Furthermore, Jiang et al. [21] predicted AGB in the Wangyedia forest farm in China using Landsat-8 and the newly launched Landsat-9 and reported RMSEs of 16.83 tha^{-1} and 17.91 tha^{-1}, respectively. The authors of [22] coupled remote sensing-derived Sentinel-2 explanatory variables with geostatistics and machine learning algorithms in Mayanmar for predicting the aboveground biomass and established accuracies of 24.91 $Mgha^{-1}$ and 34.72 $Mgha^{-1}$ for the Random Forest-based ordinary kriging and the Random Forest-based co-kriging, respectively.

As demonstrated by the results in [21], Landsat-8 built AGB estimation models can still be superior to AGB models built using the successor Landsat-9 sensor, despite the relative spectral and radiometric improvements in the latest Landsat-9 sensor. This is also justified by the results of the present study as the results from both the Landsat-8 and Sentinel-2-based Bayesian and frequentist approaches are not significantly different from each other. From a practitioner's point of view, the recommendation for the use and adoption of a particular sensor and statistical approach largely depends on the policy problem being addressed. Furthermore, the costs involved and ease of use and application of the sensors and methodology also play a bigger role. Despite its lack of simplicity and difficulties in implementation, the Bayesian approach is more appealing and pragmatic for natural resources monitoring and reporting, as different studies of a particular region would eventually constitute a database that subsequent studies on carbon assessment would rely on for the updating of priors and posteriors in the prediction process.

5. Conclusions

This study set out to compare the prediction performances of two inferential geostatistical frameworks, one making use of a hierarchical Bayesian geostatistical approach, and the other one utilizing a frequentist geostatistical approach, for C stock prediction in a managed plantation forest ecosystem in the eastern highlands of Zimbabwe. Broadband spectral indices from two multispectral remote sensing platforms of Landsat-8 and Sentinel-2 were employed as prediction aids for the geostatistical methodologies. We established notable differences between the two inferential approaches in the prediction of C stock, with the Sentinel-2-based hierarchical Bayesian geostatistical approach yielding lower ($0.4 \leq MgCha^{-1} \leq 1.8$) prediction uncertainty values than its frequentist geostatistical KED ($2 \leq MgCha^{-1} \leq 11$) modelling counterpart did. In both geostatistical methods, the Sentinel-2-driven C stock prediction models outperformed the Landsat-8-driven C stock prediction model counterparts.

The bigger policy problem pertaining to climate change mitigation and climate change action being addressed in this study requires accurate and sustainable carbon accounting and verification tools. In that regard, despite the ease of application and use of the frequentist inferential methodology, we conclude that the Bayesian-based technique, coupled with high-quality remote sensing information, is a better method for predicting C stock. This is a critical steppingstone in accounting for carbon to achieve climate change adaptation and mitigation under the United Nations Framework Convention on Climate Change (UNFCCC). The Bayesian-based Sentinel-2 C stock predictive model is preferable to its Sentinel-2-based frequentist counterpart for assisting natural resources managers and other forest practitioners in providing advice to governments on decision for afforestation and reforestation.

Using the Bayesian concept that "today's posterior is tomorrow's prior", we are able to construct a meaningful database of forest parameters, specifically C stock parameters, that will aid future estimation and prediction, even with limited datasets, thereby eliminating the need for costly sampling campaigns. The sustainability of such databases is further made better by the availability of freely available satellite data that are continuously being improved in terms of quality and scale of coverage. Conversely, the frequentist geostatistical approach cannot work optimally under limited sample sizes, and therefore, this makes it a costly alternative in the long term for updating C stock databases kept for C stock monitoring and accounting. We therefore recommend and conclude that the Bayesian-based C stock prediction method, coupled with high-quality remote sensing information such as that of Sentinel-2, is a useful inferential statistical methodology for reporting C stock in managed plantation forest ecosystems.

Author Contributions: Writing—original draft, T.S.C.; Writing—review & editing, T.D.; Project administration, O.M. All authors have read and agreed to the published version of the manuscript.

Funding: This work is based on the research supported in part by the National Research Foundation of South Africa (Grant Number: 84157).

Data Availability Statement: Not applicable.

Acknowledgments: The authors extend their gratitude to Kutsaranga, Matowanyika and Mukwekwe for giving us access to the plantation forest of the sampled region at Lot 75A of Nyanga Downs in Manicaland Province. This research received support from the University of KwaZuluNatal, college of Agricultural, Earth and Environmental Science.

Conflicts of Interest: The authors declare that they have no financial interest nor personal relationships that could otherwise sway the work reported and presented in this article.

References

1. Araki, S.; Yamamoto, K.; Kondo, A. Application of regression kriging to air pollutant concentrations in Japan with high spatial resolution. *Aerosol. Air Qual. Res.* **2015**, *15*, 234–241. [CrossRef]
2. Shoko, C.; Gara, T. Remote sensing of aboveground grass biomass between protected and non-protected areas in savannah rangelands. *Afr. J. Ecol.* **2021**, *59*, 687–695. [CrossRef]
3. Gibbs, H.K.; Brown, S.; Niles, J.O.; Foley, J.A. Monitoring and estimating tropical forest carbon stocks: Making REDD a reality. *Environ. Res. Lett.* **2007**, *2*, 45023. [CrossRef]
4. Van Amstel, A. IPCC 2006 Guidelines for National Greenhouse Gas Inventories. 2006. Available online: https://www.ipcc-nggip.iges.or.jp/public/2006gl/ (accessed on 13 February 2023).
5. European-Commission. *Timber Trade Flows within, to and from Eastern and Southern African Countries*; European-Commission: Brussels, Belgium, 2017.
6. Government of Zimbabwe. Zimbabwe Revised Nationally Determined Contribution. 2021. Available online: https://unfccc.int/sites/default/files/NDC/2022-06/Zimbabwe%20Revised%20Nationally%20Determined%20Contribution%202021%20Final.pdf (accessed on 13 February 2023).
7. Brown, S. Estimating biomass and biomass change of tropical forests: A primer. *FAO For. Pap.* **1997**, *134*.
8. Initiative For Climate Action Transparency (ICAT). Zimbabwe on Track to Better Climate Action Transparency. 2022. Available online: https://climateactiontransparency.org/zimbabwe-on-track-to-better-climate-action-transparency/ (accessed on 28 December 2022).
9. Shi, L.; Liu, S. *Methods of Estimating Forest Biomass: A Review*; Tumuluru, J.S., Ed.; IntechOpen: Rijeka, Croatia, 2017; Chapter 2.
10. Gelman, A.; John, B.C.; Hals, S.S.; Rubin, D.B. *Bayesian Data Analysis*, 2nd ed.; Chapman and Hall/CRC: London, UK, 2004.
11. Cameletti, M.; Biondi, F. Hierarchical modeling of space-time dendroclimatic fields: Comparing a frequentist and a Bayesian approach. *Arct. Antarct. Alp. Res.* **2019**, *51*, 115–127. [CrossRef]
12. Ghosh, G.; Carriazo, F. *Bayesian and Frequentist Approaches to Hedonic Modeling in a Geo-Statistical Framework*; Agricultural and Applied Economics Association (AAEA): San Antonio, TX, USA, 2007.

13. Finley, A.O.; Banerjee, S.; McRoberts, R.E. A Bayesian approach to multi-source forest area estimation. *Environ. Ecol. Stat.* **2008**, *15*, 241–258. [CrossRef]
14. Babcock, C.; Finley, A.O.; Bradford, J.B.; Kolka, R.; Birdsey, R.; Ryan, M.G. LiDAR based prediction of forest biomass using hierarchical models with spatially varying coefficients. *Remote Sens. Environ.* **2015**, *169*, 113–127. [CrossRef]
15. Green, E.J.; Finley, A.O.; Strawderman, W.E. *Introduction to Bayesian Methods in Ecology and Natural Resources*; Springer: Cham, Switzerland, 2020.
16. Chinembiri, T.S.; Mutanga, O.; Dube, T. Hierarchical Bayesian geostatistics for C stock prediction in disturbed plantation forest in Zimbabwe. *Ecol. Inform.* **2023**, *73*, 101934. [CrossRef]
17. Do, A.N.T.; Tran, H.D.; Ashley, M.; Nguyen, A.T. Monitoring landscape fragmentation and aboveground biomass estimation in Can Gio mangrove biosphere reserve over the past 20 years. *Ecol. Inform.* **2022**, *70*, 101743. [CrossRef]
18. Hudson, G.; Wackernagel, H. Mapping temperature using kriging with external drift: Theory and an example from scotland. *Int. J. Climatol.* **1994**, *14*, 77–91. [CrossRef]
19. Wheeler, D.C.; Waller, L.A. Comparing spatially varying coefficient models: A case study examining violent crime rates and their relationships to alcohol outlets and illegal drug arrests. *J. Geogr. Syst.* **2009**, *11*, 1–22. [CrossRef]
20. Sales, M.H.; Souza, C.M.; Kyriakidis, P.C.; Roberts, D.A.; Vidal, E. Improving spatial distribution estimation of forest biomass with geostatistics: A case study for Rondônia, Brazil. *Ecol. Modell.* **2007**, *205*, 221–230. [CrossRef]
21. Jiang, F.; Sun, H.; Chen, E.; Wang, T.; Cao, Y.; Liu, Q. Above-ground biomass estimation for coniferous forests in Northern China using regression kriging and landsat 9 images. *Remote Sens.* **2022**, *14*, 5734. [CrossRef]
22. Wai, P.; Su, H.; Li, M. Estimating aboveground biomass of two different forest types in myanmar from sentinel-2 data with machine learning and geostatistical algorithms. *Remote Sens.* **2022**, *14*, 2146. [CrossRef]
23. Korhonen, L.; Hadi; Packalen, P.; Rautiainen, M. Comparison of sentinel-2 and landsat 8 in the estimation of boreal forest canopy cover and leaf area index. *Remote Sens. Environ.* **2017**, *195*, 259–274. [CrossRef]
24. Astola, H.; Häme, T.; Sirro, L.; Molinier, M.; Kilpi, J. Comparison of sentinel-2 and landsat 8 imagery for forest variable prediction in boreal region. *Remote Sens. Environ.* **2019**, *223*, 257–273. [CrossRef]
25. Jha, N.; Tripathi, N.K.; Barbier, N.; Virdis, S.G.P.; Chanthorn, W.; Viennois, G.; Brockelman, W.Y.; Nathalang, A.; Tongsima, S.; Sasaki, N.; et al. The real potential of current passive satellite data to map aboveground biomass in tropical forests. *Remote Sens. Ecol. Conserv.* **2021**, *7*, 504–520. [CrossRef]
26. Finley, A.O.; Banerjee, S.; McRoberts, R.E. Hierarchical spatial models for predicting tree species assemblages across large domains. *Ann. Appl. Stat.* **2009**, *3*, 1052–1079. [CrossRef]
27. Banerjee, S.; Finley, A.O.; Waldmann, P.; Ericsson, T. Hierarchical spatial process models for multiple traits in large genetic trials. *J. Am. Stat. Assoc.* **2010**, *105*, 506–521. [CrossRef]
28. Babcock, C.; Finley, A.O.; Andersen, H.-E.; Pattison, R.; Cook, B.D.; Morton, D.C.; Alonzo, M.; Nelson, R.; Gregoire, T.; Ene, L.; et al. Geostatistical estimation of forest biomass in interior Alaska combining landsat-derived tree cover, sampled airborne lidar and field observations. *Remote Sens. Environ.* **2018**, *212*, 212–230. [CrossRef]
29. Neyman, J. Outline of a theory of statistical estimation based on the classical theory of probability. *Philos. Trans. R. Soc. A* **1937**, *236*, 333–380. [CrossRef]
30. Murphy, K.P. A Probabilistic Perspective. *Text B*. 2012. Available online: https://d1wqtxts1xzle7.cloudfront.net/55735470/Machine_learning_A_Probabilistic_Perspective.pdf?1517974187=&response-content-disposition=inline%3B+filename%3DMachine_Learning_A_Probabilistic_Perspec.pdf&Expires=1679287483&Signature=SEV-l8rcLLC3o8k0iRZX9fOoWoZyp82ssxglfGtk0vQxpatA4vLCM8nN-HADoVT8IzBf631g3xykOibpqa4vc2nNoievSdbei8VU-xjSNRe0cS0w6r58QVkyRnmE7tgpLWh8-6dRDE-x-x88aY84sbUQQOxIgzn1ZjIQT2ifMVBRXogQHsYEtdp04qL5umm-KJ9iqeyV3SpZO0rLLEaXArtn6ALLV2PXVBy-uWeLAWsvMloCuxxXAIyoCSHaf32VWrL8tICMlM2bvMWW0r62FtRbd1d7jz3dNvL-ENGFXJOOgFjrjBwGY~Xa3u2QNBjcTsTVMRn-M9LL7AjPf9oNUQ__&Key-Pair-Id=APKAJLOHF5GGSLRBV4ZA (accessed on 13 February 2023).
31. Hazra, A. Using the confidence interval confidently. *J. Thorac. Dis.* **2017**, *9*, 4124–4129. [CrossRef]
32. Edwards, W.; Lindman, H.; Savage, L.J. Bayesian statistical inference for psychological research. *Psychol. Rev.* **1963**, *70*, 193–242. [CrossRef]
33. Box, G.E.P.; Tiao, G.C. *Bayesian Inference in Statistical Analysis*; Wiley: New York, NY, USA, 1992.
34. Goulard, M.; Voltz, M. Linear coregionalization model: Tools for estimation and choice of cross-variogram matrix. *Math. Geol.* **1992**, *24*, 269–286. [CrossRef]
35. Forestry-Commission. *Zimbabwe Land and Vegetation Cover Area Estimates*; Forestry-Commission: Harare, Zimbabwe, 2021.
36. Zvobgo, L.; Tsoka, J. Deforestation rate and causes in upper manyame sub-catchment, Zimbabwe: Implications on achieving national climate change mitigation targets. *Trees For. People* **2021**, *5*, 100090. [CrossRef]
37. Whitlow, T. *Land Degradation in Zimbabwe. A Geographical Study*; University of Zimbabwe (UZ): Harare, Zimbabwe, 1998.
38. FAO. *Forestry Outlook Study for Africa. Regional Report—Opportunities and Challenges towards 2020*; FAO: Rome, Italy, 2003.
39. Drusch, M.; Del Bello, U.; Carlier, S.; Colin, O.; Fernandez, V.; Gascon, F.; Hoersch, B.; Isola, C.; Laberinti, P.; Martimort, P.; et al. Sentinel-2: ESA's optical high-resolution mission for GMES operational services. *Remote Sens. Environ.* **2012**, *120*, 25–36. [CrossRef]
40. Frampton, W.J.; Dash, J.; Watmough, G.; Milton, E.J. Evaluating the capabilities of sentinel-2 for quantitative estimation of biophysical variables in vegetation. *ISPRS J. Photogramm. Remote Sens.* **2013**, *82*, 83–92. [CrossRef]

41. Ranghetti, L.; Boschetti, M.; Nutini, F.; Busetto, L. 'sen2r': An R toolbox for automatically downloading and preprocessing sentinel-2 satellite data. *Comput. Geosci.* **2020**, *139*, 104473. [CrossRef]
42. Walvoort, D.; Brus, D.; de Gruijter, J. An R package for spatial coverage sampling and random sampling from compact geographical strata by k-means. *Comput. Geosci.* **2010**, *36*, 1261–1267. [CrossRef]
43. Brus, D.; de Gruijter, J.; van Groenigen, J. Chapter 14 designing spatial coverage samples using the k-means clustering algorithm. *Dev. Soil Sci.* **2006**, *31*, 183–192.
44. Li, C.; Li, X. Hazard rate and reversed hazard rate orders on extremes of heterogeneous and dependent random variables. *Stat. Probab. Lett.* **2019**, *146*, 104–111. [CrossRef]
45. Bordoloi, R.; Das, B.; Tripathi, O.; Sahoo, U.; Nath, A.; Deb, S.; Das, D.; Gupta, A.; Devi, N.; Charturvedi, S.; et al. Satellite based integrated approaches to modelling spatial carbon stock and carbon sequestration potential of different land uses of Northeast India. *Environ. Sustain. Indic.* **2022**, *13*, 100166. [CrossRef]
46. Ravindranath, N.H.; Ostwald, M. Carbon inventory methods handbook for greenhouse gas inventory, carbon mitigation and roundwood production projects. *Adv. Glob. Chang. Res. Vol.* **2008**, *29*. [CrossRef]
47. Zunguze, A.X. *Quantificação de Carbono Sequestrado em Povoamentos de Eucalyptus Spp na Floresta de Inhamacari-Manica*; Universidade Eduardo Mondlane: Maputo, Mozambique, 2012.
48. Finley, A.O.; Banerjee, S.; MacFarlane, D.W. A hierarchical model for quantifying forest variables over large heterogeneous landscapes with uncertain forest areas. *J. Am. Stat. Assoc.* **2011**, *106*, 31–48. [CrossRef]
49. Finley, A.; Sudipto, B.; Carlin, B. spBayes: An R package for univariate and multivariate hierarchical point-referenced spatial models. *J. Stat. Softw.* **2007**, *19*. [CrossRef]
50. R Core Development, T. *A Language and Environment for Statistical Computing*; R Foundation for Statistical Computing: Vienna, Austria, 2008.
51. Gelfand, A.E. Hierarchical modeling for spatial data problems. *Spat. Stat.* **2012**, *1*, 30–39. [CrossRef]
52. Demirhan, H.; Kalaylioglu, Z. Joint prior distributions for variance parameters in Bayesian analysis of normal hierarchical models. *J. Multivar. Anal.* **2015**, *135*, 163–174. [CrossRef]
53. Duchêne, S.; Duchêne, D.A.; Di Giallonardo, F.; Eden, J.-S.; Geoghegan, J.L.; Holt, K.E.; Ho, S.Y.W.; Holmes, E.C. Cross-validation to select Bayesian hierarchical models in phylogenetics. *BMC Evol. Biol.* **2016**, *16*, 115. [CrossRef]
54. Spiegelhalter, D.J.; Best, N.G.; Carlin, B.P.; Van Der Linde, A. Bayesian measures of model complexity and fit. *J. R. Stat. Soc. Ser. B Stat. Methodol.* **2002**, *64*, 583–639. [CrossRef]
55. Jackman, S. Estimation and inference via bayesian simulation: An introduction to markov chain monte carlo. *Am. J. Pol. Sci.* **2000**, *44*, 375–404. [CrossRef]
56. Odeh, I.O.A.; McBratney, A.B.; Chittleborough, D.J. Further results on prediction of soil properties from terrain attributes: Heterotopic cokriging and regression-kriging. *Geoderma* **1995**, *67*, 215–226. [CrossRef]
57. Kupfersberger, H.; Deutsch, C.V.; Journel, A.G. Deriving constraints on small-scale variograms due to variograms of large-scale data. *Math. Geol.* **1998**, *30*, 837–852. [CrossRef]
58. Hengl, T.; Walvoort, D.J.J.; Brown†, A.; Rossiter, D.G. A double continuous approach to visualization and analysis of categorical maps. *Int. J. Geogr. Inf. Sci.* **2004**, *18*, 183–202. [CrossRef]
59. Webster, R.; Welham, S.J.; Potts, J.M.; Oliver, M.A. Estimating the spatial scales of regionalized variables by nested sampling, hierarchical analysis of variance and residual maximum likelihood. *Comput. Geosci.* **2006**, *32*, 1320–1333. [CrossRef]
60. Stoyan, D. *Statistical Analysis of Spatial and Spatio-Temporal Point Patterns*, 3rd ed.; CRC Press: New York, NY, USA. [CrossRef]
61. Sahu, S.K. *Bayesian Modeling of Spatio Temporal Data with R*, 1st ed.; Chapman and Hall/CRC: Southampton, UK, 2022.
62. Pascual, A.; Tupinambá-Simões, F.; de Conto, T. Using multi-temporal tree inventory data in eucalypt forestry to benchmark global high-resolution canopy height models. A showcase in Mato Grosso, Brazil. *Ecol. Inform.* **2022**, *70*, 101748. [CrossRef]
63. Box, G.E.P.; Cox, D.R. An analysis of transformations revisited, rebutted. *J. Am. Stat. Assoc.* **1982**, *77*, 209–210. [CrossRef]
64. El-Askary, H.; Abd El-Mawla, S.H.; Li, J.; El-Hattab, M.M.; El-Raey, M. Change detection of coral reef habitat using Landsat-5 TM, Landsat 7 ETM+ and Landsat 8 OLI data in the Red Sea (Hurghada, Egypt). *Int. J. Remote Sens.* **2014**, *35*, 2327–2346. [CrossRef]
65. Jia, K.; Wei, X.; Gu, X.; Yao, Y.; Xie, X.; Li, B. Land cover classification using landsat 8 operational land imager data in Beijing, China. *Geocarto Int.* **2014**, *29*, 941–951. [CrossRef]
66. Isaaks, E.H.; Srivastava, R.M. *Applied Geostatistics*; Oxford University Press: Oxford, UK, 1989.
67. Diggle, P.; Ribeiro, P.J. Model-Based Geostatistics. *J. R. Stat. Soc. C* **2007**, *846*, 15–27.
68. Baloloy, A.B.; Blanco, A.; Candido, C.G.; Argamosa, R.J.L.; Dumalag, J.B.L.C.; Dimapilis, L.L.C.; Paringit, E. Estimation of mangrove forest aboveground biomass using multispectral bands, vegetation indices and biophysical variables derived from optical satellite imageries: Rapideye, planetscope and sentinel-2. *ISPRS Ann. Photogramm. Remote Sens. Spat. Inf. Sci.* **2018**, *IV-3*, 29–36. [CrossRef]
69. Sovdat, B.; Kadunc, M.; Batič, M.; Milčinski, G. Natural color representation of Sentinel-2 data. *Remote Sens. Environ.* **2019**, *225*, 392–402. [CrossRef]
70. Wang, Q.; Li, J.; Jin, T.; Chang, X.; Zhu, Y.; Li, Y.; Sun, J.; Li, D. Comparative analysis of landsat-8, sentinel-2, and GF-1 data for retrieving soil moisture over wheat farmlands. *Remote Sens.* **2020**, *12*, 2708. [CrossRef]

71. Jiang, F.; Kutia, M.; Ma, K.; Chen, S.; Long, J.; Sun, H. Estimating the aboveground biomass of coniferous forest in Northeast China using spectral variables, land surface temperature and soil moisture. *Sci. Total Environ.* **2021**, *785*, 147335. [CrossRef] [PubMed]
72. Dang, A.T.N.; Nandy, S.; Srinet, R.; Luong, N.V.; Ghosh, S.; Senthil Kumar, A. Forest aboveground biomass estimation using machine learning regression algorithm in Yok Don National Park, Vietnam. *Ecol. Inform.* **2019**, *50*, 24–32. [CrossRef]
73. Takagi, K.; Yone, Y.; Takahashi, H.; Sakai, R.; Hojyo, H.; Kamiura, T.; Nomura, M.; Liang, N.; Fukazawa, T.; Miya, H.; et al. Forest biomass and volume estimation using airborne LiDAR in a cool-temperate forest of northern Hokkaido, Japan. *Ecol. Inform.* **2015**, *26*, 54–60. [CrossRef]
74. Kramer, C.Y. Extension of multiple range tests to group means with unequal numbers of replications. *Biometrics* **1956**, *12*, 307–310. [CrossRef]
75. Chinembiri, T.S.; Bronsveld, M.C.; Rossiter, D.G.; Dube, T. The precision of C stock estimation in the ludhikola watershed using model-based and design-based approaches. *Nat. Resour. Res.* **2013**, *22*, 297–309. [CrossRef]
76. Tveito, O.E.; Wegehenkel, M.; van der Wel, F. The use of geographic information systems in climatology and meteorology. *SAGE* **2003**, *27*. [CrossRef]
77. Gupta, A.; Kamble, T.; Machiwal, D. Comparison of ordinary and Bayesian kriging techniques in depicting rainfall variability in arid and semi-arid regions of north-west India. *Environ. Earth Sci.* **2017**, *76*, 512. [CrossRef]
78. Dube, T.; Mutanga, O. Evaluating the utility of the medium-spatial resolution Landsat 8 multispectral sensor in quantifying aboveground biomass in uMgeni catchment, South Africa. *ISPRS J. Photogramm. Remote Sens.* **2015**, *101*, 36–46. [CrossRef]
79. Li, S.; Ganguly, S.; Dungan, J.L.; Wang, W.; Nemani, R.R. Sentinel-2 MSI radiometric characterization and cross-calibration with landsat-8 OLI. *Adv. Remote Sens.* **2017**, *6*, 147–159. [CrossRef]
80. Ver Hoef, J. Sampling and geostatistics for spatial data. *Écoscience* **2002**, *9*, 152–161. [CrossRef]
81. Babcock, C.; Finley, A.O.; Cook, B.D.; Weiskittel, A.; Woodall, C.W. Modeling forest biomass and growth: Coupling long-term inventory and LiDAR data. *Remote Sens. Environ.* **2016**, *182*, 1–12. [CrossRef]
82. FAO. *Global Forest Resources Assessment Country Report, Zimbabwe*; FAO: Rome, Italy, 2005.
83. Wu, C.; Shen, H.; Shen, A.; Deng, J.; Gan, M.; Zhu, J.; Xu, H.; Wang, K. Comparison of machine-learning methods for above-ground biomass estimation based on Landsat imagery. *J. Appl. Remote Sens.* **2016**, *10*, 35010. [CrossRef]
84. Xiong, Y.; Wang, H. Spatial relationships between NDVI and topographic factors at multiple scales in a watershed of the Minjiang River, China. *Ecol. Inform.* **2022**, *69*, 101617. [CrossRef]
85. Meyer, H.L.; Marco, H.; Burkhard, B.; Joseph, P.; Dirk, P. Comparison of landsat-8 and sentinel-2 data for estimation of leaf area index in temperate forests. *Remote Sens.* **2019**, *11*, 1160. [CrossRef]

remote sensing

MDPI

Article

Integrating Semi-Supervised Learning with an Expert System for Vegetation Cover Classification Using Sentinel-2 and RapidEye Data

Nasir Farsad Layegh [1,*], Roshanak Darvishzadeh [1], Andrew K. Skidmore [1], Claudio Persello [1] and Nina Krüger [2]

[1] Faculty of Geo-Information Science and Earth Observation (ITC), University of Twente, 7500 AE Enschede, The Netherlands; r.darvish@utwente.nl (R.D.); a.k.skidmore@utwente.nl (A.K.S.); c.persello@utwente.nl (C.P.)

[2] M.O.S.S. Computer Grafik Systeme GmbH, Hohenbrunner Weg 13, 82024 Munich, Germany; nkrueger@moss.de

* Correspondence: farsad_layegh@yahoo.com

Abstract: In complex classification tasks, such as the classification of heterogeneous vegetation covers, the high similarity between classes can confuse the classification algorithm when assigning the correct class labels to unlabelled samples. To overcome this problem, this study aimed to develop a classification method by integrating graph-based semi-supervised learning (SSL) and an expert system (ES). The proposed method was applied to vegetation cover classification in a wetland in the Netherlands using Sentinel-2 and RapidEye imagery. Our method consisted of three main steps: object-based image analysis (OBIA), integration of SSL and an ES (SSLES), and finally, random forest classification. The generated image objects and the related features were used to construct the graph in SSL. Then, an independently developed and trained ES was used in the labelling stage of SSL to reduce the uncertainty of the process, before the final classification. Different spectral band combinations of Sentinel-2 were then considered to improve the vegetation classification. Our results show that integrating SSL and an ES can result in significantly higher classification accuracy (83.6%) compared to a supervised classifier (64.9%), SSL alone (71.8%), and ES alone (69.5%). Moreover, utilisation of all Sentinel-2 red-edge spectral band combinations yielded the highest classification accuracy (overall accuracy of 83.6% with SSLES) compared to the inclusion of other band combinations. The results of this study indicate that the utilisation of an ES in the labelling process of SSL improves the reliability of the process and provides robust performance for the classification of vegetation cover.

Keywords: semi-supervised learning; expert system; object-based image analysis; Sentinel-2; vegetation cover; image classification

1. Introduction

Accurate mapping of vegetation cover in an ecosystem can help to initiate its protection and restoration programmes efficiently [1]. It is, therefore, necessary to acquire accurate and up-to-date information about the status of the vegetation cover of an ecosystem through regular monitoring. Traditional field inventory monitoring methods of vegetation covers are usually expensive and time-consuming [2]. Thus, a viable approach can be the use of satellite remote sensing data. Satellite data provide advantages such as large area coverage, ongoing data collection, and cost-effectiveness for monitoring and mapping purposes [3,4]. The advent of remote sensing technologies, with high-resolution multispectral satellite sensors, has provided new opportunities to monitor vegetation cover at different spatial and temporal scales. For instance, Sentinel-2 MSI, covering a wide spectral range (400–2400 nm) including three red-edge bands, may allow more efficient discrimination of different vegetation types.

A common approach to extracting information from satellite data, to distinguish different vegetation types, is using image classification algorithms [5,6]. Usually, there are three common challenges involved in image classification that could affect its accuracy: (1) collecting a sufficient number of training samples, (2) creating a balanced training and test set, and (3) fine-tuning the algorithm parameters to obtain the optimum performance [7–9]. A relevant solution to these challenges is the introduction of the semi-supervised learning (SSL) technique [10], which uses relatively few labelled samples and a large number of unlabelled data to train a model [11].

The existing paradigms of SSL for the classification of remote sensing data can be divided into four major categories: (1) generative mixture models, such as expectation–maximisation algorithms [12,13], (2) low-density separation algorithms, such as transductive support vector machines (TSVMs) [14,15], (3) self-learning methods [16–19], and (4) graph-based methods [20–23]. Among the SSL methods, graph-based approaches have recently received significant attention due to their ability to provide a relatively high classification accuracy while retaining computational simplicity [24–27]. While graph-based algorithms can improve the classification performance by using the distribution of unlabelled samples, they have some limitations [9]. One of these limitations occurs in complex classification tasks, such as identifying different classes in heterogeneous vegetation covers. In such a case, samples from the same vegetation class may show low similarity (i.e., high intra-class variability), and two samples from two different vegetation classes show high similarity (i.e., low inter-class variability). This "similarity" problem can confuse the graph-based algorithm, and the semi-labelled samples may not have the correct label. In this case, unlabelled samples can be detrimental to the graph-based algorithm, as they may degrade the accuracy by misguiding the classifier [11]. One of the common approaches to tackling the similarity problem is using non-parametric classifiers since they do not make any underlying assumptions about the distribution of data [7]. However, these classifiers require a representative amount of training data to estimate the mapping function, and they are also subject to overfitting. Consequently, in studies such as vegetation cover classification where there might be an imbalanced distribution of features and training samples, these classifiers would underperform [28]. To help solve the mentioned problem of SSL, this study aimed to use expert knowledge within an independently developed and trained expert system (ES) in the labelling process of a graph-based algorithm. The contribution of an ES can help the problem by refining the semi-labelled samples. In this context, expert knowledge is defined as the experience and existing knowledge of the expert in the specific domains of study, technical practices, and prior information on the study area [29,30]. The developed ES should have the ability to classify the unlabelled samples using expert knowledge, independently from SSL. This ability of the ES can help SSL to assign the most certain class label to the unlabelled samples by filtering out samples with less certain labels.

Motivated by the above insights, in this study, a novel classification approach was proposed for the classification of satellite images by integrating graph-based SSL and an ES (SSLES). The main idea of the proposed approach was to construct a graph in SSL, based on image features, and use an ES in the labelling process of SSL to assign the most probable class labels to the selected unlabelled samples and then perform the classification using a standard supervised classifier. The study specifically aimed to address two objectives, as follows: (1) to investigate the performance of SSLES for vegetation cover classification and (2) to investigate the potential of Sentinel-2 spectral data for vegetation cover classification.

2. Study Area and Materials

2.1. Study Area

The study area was Schiermonnikoog Island in the Netherlands, located between 53°27′20″N–53°30′40″N latitude and 06°06′35″E–06°20′56″E longitude, with an area of 199.1 km^2 (Figure 1). The vegetation cover on the south and south-east shore of the island has adapted to the regular inundation of seawater and has formed a salt marsh [31].

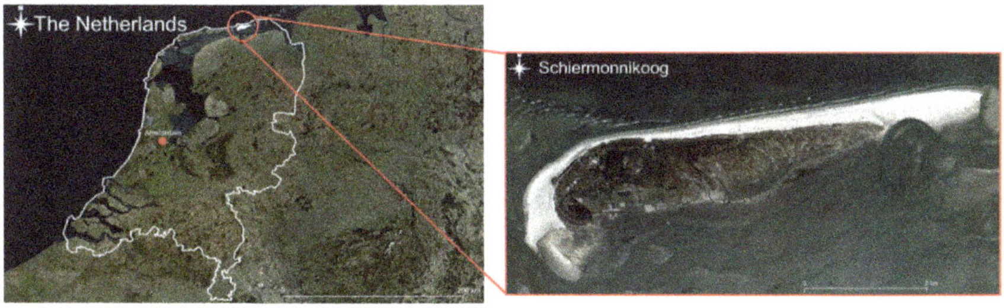

Figure 1. Location of Schiermonnikoog national park.

In this study, vegetation species of the island were categorised into 10 functional groups for classification based on the reference vegetation map [32], namely: high matted grass, low matted grass, agriculture, forest, green beach, tussock grass, high shrub, herbs, low salix shrub, low hippopahe shrub. The natural vegetation cover has a large spatial and temporal variability, due to the dynamic influences of the tide, wind, and grazing [33,34]. This area was chosen to test the proposed classification methodology as it is representative of a diverse and mixed vegetation cover.

2.2. Materials

2.2.1. Sentinel-2 Data

The main satellite imagery used in this study was the standard Sentinel-2 Level-1C product, which is in UTM/WGS84 projection, and its per-pixel radiometric measurements are provided in top of atmosphere (TOA) reflectance [35]. The Sentinel-2 image of the study area was acquired on 17 July 2016 belonging to the relative orbit of R008 and was downloaded from ESA Sentinel-2 Pre-operation Hub (https://scihub.copernicus.eu/, accessed on 30 July 2016). The image was chosen from July to obtain a cloud-free image with vigorously growing vegetation. The atmospheric correction of the image was performed using Sen2Cor software [35], and the top of canopy (TOC) reflectance was calculated for further analysis.

Sentinel-2 offers a multispectral sensor in 13 bands from 443 to 2190 nm with three different geometric resolutions as follows:

- 10 m resolution bands: blue (490 nm), green (560 nm), red (665 nm), and near-infrared (842 nm).
- 20 m resolution bands: four red-edge/NIR bands with central wavelength at 705 nm, 740 nm, 783 nm, and 865 nm, respectively, and shortwave infrared-1 and -2 (1610 nm and 2190 nm).
- 60 m resolution bands: coastal (443 nm), water vapour (1375 nm), and cirrus (1376).

In this study, all the Sentinel-2 bands were resampled to 5-meter resolution for processing so that they had the same resolution as the RapidEye image.

To achieve a higher classification accuracy and assess the capability of Sentinel-2 data in classifying the vegetation types, its spectral bands were combined into different groups for subsequent assessment to find the most informative band combination for vegetation classification accuracy. Based on previous studies, the most important regions of the spectrum to study vegetation cover are the red-edge, shortwave, and red–infrared regions [4,36–40]. Consequently, six groups of band combinations were considered to classify vegetation cover, as follows:

- Group 1: All spectral bands;
- Group 2: Red and infrared bands;
- Group 3: All shortwave infrared bands;
- Group 4: All red-edge bands;
- Group 5: Red, infrared, and red-edge bands;
- Group 6: Red-edge and shortwave infrared bands.

2.2.2. RapidEye Data

In this study, a RapidEye image was also acquired for Schiermonnikoog Island on 18 July 2015. The pre-processed data were obtained at level 3A, which means radiometric and geometric corrections, as well as geo-referencing, were applied. The image covers 25 km × 25 km with the orthorectified pixel size of 5 m × 5 m. Due to clear weather conditions during the image acquisition, no further atmospheric correction was applied. Both Sentinel-2 and RapidEye images were chosen from a similar time of year to ensure that in both, the vegetation cover is as alike as possible. In this study, the RapidEye image was used for segmentation only, and the features were extracted from Sentinel-2 data.

2.2.3. Reference Data and Sampling

The reference data used in this study included field observations of dominant vegetation species for 30 vegetation plots (30 m × 30 m) collected in July 2015 and a vegetation map belonging to 2010 [32]. This map was obtained from experts' visual interpretation of aerial photographs (1:10,000) combined with extensive field inventory, and it included the same vegetation classes as this study.

To select the training and test samples, stratified random sampling was implemented, where each vegetation class was considered a stratum [41]. The resulting training sample size became 650 for the 10 vegetation classes, and the samples were extracted from the vegetation plots. Table 1 reports the number of samples per stratum. In addition, 434 more samples were identified as an independent test set (2/3 of the number of the training samples). A sample in the context of this study is referred to as an image object (as the result of image segmentation), representing a vegetation patch on the ground.

Table 1. The vegetation classes in Schiermonnikoog Island and the number of their collected samples.

Class Name	Number of Training Samples	Number of Test Samples
High matted grass	160	107
Low matted grass	142	95
Agriculture	71	47
Forest	58	39
Green beach	58	39
Tussock grass	45	30
High shrub	45	30
Herbs	35	23
Low salix shrub	25	17
Low hippopahe shrub	11	7
Sum	650	434

2.2.4. Knowledge Sources

In this study, three main sources of knowledge were identified to be used as input to build the ES knowledge base. These sources were different and separate from the reference maps used for the sampling process:

- A reference vegetation map of the study area, generated in 2010 [32]. As this map was generated with experts' visual interpretation of aerial photographs and extensive fieldwork, it contained some level of experts' knowledge.
- Ancillary data, including field records of dominant vegetation types for 30 vegetation plots, NDVI data from the Sentinel-2 image, and a digital elevation model (DEM) of the island (produced from laser altimetry by the Dutch ministry of public works, Rijkswaterstaat) to generate height, slope, aspect, etc.
- Published resources about vegetation cover in Schiermonnikoog and its ecology [31,42–44].

3. Methods

The architecture of the proposed classification approach contained three parts: object-based image analysis (OBIA), semi-supervised learning and expert system (SSLES), and classification. Using OBIA, the satellite image was segmented to generate image objects and features. Using SSLES, the number of training samples was increased by labelling a set of most informative unlabelled samples. In the final step, classification was performed on the datasets using a standard supervised classifier.

3.1. Object-Based Image Analysis (OBIA)

The segmentation method of the mean shift was used to generate image objects [45]. This algorithm requires two parameters to be tuned to obtain an optimal segmentation result:

- Spatial radius h_s (spatial distance between classes);
- Range radius h_r (the spectral difference between classes).

The segmentation was performed using a 5 m × 5 m RapidEye image, due to its finer spatial resolution compared to Sentinel-2. This could result in a higher spatial accuracy of image objects [46]. For a quantitative evaluation of the segmentation results, the method proposed by [47,48] was used which measures both topological and geometric similarity between the segmented and the reference objects. The method relies on the ratio of the intersected area of the segments and the reference objects, and depending on the size of overlap for each segment and object, over- and under-segmentation indices are calculated.

Using the segmented map and the reference vegetation map, labelled and unlabelled objects were generated. For this, the segmented map was overlaid with the reference samples' layer (obtained from the sampling process that contained training and test samples). Any image object that contained the centroid of a reference sample and had more than 50% overlap with it was treated as a labelled object. The rest were considered unlabelled objects.

Sentinel-2 data were resampled to 5 m resolution to match the spatial resolution of RapidEye data and then be able to extract the features using the segmentation results. For mapping vegetation, three categories of features were considered, which have been recognised as important in previous studies [37,49–53], as follows: (a) a set of spectral features consisting of the mean, standard deviation, median, and minimum and maximum values of pixels within an image object, (b) a set of textural features including GLCM (grey-level co-occurrence matrix) and GLDV (grey-level difference vector), and (c) a set of geometrical features representing the area and perimeter of the image objects.

The final results of OBIA are image objects with a corresponding informative set of image features.

3.2. Semi-Supervised Learning

The graph-based semi-supervised learning (SSL) method was used to increase the training samples, before classification. This method aims to construct a graph $G = (V, E)$ connecting similar samples. V consists of N = L + U samples, where L and U are the numbers of labelled and unlabelled samples, respectively. Edges E are represented typically by a symmetric similarity weight matrix $W \in R^{N \times N}$ [26]. The k-nearest neighbour (KNN) approach was used to construct the weight matrix, denoted as W^W:

$$W_{i,j}^W \begin{cases} exp^{-\frac{\|x_i - x_j\|^2}{2\sigma^2}} & if \quad x_j \in NB_K^W(x_i) \\ 0 & Otherwise \\ & i,j \in \{1,2,\ldots,l+u\} \end{cases} \tag{1}$$

where σ is the Gaussian kernel bandwidth, $\|x_i - x_j\|$ is the similarity measure between two samples, and $NB_K^W(x_i)$ is a set of K nearest neighbours of sample x_i. In this study, the similarity measure between two samples was based on image features obtained from OBIA.

To construct the graph and assign the class labels to the unlabelled samples, the energy function proposed by [54] was used to be optimised, defined as follows:

$$\begin{aligned} &\min_f \sum_{i\in\{1,2,\ldots,l\}} (f_i - y_i)^2 + \tfrac{1}{2}\sum_{i,j\in\{1,2,\ldots,l+u\}} W^W_{i,j}(f_i - f_j)^2 \\ &= (f_l - y_l)^T (f_l - y_l) + \tfrac{1}{2} f^T \Delta f \end{aligned} \quad (2)$$

where $f = (f_L, f_U)^T$ is composed of f_L and f_U which are the predicted class labels of the labelled and unlabelled samples. y_i is the vector class label of the sample i. Δ is a graph Laplacian matrix obtained by $\Delta = D - W$, and D is the diagonal degree matrix given by $D_{ii} = \sum_j W_{ij}$.

The label propagation technique was employed at the end to propagate the information through the graph to the unlabelled samples [11,55]. For this purpose, the weight of the edges in the graph was computed, according to Equation (2), and then the probability matrix was estimated as $P = D^{-1}W^W$. The edge with the highest probability is the determiner of the label for the unlabelled sample.

3.3. *Expert System*

The expert system (ES) approach used in this study was described in detail by [56]. The ES, here, was developed to answer the question of "What vegetation type is probable to occur in a given image object?" in reference to the samples that obtained labels with SSL.

Bayes' theory was used in the ES to compute the probability of the rule that the hypothesis (H_a) occurs in an image object given a piece of evidence (E_b), i.e.,

$$P(H_a|E_b) = \frac{P(E_b|H_a)P(H_a)}{P(E_b)} \quad (3)$$

where $P(E_b | H_a)$ is the *a priori* conditional probability that there is a piece of evidence E_b (e.g., a mean slope of less than 0.1°) given a hypothesis H_a (e.g., "high shrub" class) that class C_i occurs in a specific image object. $P(H_a)$ is the probability for the hypothesis (H_a) that class C_i occurs in an object. $P(E_b)$ is the probability that an object has an item of evidence $\{E_b\}$. The following steps were taken to compute the probability rules:

- Generate a histogram population of each feature layer in the knowledge base;
- Divide each histogram into 10 quantiles, representing the frequency of the occurrence of each class at each percentile of the feature layer;
- Normalize the frequency values by fitting a normal distribution.

3.4. *SSLES Algorithm*

Algorithm 1 demonstrates the inputs, output, and steps of SSLES. In this algorithm, which is fully implemented in MATLAB_R2015b, the inputs are image objects with the respective set of features. Further, steps 1–4 are related to SSL which generates a graph based on image features and propagates the class labels to the potential unlabelled samples. In step 5, the developed ES performs an independent class label prediction on the semi-labelled samples of the SSL in step 4. Finally, if both the ES and the SSL agree on the same class label, the labelled sample is added to the training set; otherwise, it is returned to the unlabelled set. The flowchart of the algorithm is presented in Figure 2.

Algorithm 1: SSLES

Inputs:

- A set of labelled objects O_L
- A set of unlabelled objects O_U
- Extracted features f_i for each object

Outputs:

- New labelled samples to be added to the original training set

Steps:

For every $o_i \in O_L$, i = 1:L

1. Measure the similarity of all labelled and unlabelled samples in the image feature space
2. Sort the similarity values and choose the first K unlabelled samples with the highest similarity values
3. Construct the graph with the KNN method, then calculate the matrix P
4. Predict the class label of the unlabelled samples using the label propagation framework
5. Predict the class label of the unlabelled samples using the developed ES
6. If both approaches have predicted the same class label for a sample, then add the object to the training set, otherwise, let it remain with the unlabelled object set

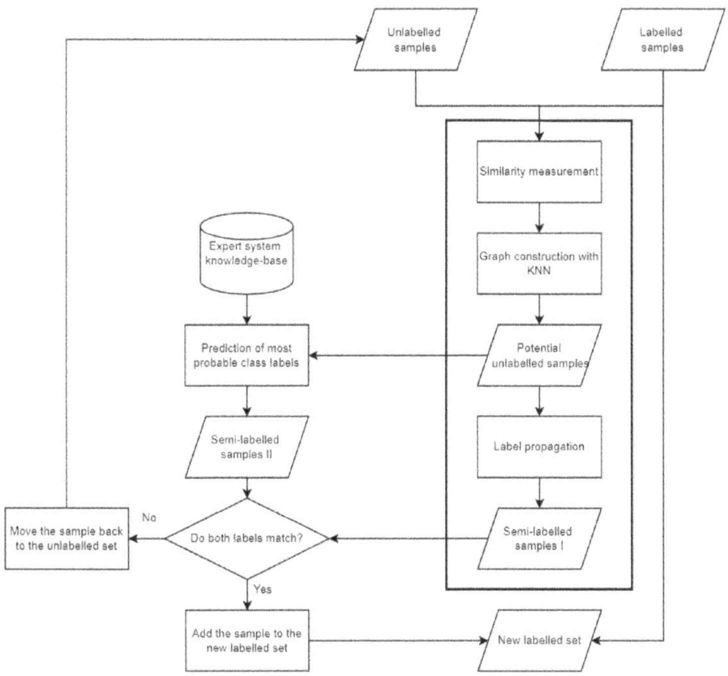

Figure 2. SSLES flowchart.

3.5. Classification and Evaluation

The final step is classification, where a standard supervised classifier is implemented. This step is performed using the random forest (RF) classifier [57,58]. This classifier has two parameters that needed to be set:

- The number of classification trees, i.e., the number of bootstrap iterations (*ntree*);
- The number of input variables used at each node (*mtry*).

Several studies have demonstrated that the default value for *mtry* can provide satisfactory results [59–61]. Therefore, this parameter is set to the default value, i.e., the square root of the total number of input features.

In this study, the RF classifier was used in three different scenarios. In the first scenario, the training set generated from SSLES was used to train the classifier; in the second scenario, the training set generated from SSL was used to train the classifier; and in the third scenario, the classifier was trained with the original training set.

For the validation of the classification results, the accuracy assessment based on the error matrix was conducted [62]. The evaluation was performed using the same test samples extracted from the reference data, and the results were evaluated in terms of overall accuracy (OA) and Cohen's kappa coefficient. To assess the statistical significance of the difference between the obtained accuracies, McNemar's test was used with a 95% confidence level and 1 degree of freedom.

4. Results

4.1. Object-Based Image Analysis

To start the image segmentation process, the parameters were tuned and evaluated. h_s and h_r were iterated in the potential range of [1:10]. Using sensitivity analysis, $h_s = 5$ and $h_r = 7$ were chosen as the optimum values. As the result of image segmentation, a total number of 5230 image objects were delineated with a mean size of 30 pixels. Figure 3 illustrates the final segmented objects on the false colour composite Sentinel-2 image of the study area.

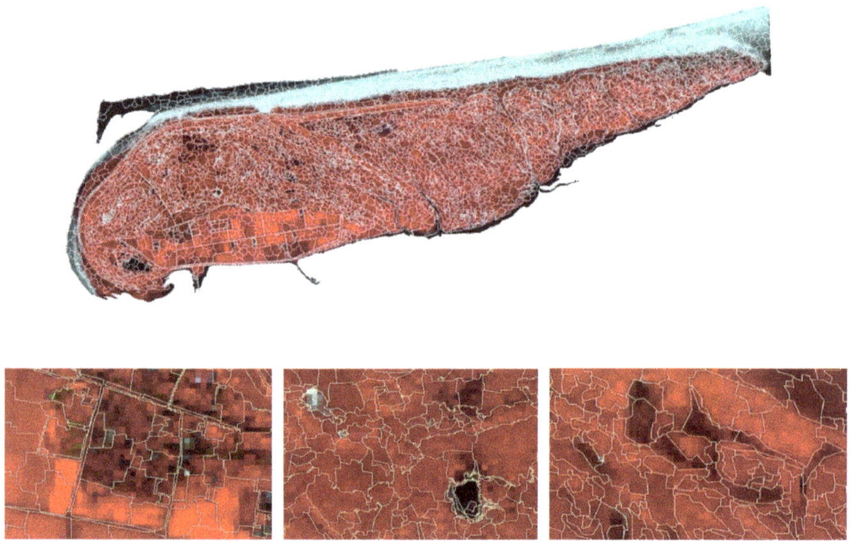

Figure 3. Segmentation result of RapidEye image with $h_s = 5$ and $h_r = 7$, on a false colour composite of Sentinel-2 (B08, B04, B03).

Next, the image objects were divided into three data sets: training, test (labelled), and unlabelled objects. The numbers of generated datasets are listed in Table 2.

Table 2. The number of image objects for three different subsets of image objects.

Training Objects	Test Objects	Unlabelled Objects	Total Number of Objects
650	434	4146	5230

4.2. Semi-Supervised Learning

To depict the idea of label propagation, a graph is shown in Figure 4 representing a set of labelled and unlabelled samples. The labels are propagated from the labelled to the unlabelled samples based on the probabilities of the arrows. In this example, although two "Herbs" samples are connected to the unlabelled sample in the centre, it is labelled as "Forest" since the probability value of the "Forest" sample is higher than "High Shrub" and two "Herbs".

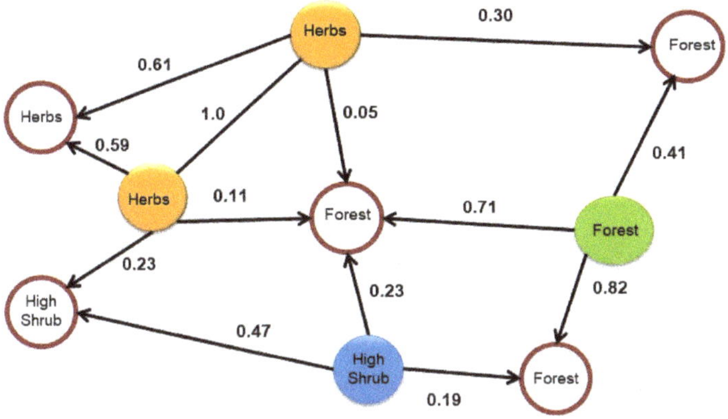

Figure 4. Illustration of label propagation procedure. Coloured circles represent four different labelled samples, and white circles represent unlabelled samples. Values indicate the probability of edges. Propagation direction is shown by the direction of the arrows, i.e., the arrow is always from a labelled sample to an unlabelled sample.

The final output of SSL is semi-labelled samples that might have incorrect labels. Following the generation of semi-labelled samples, an ES was used to classify the semi-labelled samples again, in parallel to SSL. Samples that obtained the same class label as SSL were merged with the original training set to generate a new extended training set; otherwise, they were moved back to the unlabelled pool.

4.3. Expert System a Priori Probabilities

The reference vegetation map was used to estimate the *a priori* probabilities for the vegetation classes and the initial conditional probabilities for all the feature layers (Table 3).

Table 3. A priori probability estimation of the vegetation classes. The probability of occurrence of each vegetation type in the study area is presented. HMG: High Matted Grass, LMG: Low Matted Grass, Ag: Agriculture, Fr: Forest, GB: Green Beach, TG: Tussock, HS: High Shrub, Hr: Herbs, LSS: Low Salix Scrub, and LHS: Low Hippopahe Shrub.

Class Name	HMG	LMG	Ag	Fr	GB	TG	HS	Hr	LSS	LHS
Probability	0.28	0.24	0.1	0.08	0.08	0.06	0.06	0.05	0.03	0.01

The ancillary data were used to extract training samples, and then their feature values were statistically analysed, i.e., using the methodology explained in Section 3.3, to define the probability of occurrence of a vegetation type within an image object. The result is six sets of rules for each feature layer that contains the probability values of occurrence of evidence (i.e., a set of features) given a hypothesis (i.e., specific vegetation type). These rules belong to the mean and standard deviation values of the feature layers (Figure 5).

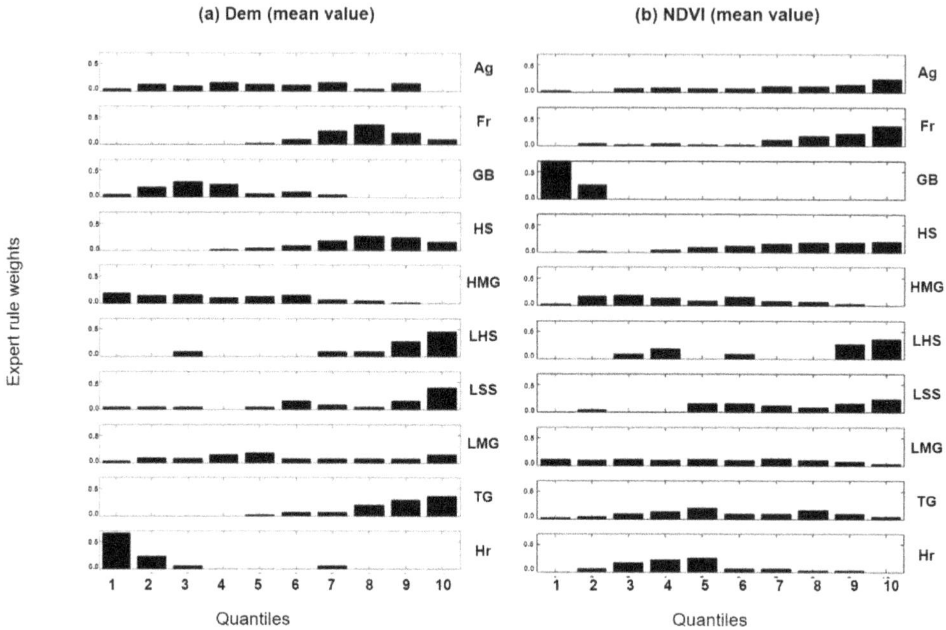

Figure 5. Example of expert rule weights for ancillary data. Y-axis shows the initial conditional probability, and the X-axis shows the 10 quantiles.

Figure 5 illustrates how rules were derived for the mean values of the ancillary data. These probability rules were applied to each input sample in order to compute the probability values. The black bars indicate the probability of that sample belonging to each class. Based on three derived facts from the published resources, three more rules were generated. The derived facts were (1) the dependency of some classes, e.g., herbs, on water availability (streams), (2) the presence/absence of some classes around the residential areas, e.g., green beach, and (3) a forestry programme adjacent to the village [63]. Using the method described before, the distances of the vegetation classes from the water streams and the residential area were analysed and then divided into three quantiles, and the probability of occurrence of vegetation classes at each distance quantile was computed (Table 4).

Table 4. Expert rules are based on the distance of the vegetation classes to streams and residential areas. Values describe the probability of occurrence of vegetation class at each distance quantile.

Vegetation Classes	Distance to Streams			Distance to the Residential Area		
	Quantile 1	Quantile 2	Quantile 3	Quantile 1	Quantile 2	Quantile 3
Ag	0.11	0.31	0.58	0.90	0.10	0.00
Fr	0.06	0.22	0.72	0.41	0.34	0.25
HMG	0.40	0.45	0.15	0.00	0.15	0.85
LMG	0.22	0.31	0.47	0.23	0.46	0.31
TG	0.09	0.21	0.70	0.27	0.55	0.18
HS	0.19	0.25	0.56	0.13	0.45	0.42
LHS	0.00	0.29	0.71	0.00	0.04	0.96
LSS	0.29	0.10	0.61	0.24	0.76	0.00
GB	0.00	0.51	0.49	0.00	0.01	0.99
Hr	0.61	0.38	0.01	0.01	0.00	0.99

According to the extracted probability rules in Table 4, each sample's feature values were examined and depending on the percentile that it lay in, a probability value was assigned. Eventually, by an iteration through the nine rule sets, samples gained nine sets of probabilities for each class. Then, these probabilities were merged into a final combined probability, and the class with the highest probability value was the class of that sample.

4.4. SSLES Results

After running SSLES, 1513 new samples were labelled. As part of the process, these newly labelled samples were combined with the original training set to generate the new training set. The table below (Table 5) summarizes the number of training samples for each class.

Table 5. The number of original training samples, newly labelled samples, and new training samples.

Class Name	Number of Original Training Samples	Number of Newly Labelled Samples	Number of New Training Samples
HMG	160	494	654
LMG	142	290	432
Ag	71	167	238
Fr	58	121	179
GB	58	106	164
TG	45	99	144
HS	45	89	134
Hr	35	67	102
LSS	25	48	73
LHS	11	32	43
Sum	650	1513	2163

Unlike most SSL implementations that exploit the labelled samples iteratively until all the samples have a label, in this study, graph-based SSL was followed by a basic supervised classifier. As shown by [64], this implementation has the advantage of reducing the computational complexity of the algorithm and can classify more new samples as well.

4.5. Classification Results and Evaluation

4.5.1. Parameter Tuning

To evaluate the performance of SSLES, three classification scenarios were conducted with SSL only, ES only, and RF methods for comparison. Before conducting the experiments, three parameters needed to be tuned to obtain the optimum results. For this, the values of the parameters were changed in a potential range, and the values yielding the highest OA were selected as the final parameters' values. The parameters were the k-nearest neighbour and kernel bandwidth belonging to the SSL and *ntree* belonging to the RF classifier, which were tuned in the range of $k = \{1, 2, \ldots, 20\}$, $\sigma = \{0.1, 0.2, \ldots, 2\}$, $ntree = \{10, 20, \ldots, 800\}$, respectively, based on previous studies. The values of parameters for graph construction were set as $k = 16$ and $\sigma = 0.2$, and regarding the RF classifier, the parameter was set to $ntree = 80$ which led to the highest OA.

4.5.2. Classification Evaluation

After tuning the parameters, the experiments were carried out. Table 6 reports the obtained classification scores for the six groups of band combinations using four methods.

Table 6. Classification results for the six groups of band combinations of Sentinel-2 data, in terms of OA and kappa coefficient. Group 1: (All spectral bands), Group 2: (Red and infrared bands), Group 3: (All shortwave infrared bands), Group 4: (All red-edge bands), Group 5: (Red, infrared, and red-edge bands), Group 6: (Red-edge and shortwave infrared bands).

Dataset	SSLES		RF		SSL Only		ES Only	
	OA (%)	Kappa	OA (%)	Kappa	OA (%)	Kappa	OA (%)	Kappa
Group 1	81.1	0.67	64.6	0.52	70.9	0.60	68.1	0.55
Group 2	73.5	0.57	58.9	0.44	62.3	0.48	60.9	0.46
Group 3	74.6	0.59	60.1	0.47	63.8	0.49	63.1	0.48
Group 4	83.6	0.70	64.9	0.56	71.8	0.61	69.5	0.57
Group 5	67.8	0.49	47.2	0.33	55.3	0.40	53.8	0.38
Group 6	79.9	0.67	64.3	0.55	58.2	0.59	68.7	0.57

As it can be observed in Table 6, SSLES produces relatively higher accuracy for all of the six datasets when compared to SSL alone, ES alone, and RF. The highest accuracy was achieved using only the red-edge bands. Figure 6 illustrates the final classified map of the study area with Group 4 of band combination, using the SSLES for classification.

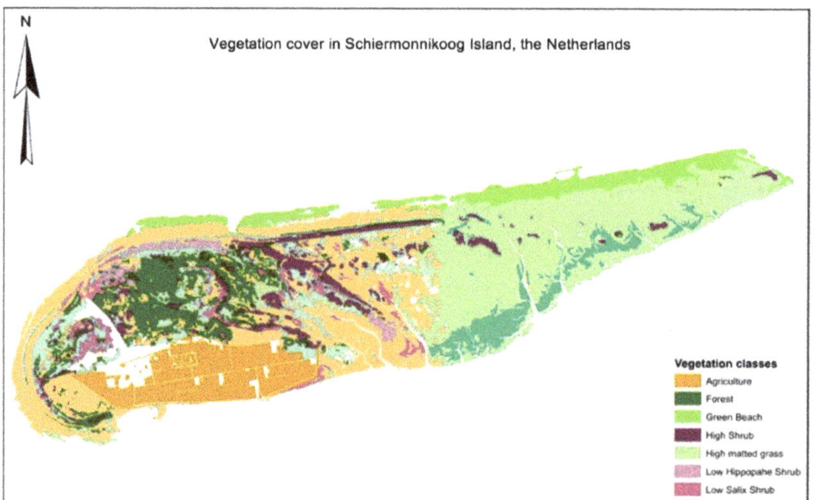

Figure 6. Classified map of the study area using the Group 4 of band combinations, red-edge spectral bands (as this group achieved the highest accuracy) of Sentinel-2 data, and the SSLES method.

Classification with Groups 1 and 6 also yielded noticeably high OA, compared to Group 4. Examining the results of these three groups with McNemar's test revealed that there are no statistically significant differences between the obtained accuracies from Groups 1, 4, and 6. The figure below (Figure 7) shows an example of an area (i.e., east of the island) where the four classification methods provided different results.

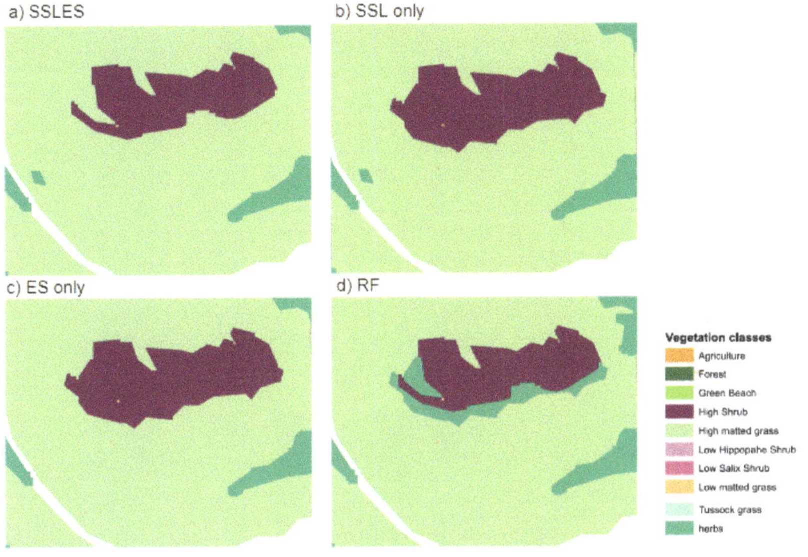

Figure 7. Example of classification result using (**a**) SSLES, (**b**) SSL only, (**c**) ES only, and (**d**) RF.

The confusion matrix of SSL is shown in Figure 8 to report the producers' accuracies of vegetation classes. Results are presented as a 2D plot where colours represent the accuracy of each class.

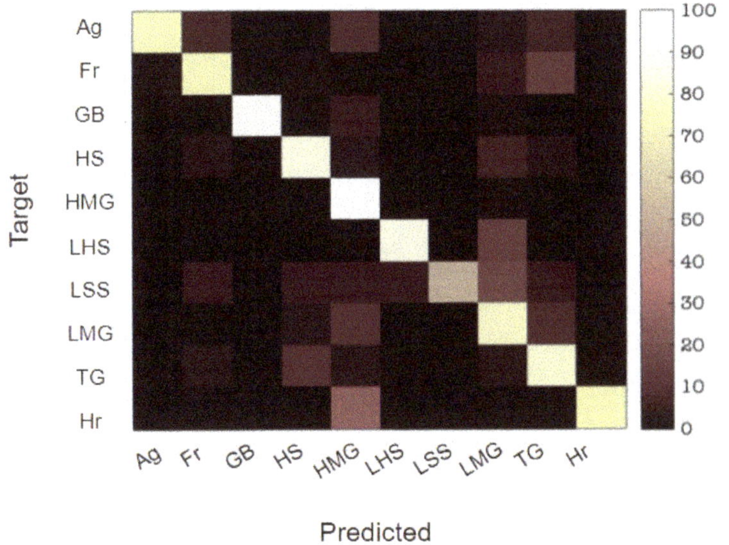

Figure 8. Producer's accuracy obtained with SSLES. The colours illustrate the accuracy, where white colour represents 100% accuracy, and black represents 0% accuracy.

In Figure 8, the highest accuracy classifications have a light/white colour in the main diagonal and dark/black colour in the other cells, which means no misclassification.

5. Discussion

The obtained results in Table 6 show that SSLES can yield higher classification accuracy than SSL. Furthermore, using the red-edge bands provided the highest accuracy, which confirmed the findings of [3] where the importance of the Sentinel-2 red-edge bands for vegetation classification was highlighted. As is shown in Figure 8, a considerable number of "Herbs" are classified as "High matted grass", and the classifier has confused the "Low matted grass" and "High matted grass" with other classes. This can probably be justified by (i) the high diversity of these two classes in the study area and (ii) the low diversity of training samples belonging to these classes. To gain better insight into SSLES's performance and the advantage of using an ES integrated with SSL, the confusion matrix obtained from SSLES was subtracted from the one from the SSL, and the results are presented in a new matrix. In the resulting matrix, if the value of a cell increased, it is labelled "**positive change**"; if the value decreased, it is labelled as "**negative change**"; otherwise, the cell was given a "**no change**" label (Figure 9).

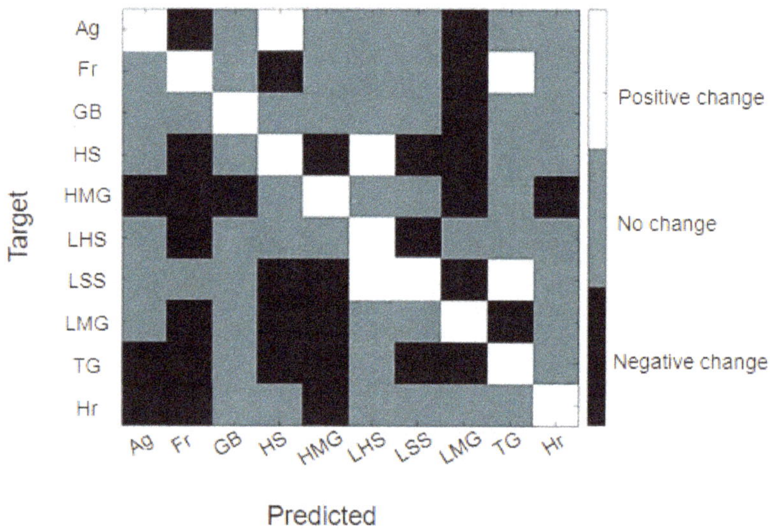

Figure 9. A detailed comparison of two classification methods, the matrix is the result of subtracting the confusion matrix of the SSLES from the SSL. The white colour represents an increase in the cell's value, black indicates a decrease, and if the value was unchanged, it was coloured cyan.

Comparing the obtained matrices reveals that the classification accuracies of all vegetation classes improved, except for five off-diagonal cells. In an ideal situation, it is expected that all the positive changes happen in the main diagonal elements and not in the off-diagonal elements, which means a decrease in misclassification. This result implies that the contribution of the ES to the labelling process of SSL has the advantages of removing the less reliable semi-labelled samples and increasing the overall quality of training samples obtained by SSL.

The performance of SSLES can be discussed by considering two perspectives. The first lies in constructing the graph in SSL, which is based on image features rather than spatial neighbourhood and spectral similarity. This could help to obtain a better estimation of the underlying class label of the potential unlabelled image objects. The second perspective is related to the role of the ES in increasing the certainty of labelling in SSL. The ES handled this through the use of probability rules that aimed to link environmental parameters and the location of vegetation types, where it is most likely that a vegetation type may occur.

To have a benchmark to evaluate and compare the performance of SSLES, it is assumed that the number of reference training samples is representative and sufficient to train a standard supervised classifier. Since SSLES increased the number of samples in the training set, there could be the risk of overfitting the classifier due to the high number of training samples. Therefore, to test the robustness of SSLES regarding the number of initial training samples, a new test was conducted using the Group 4 dataset. For this, only 50% of the original reference training samples were used for SSLES, and the result was evaluated by the same test set that had been used previously. This resulted in an OA of 80% and compared to the case of using all the training samples, no statistically significant difference was observed. The classification accuracies could have been negatively affected by various factors. These include the uncertainties associated with the samples obtained from OBIA. In OBIA, if segmentation has low quality, it may not be able to separate two different classes properly in the image; hence, the extracted features for an image object will be the mix of features of two different classes [65,66]. To avoid this problem in the current study, the segmentation results were compared to the reference sample polygons. Nevertheless, in the case of having uncertainty in the reference vegetation map, some level of uncertainty may be found in the segmentation. The segmentation was performed using a RapidEye 5 m image while the features were extracted from the Sentinel-2 image. Although there was a one-year gap between the two images, they were acquired at the same time of year. In such a relatively short time period, no changes are expected in the vegetation structure of the study area except for the agriculture class. This was further confirmed by the high-resolution satellite imagery of GoogleEarth. The limitations of SSLES can be discussed considering the restrictions of SSL and the ES. In terms of the ES, assigning the quantitative values for the *a priori* probabilities has some uncertainties since it is based on the reference vegetation map mainly. Using the knowledge and experience of specialised experts such as ecologists could result in a stronger knowledge base, but due to time constraints, it was not possible. However, the assumption of this study was to use any available source of (prior) knowledge and expertise. The sources of knowledge could be either in the form of a human expert or other resources such as scientific research and published works. Regarding SSL limitations, the KNN method was used for graph construction. A recent study by [67] showed that the KNN may result in irregular graph construction where each node is connected to more than K neighbours. In this case, the algorithm may end up assigning the incorrect class label to the connected nodes. This problem might be more pronounced in the present study because of the high similarity between the vegetation classes.

6. Conclusions

In this study, a developed approach for semi-supervised classification of satellite images was proposed and applied for vegetation cover classification. The algorithm constructs a graph based on image features from OBIA. It uses Euclidean distance to compute the similarity between samples, where KNNs are selected for labelling. Using an ES to supervise the labelling process of the graph-based SSL algorithm was the key point in this study.

The capability of OBIA, SSL, and the ES for classification, particularly in the field of remote sensing, has already been investigated in the literature. The novel contribution of this study was the integration of these three into a single algorithm. Results prove the effectiveness of the proposed algorithm for the challenging problem of vegetation cover classification where some vegetation classes show similar characteristics. The capability of Sentinel-2 spectral data in vegetation classification was assessed, and the results prove that the red-edge band's combination could yield the highest overall accuracy for vegetation cover classification.

In a future study, linking the vegetation classification levels to the concept of Anderson levels for land cover mapping could be considered to adjust the vegetation classes where the highest level has the highest accuracy [68]. From an algorithm perspective, the potential applicability of the SSLES method on different land covers and biomes using different remote sensing data could be analysed. Concerning SSL, different strategies need to be investigated for selecting the unlabelled samples in a more informative and reliable way.

Using alternative approaches for graph construction instead of KNN, as well as using a different similarity measurement method such as JM distance, should be of high priority. Finally, the possibility of applying the presented method for the classification of large-scale or big data in remote sensing may be investigated.

Author Contributions: Conceptualization, N.F.L. and R.D.; Formal analysis, N.F.L.; Methodology, N.F.L. and R.D.; Project administration, R.D. and A.K.S.; Resources, R.D.; Software, N.F.L.; Supervision, R.D. and A.K.S.; Validation, N.F.L.; Visualization, N.F.L.; Writing—original draft, N.F.L.; Writing—review & editing, N.F.L., R.D., A.K.S., C.P. and N.K. All authors have read and agreed to the published version of the manuscript.

Funding: This research was funded by the Erasmus+ programme of Education, Audio-visual and Culture Executive Agency (EACEA) of the European Union.

Conflicts of Interest: The authors declare no conflict of interest.

References

1. Egbert, S.L.; Park, S.; Price, K.P.; Lee, R.-Y.; Wu, J.; Duane Nellis, M. Using conservation reserve program maps derived from satellite imagery to characterize landscape structure. *Comput. Electron. Agric.* **2002**, *37*, 141–156. [CrossRef]
2. Xie, Y.; Sha, Z.; Yu, M. Remote sensing imagery in vegetation mapping: A review. *J. Plant Ecol.* **2008**, *1*, 9–23. [CrossRef]
3. Immitzer, M.; Vuolo, F.; Atzberger, C. First experience with Sentinel-2 data for crop and tree species classifications in central Europe. *Remote Sens.* **2016**, *8*, 166. [CrossRef]
4. Mui, A.; He, Y.; Weng, Q. An object-based approach to delineate wetlands across landscapes of varied disturbance with high spatial resolution satellite imagery. *ISPRS J. Photogramm. Remote Sens.* **2015**, *109*, 30–46. [CrossRef]
5. Richards, J.A. *Remote Sensing Digital Image Analysis*; Springer: Berlin/Heidelberg, Germany, 2013; ISBN 978-3-642-30061-5.
6. Bhatnagar, S.; Gill, L.; Regan, S.; Naughton, O.; Johnston, P.; Waldren, S.; Ghosh, B. Mapping vegetation communities inside wetlands using Sentinel-2 imagery in Ireland. *Int. J. Appl. Earth Obs. Geoinf.* **2020**, *88*, 102083. [CrossRef]
7. Skidmore, A.K.; Forbes, G.W.; Carpenter, D.J. Technical note non-parametric test of overlap in multispectral classification. *Int. J. Remote Sens.* **1988**, *9*, 777–785. [CrossRef]
8. Mellor, A.; Boukir, S.; Haywood, A.; Jones, S. Exploring issues of training data imbalance and mislabelling on random forest performance for large area land cover classification using the ensemble margin. *ISPRS J. Photogramm. Remote Sens.* **2015**, *105*, 155–168. [CrossRef]
9. Persello, C.; Bruzzone, L. Active and semisupervised learning for the classification of remote sensing images. *IEEE Trans. Geosci. Remote Sens.* **2014**, *52*, 6937–6956. [CrossRef]
10. Board, R.; Pitt, L. Semi-supervised learning. *Mach. Learn.* **1989**, *4*, 41–65. [CrossRef]
11. Chapelle, O.; Schölkopf, B.; Zien, A. *Semi-Supervised Learning*; The MIT Press: London, UK, 2010; ISBN 9780262255899.
12. Jackson, Q.; Landgrebe, D.A. An adaptive classifier design for high-dimensional data analysis with a limited training data set. *IEEE Trans. Geosci. Remote Sens.* **2001**, *39*, 2664–2679. [CrossRef]
13. Prabukumar, M.; Shrutika, S. Band clustering using expectation–maximization algorithm and weighted average fusion-based feature extraction for hyperspectral image classification. *J. Appl. Remote Sens.* **2018**, *12*, 046015. [CrossRef]
14. Dalponte, M.; Ene, L.T.; Marconcini, M.; Gobakken, T.; Næsset, E. Semi-supervised SVM for individual tree crown species classification. *ISPRS J. Photogramm. Remote Sens.* **2015**, *110*, 77–87. [CrossRef]
15. Bruzzone, L.; Chi, M.; Marconcini, M. A novel transductive SVM for semisupervised classification of remote-sensing images. *IEEE Trans. Geosci. Remote Sens.* **2006**, *44*, 3363–3373. [CrossRef]
16. Maulik, U.; Chakraborty, D. A self-trained ensemble with semisupervised SVM: An application to pixel classification of remote sensing imagery. *Pattern Recognit.* **2011**, *44*, 615–623. [CrossRef]
17. Dopido, I.; Li, J.; Marpu, P.R.; Plaza, A.; Bioucas Dias, J.M.; Benediktsson, J.A. Semisupervised self-learning for hyperspectral image classification. *IEEE Trans. Geosci. Remote Sens.* **2013**, *51*, 4032–4044. [CrossRef]
18. Geiß, C.; Aravena Pelizari, P.; Blickensdörfer, L.; Taubenböck, H. Virtual support vector machines with self-learning strategy for classification of multispectral remote sensing imagery. *ISPRS J. Photogramm. Remote Sens.* **2019**, *151*, 42–58. [CrossRef]
19. Lu, X.; Zhang, J.; Li, T.; Zhang, Y. Incorporating diversity into self-learning for synergetic classification of hyperspectral and panchromatic images. *Remote Sens.* **2016**, *8*, 804. [CrossRef]
20. Camps-Valls, G.; Bandos Marsheva, T.V.; Zhou, D. Semi-supervised graph-based hyperspectral image classification. *IEEE Trans. Geosci. Remote Sens.* **2007**, *45*, 3044–3054. [CrossRef]
21. Gu, Y.; Feng, K. L1-graph semisupervised learning for hyperspectral image classification. In Proceedings of the 2012 IEEE International Geoscience and Remote Sensing Symposium, Munich, Germany, 22–27 July 2012; pp. 1401–1404.
22. Ma, L.; Crawford, M.M.; Yang, X.; Guo, Y. Local-manifold-learning-based graph construction for semisupervised hyperspectral image classification. *IEEE Trans. Geosci. Remote Sens.* **2015**, *53*, 2832–2844. [CrossRef]
23. Zhao, Y.; Su, F.; Yan, F. Novel Semi-supervised hyperspectral image classification based on a superpixel graph and discrete potential method. *Remote Sens.* **2020**, *12*, 1528. [CrossRef]

24. Zhu, X.; Ghahramani, Z.; Lafferty, J. Semi-supervised learning using gaussian fields and harmonic functions. In Proceedings of the 20th International Conference on Machine Learning, Washington, DC, USA, 21 August 2003; pp. 912–919.
25. Kim, K.-H.; Choi, S. Label propagation through minimax paths for scalable semi-supervised learning. *Pattern Recognit. Lett.* **2014**, *45*, 17–25. [CrossRef]
26. Ma, L.; Ma, A.; Ju, C.; Li, X. Graph-based semi-supervised learning for spectral-spatial hyperspectral image classification. *Pattern Recognit. Lett.* **2016**, *83*, 133–142. [CrossRef]
27. Chong, Y.; Ding, Y.; Yan, Q.; Pan, S. Graph-based semi-supervised learning: A review. *Neurocomputing* **2020**, *408*, 216–230. [CrossRef]
28. Skidmore, A.K.; Turner, B.J. Forest mapping accuracies are improved using a supervised nonparametric classifier with SPOT data. *Photogramm. Eng. Remote Sens. PERS* **1988**, *54*, 1415–1421.
29. Hayes-Roth, F.; Waterman, D.; Lenat, D. *Building Expert Systems*; Addison-Wesley, Reading: Boston, MA, USA, 1983; ISBN 0-201-10686-8.
30. Booker, J.M.; McNamara, L.A. Solving black box computation problems using expert knowledge theory and methods. *Reliab. Eng. Syst. Saf.* **2004**, *85*, 331–340. [CrossRef]
31. Schmidt, K.S.; Skidmore, A.K. Spectral discrimination of vegetation types in a coastal wetland. *Remote Sens. Environ.* **2003**, *85*, 92–108. [CrossRef]
32. Pranger, D.P.; Tolman, M.E. *Toelichting Bij De Vegetatiekartering Schiermonnikoog Op Basis Van False Colour-Luchtfoto's 1:10.000 [Explanation to the Vegetation Mapping Schiermonnikoog 2010 on the Basis of False Colour Aerial Photographs 1:10.000, in Dutch]*; Rijkswaterstaat: Delft, The Netherlands, 2012.
33. Vrieling, A.; Skidmore, A.K.; Wang, T.; Meroni, M.; Ens, B.J.; Oosterbeek, K.; O'Connor, B.; Darvishzadeh, R.; Heurich, M.; Shepherd, A.; et al. Spatially detailed retrievals of spring phenology from single-season high-resolution image time series. *Int. J. Appl. Earth Obs. Geoinf.* **2017**, *59*, 19–30. [CrossRef]
34. Darvishzadeh, R.; Wang, T.; Skidmore, A.; Vrieling, A.; O'Connor, B.; Gara, T.W.; Ens, B.J.; Paganini, M. Analysis of Sentinel-2 and RapidEye for Retrieval of Leaf Area Index in a Saltmarsh Using a Radiative Transfer Model. *Remote Sens.* **2019**, *11*, 671. [CrossRef]
35. ESA SENTINEL-2 User Handbook. Available online: https://sentinels.copernicus.eu/web/sentinel/user-guides/document-library/-/asset_publisher/xlslt4309D5h/content/sentinel-2-user-handbook (accessed on 24 July 2015).
36. Atzberger, C.; Darvishzadeh, R.; Schlerf, M.; le Maire, G. Suitability and adaptation of prosail radiative transfer model for hyperspectral grassland studies. *Remote Sens. Lett.* **2013**, *4*, 55–64. [CrossRef]
37. Tigges, J.; Lakes, T.; Hostert, P. Urban vegetation classification: Benefits of Multitemporal rapideye satellite data. *Remote Sens. Environ.* **2013**, *136*, 66–75. [CrossRef]
38. Darvishzadeh, R.; Atzberger, C.; Skidmore, A.K.; Abkar, A.A. Leaf area index derivation from hyperspectral vegetation indicesand the red edge position. *Int. J. Remote Sens.* **2009**, *30*, 6199–6218. [CrossRef]
39. Gilmore, M.S.; Wilson, E.H.; Barrett, N.; Civco, D.L.; Prisloe, S.; Hurd, J.D.; Chadwick, C. Integrating multi-temporal spectral and structural information to map wetland vegetation in a lower connecticut river tidal marsh. *Remote Sens. Environ.* **2008**, *112*, 4048–4060. [CrossRef]
40. Macintyre, P.; van Niekerk, A.; Mucina, L. Efficacy of multi-season Sentinel-2 imagery for compositional vegetation classification. *Int. J. Appl. Earth Obs. Geoinf.* **2020**, *85*, 101980. [CrossRef]
41. Cochran, W.G. *Sampling Techniques*; John Wiley and Sons: New York, NY, USA, 1977; p. 428.
42. Bird, E. *Coastal Geomorphology. An Introduction*; John Wiley and Sons Ltd.: Chichester, UK, 2008; ISBN 9780874216561.
43. Rundquist, D.C.; Narumalani, S.; Narayanan, R.M. A Review of Wetlands Remote Sensing and Defining New Considerations. *Remote Sens. Rev.* **2001**, *20*, 207–226. [CrossRef]
44. Schmidt, K.S.; Skidmore, K.; Kloosterman, E.H.; van Oosten, H.; Kumar, L.; Janssen, J.M. Mapping Coastal Vegetation Using an Expert System and Hyperspectral Imagery. *Photogramm. Eng. Remote Sens.* **2004**, *70*, 703–715. [CrossRef]
45. Comaniciu, D.; Meer, P. Mean Shift: A robust approach toward feature space analysis. *IEEE Trans. Pattern Anal. Mach. Intell.* **2002**, *24*, 603–619. [CrossRef]
46. Laurent, V.C.E.; Schaepman, M.E.; Verhoef, W.; Weyermann, J.; Chavez, R.O. Bayesian object-based estimation of LAI and chlorophyll from a simulated Sentinel-2 top-of-atmosphere radiance image. *Remote Sens. Environ.* **2014**, *140*, 318–329. [CrossRef]
47. Clinton, N. An accuracy assessment measure for object based image segmentation an accuracy assessment measure for object based image. *Int. Arch. Photogramm. Remote Sens. Spat. Inf. Sci.* **2008**, *37*, 1189–1194.
48. Möller, M.; Lymburner, L.; Volk, M. The comparison index: A tool for assessing the accuracy of image segmentation. *Int. J. Appl. Earth Obs. Geoinf.* **2007**, *9*, 311–321. [CrossRef]
49. Haralick, R.M.; Shanmugam, K.; Dinstein, I. Textural features for image classification. *IEEE Trans. Syst. Man Cybern.* **1973**, *3*, 610–621. [CrossRef]
50. Yu, Q.; Gong, P.; Clinton, N.; Biging, G.; Kelly, M.; Schirokauer, D. Object based detailed vegetation classification with airborne high spatial resolution remote sensing imagery. *Photogramm. Eng. Remote Sens.* **2006**, *72*, 799–811. [CrossRef]
51. Mathieu, R.; Aryal, J.; Chong, A.K. Object-based classification of ikonos imagery for mapping large-scale vegetation communities in urban areas. *Sensors* **2007**, *7*, 2860–2880. [CrossRef]

52. Pham, L.T.H.; Brabyn, L.; Ashraf, S. Combining QuickBird, LiDAR, and GIS topography indices to identify a single native tree species in a complex landscape using an object-based classification approach. *Int. J. Appl. Earth Obs. Geoinf.* **2016**, *50*, 187–197. [CrossRef]
53. Fu, B.; Xie, S.; He, H.; Zuo, P.; Sun, J.; Liu, L.; Huang, L.; Fan, D.; Gao, E. Synergy of multi-temporal polarimetric SAR and optical Image satellite for mapping of marsh vegetation using object-based random forest algorithm. *Ecol. Indic.* **2021**, *131*, 108173. [CrossRef]
54. Rohban, M.H.; Rabiee, H.R. Supervised neighborhood graph construction for semi-supervised classification. *Pattern Recognit.* **2012**, *45*, 1363–1372. [CrossRef]
55. Szummer, M.; Jaakkola, T. Partially labelled classification with Markov random walks. *Adv. Neural Inf. Processing Syst.* **2001**, *14*, 945–952.
56. Skidmore, A. An expert system classifies eucalypt forest types using thematic mapper data and a digital terrain model. *Photogramm. Eng. Remote Sens.* **1989**, *55*, 1449–1464.
57. Breiman, L. *Random Forests—Random Features;* Technical report 567; Statistics department of University of California: Berkeley, CA, USA, 1999.
58. Breiman, L. Random forests. *Mach. Learn.* **2001**, *45*, 5–32. [CrossRef]
59. Immitzer, M.; Atzberger, C.; Koukal, T. Tree species classification with random forest using very high spatial resolution 8-band WorldView-2 Satellite data. *Remote Sens.* **2012**, *4*, 2661–2693. [CrossRef]
60. Liaw, A.; Wiener, M. Classification and Regression by RandomForest. *R News* **2002**, *2*, 18–22.
61. Gislason, P.O.; Benediktsson, J.A.; Sveinsson, J.R. Random forests for land cover classification. *Pattern Recognit. Lett.* **2006**, *27*, 294–300. [CrossRef]
62. Congalton, R.G.; Green, K. *Assessing the Accuracy of Remotely Sensed Data: Principles and Practices*, 2nd ed.; CRC Press: Boca Raton, FL, USA, 2008; ISBN 9781420055122-CAT# 55127.
63. Beukeboom, T.J. *The Hydrology of the Frisian Islands;* Rodopi bv Editions: Amsterdam, The Netherlands, 1976; ISBN 9062034195.
64. Wang, L.; Hao, S.; Wang, Q.; Wang, Y. Semi-supervised classification for hyperspectral imagery based on spatial-spectral label propagation. *ISPRS J. Photogramm. Remote Sens.* **2014**, *97*, 123–137. [CrossRef]
65. Song, M.; Civco, D.L.; Hurd, J.D. A competitive pixel-object approach for land cover classification. *Int. J. Remote Sens.* **2005**, *26*, 4981–4997. [CrossRef]
66. Liu, D.; Xia, F. Assessing object-based classification: Advantages and limitations. *Remote Sens. Lett.* **2010**, *1*, 187–194. [CrossRef]
67. Jebara, T.; Wang, J.; Chang, S.-F. Graph construction and b-matching for semi-supervised learning. *Int. Conf. Mach. Learn. ICML* **2009**, 1–8. [CrossRef]
68. Anderson, J.; Hardy, E.; Roach, J.; Witmer, R. *A Land Use and Land Cover Classification Systems for Use with Remote Sensor Data;* Geological Survey Professional Paper 964; US Government Printing Office: Washington, DC, USA, 1976.

remote sensing

Review

A Systematic Review on Advancements in Remote Sensing for Assessing and Monitoring Land Use and Land Cover Changes Impacts on Surface Water Resources in Semi-Arid Tropical Environments

Makgabo Johanna Mashala [1,2,*], Timothy Dube [3], Bester Tawona Mudereri [4,5], Kingsley Kwabena Ayisi [2] and Marubini Reuben Ramudzuli [1]

1 Department of Geography and Environmental Studies, University of Limpopo, Private Bag X1106, Sovenga 0727, South Africa
2 Risk and Vulnerability Science Centre (RVSC), University of Limpopo, Private Bag X1106, Sovenga 0727, South Africa
3 Institute of Water Studies, Department of Earth Science, University of the Western Cape, Private Bag X17, Bellville 7535, South Africa
4 International Centre of Insect Physiology and Ecology (icipe), Nairobi P.O. Box 30772-00100, Kenya
5 School of Animal, Plant and Environmental Sciences, University of the Witwatersrand, Private Bag 3, Johannesburg 2050, South Africa
* Correspondence: 201412998@keyaka.ul.ac.za; Tel.: +27-786252362

Abstract: This study aimed to provide a systematic overview of the progress made in utilizing remote sensing for assessing the impacts of land use and land cover (LULC) changes on water resources (quality and quantity). This review also addresses research gaps, challenges, and opportunities associated with the use of remotely sensed data in assessment and monitoring. The progress of remote sensing applications in the assessment and monitoring of LULC, along with their impacts on water quality and quantity, has advanced significantly. The availability of high-resolution satellite imagery, the integration of multiple sensors, and advanced classification techniques have improved the accuracy of land cover mapping and change detection. Furthermore, the study highlights the vast potential for providing detailed information on the monitoring and assessment of the relationship between LULC and water resources through advancements in data science analytics, drones, web-based platforms, and balloons. It emphasizes the importance of promoting research efforts, and the integration of remote sensing data with spatial patterns, ecosystem services, and hydrological models enables a more comprehensive evaluation of water quantity and quality changes. Continued advancements in remote sensing technology and methodologies will further improve our ability to assess and monitor the impacts of LULC changes on water quality and quantity, ultimately leading to more informed decision making and effective water resource management. Such research endeavors are crucial for achieving the effective and sustainable management of water quality and quantity.

Keywords: arid environment; land cover assessment and monitoring; machine learning; satellite data; water quality and quantity; water resources management

1. Introduction

Freshwater is a valuable natural resource that sustains biodiversity, carbon and nutrient cycles, food provision, and ecological functions [1]. Globally, freshwater resources offer significant socio-economic and ecological benefits, serving industrial, agricultural, and domestic needs. Semi-arid regions, covering about 15% of the Earth's surface [2], are characterized by unpredictable weather, long dry seasons, and erratic rainfall [3]. In the global semi-arid tropical regions, particularly Southern Africa, renewable freshwater resources are estimated to be around 2300 cubic kilometers [4]. Around seventy percent of the available water resources are in transboundary rivers, with the remaining thirty percent distributed between lakes and groundwater [5]. Any modifications or fluctuations in the water supply can have significant consequences for vital economic sectors, especially agriculture, and the overall natural capital. Scientific research has demonstrated that alterations in seasonal and inter-annual rainfall patterns, climate change effects, droughts, floods, and human activities have collectively influenced water systems, increasing their vulnerability [5,6].

Thus, it is crucial to manage these delicate freshwater resources carefully. Sub-Saharan African countries heavily rely on freshwater resources for agriculture and residential use, intensifying the pressure on ensuring water security and sustainable management [7]. Uneven rainfall distribution and external influences on water resources highlight the importance of water conservation and effective management. Extensive research had revealed that changes in land use and land cover (LULC), climate change impacts, and the proliferation of invasive alien species make water resources vulnerable [8,9]. The study of LULC change gained prominence in the mid-20th century [10], influencing water resources, human livelihoods, and ecosystem health. Human activities, such as agriculture, mining, and urbanization, significantly drive LULC transformation. These changes affect hydrological processes, climate variability, ecosystem services, drainage systems, and increase vulnerability to floods [11–13] Consequently, they directly impact the quantity and quality of water resources. For instance, studies have demonstrated how LULC changes contribute to a declining water quality in rivers and lakes, adversely affecting ecology and water quality [7,14,15]. Therefore, identifying effective methods to assess and monitor the impacts of LULC dynamics on water quality and quantity is vital for efficient water resource management.

Remote sensing has proven to be a cost-effective and efficient tool for providing spatially explicit data on various ecosystems, including surface water resources [16]. Earth observation techniques, such as modern UAVs, balloons, multispectral, and hyperspectral sensors, can help monitor semi-arid tropical environments and potentially address water scarcity, pollution, and the conservation of water quality. Understanding the relationship between LULC and water resources is essential for effective watershed management, policymaking, future LULC development considerations, and freshwater protection. Giri and Qui [17] provide a detailed review addressing land use and water quality in the 21st century. The study also provides insights into the factors that contribute to water quality problems, the indices used to evaluate water quality, techniques for identifying suitable explanatory variables for water quality, and the processing methods needed to capture spatial effects. Moreover, the study explores the modelling of water quality, using the identified explanatory variables to gain insights. Ullah et al. [18] reviewed the impacts of land use on surface water quality using a statistical approach. They indicated that each statistical method has a unique purpose, application and assumptions aimed at providing solutions to different problems. Meanwhile, Ozbay et al. [19] reviewed the relationship between land use and water quality and its assessment, using hyperspectral remote sensing in the mid-Atlantic estuary. Their main goal was to provide research findings on the application of hyperspectral remote sensing in order to monitor specific LULC and water quality. Meanwhile, previous studies have explored this relationship and the application of remote sensing in monitoring water quality, quantity, and specific LULC changes. However, bibliometric analyses of comprehensive systematic reviews on the use of remote sensing in global semi-arid tropical environments to understand LULC changes and their impacts on water resources are lacking.

This study aims to bridge this research gap by providing a comprehensive systematic overview of the progress, challenges, and opportunities related to the use of remote sensing applications in order to assess and monitor LULC changes and their effects on water quality and quantity in semi-arid tropical environments. This study aims to address the following key questions: (i) Which water quality and quantity parameters can be detected using remote sensing? (ii) What role does remote sensing play in understanding the relationship between LULC changes and water quality and quantity? (iii) Which methods have been utilized to assess and monitor these changes? (iv) What challenges have been encountered in these endeavors? (v) What can be achieved in the future to improve our understanding and monitoring of LULC changes and their impacts on water resources? Through these efforts, the study seeks to offer valuable insights into the potential of remote sensing technologies and provide suggestions with which to better assess and monitor LULC changes and their impacts on water resources in semi-arid tropical regions.

2. Research Method and Literature Search

This study conducted a systematic literature review that aimed to establish progress and identify existing gaps, using remotely sensed data to map and monitor LULC changes and their effects on water quality. This study also aimed to further outline the challenges and opportunities associated with remote sensing applications for assessing and monitoring the impacts of LULC change on surface water resources.

Literature Search and Data Extraction

The literature searches for this study utilized the Google Scholar, Scopus, and Web of Science databases, targeting peer-reviewed international journals related to remote sensing, hydrology, ecology, geographical information systems (GIS), and water resources. The search strategy involved defining appropriate search strings and identifying relevant keywords, phrases, and terms. To identify relevant keywords, phrases, and terms, we used the most cited literature reviews. Initial searches included terms such as "land use and land cover change", "impacts", and "water quality and quantity", resulting in a total of 18,187 publications being retrieved (17,500 from Google Scholar, 411 from Scopus, and 276 from Web of Science).

The retrieved articles underwent further screening using level 2 search criteria, including keywords such as "remote sensing", "tropical semi-arid", and the years 2001–2021. This process yielded a total of 1248 articles from Google Scholar, 121 articles from Scopus, and 95 articles from Web of Science. In the level three screening, additional keywords such as "catchment scale", "sub-catchment scale", "algorithms", "riparian buffers", "land cover land use classifications", "land use land cover monitoring challenges", "machine learning", "freshwater resources", "deep learning", "hydrological model", "spatial pattern", "ecosystem services", "change detection", "multi-spatial scale", "catchment management", "buffer zone" and "water pollution" were used. This resulted in a final compilation of 197 articles in EndNote for further screening, eliminating duplications and excluding non-English papers, gray literature, extended abstracts, conference proceedings, fee articles, and those not published between 2001 and 2021 (Figure 1). The remaining 197 articles were captured in Microsoft Excel and used to comprehensively outline the progress, gaps, challenges, and opportunities related to using remote sensing to assess and monitor land use and land cover changes and their impacts on surface water resources in semi-arid environments. Bibliometric analysis was employed to assess the published articles and identify key terms related to mapping and monitoring LULC changes and their impacts on surface water quality.

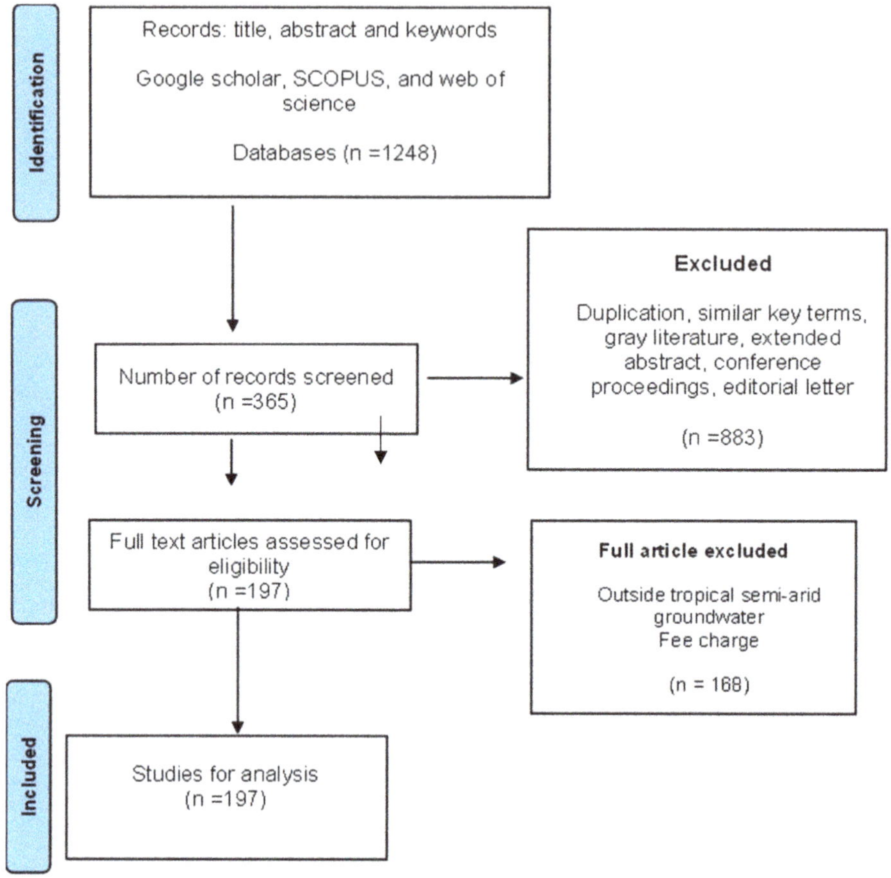

Figure 1. Methodology undertaken for selection of articles considered in the review.

3. Results

3.1. Progress of Remote Sensing in Assessment and Monitoring Land Use and Land Cover Changes

Land use and land cover (LULC) change examines the transformation and alteration of Earth's land surface, including changes in land use patterns and the conversion of natural land cover to human-modified landscapes. The assessment and monitoring of LULC changes using remote sensing have made significant progress over the years. Remote sensing technologies have advanced in terms of data acquisition, spatial resolution, spectral coverage, and temporal frequency, enabling the more accurate and detailed analysis of LULC dynamics. Some key areas of progress in the remote sensing assessment and monitoring of LULC changes include data availability, fine-scale mapping, the integration of multisource data, classification algorithms, change detection techniques, and web-based platforms and open data initiatives.

3.1.1. Data Availability and Integration of Multisource Data

The availability of satellite imagery data from platforms such as Landsat, Sentinel, and other commercial satellites has greatly improved. These datasets provide consistent seasonal and long-term coverage, allowing for the analysis of LULC changes over time. Advances in remote sensing have enabled the mapping and monitoring of LULC changes at finer spatial scales. High-resolution imagery and data fusion techniques have enhanced our ability to capture detailed LULC information, including urban areas, agricultural fields, and small-scale changes [20]. The integration of multisource remote sensing data, such as optical, radar, and LiDAR, has expanded the capabilities of LULC change assessment. Combining data from different sensors enhances our understanding of LULC dynamics and provides valuable information regarding vegetation structure, terrain characteristics, and three-dimensional mapping. Remote sensing has progressed in the development of change detection techniques, enabling the identification and quantification of LULC changes. These techniques include image differencing, spectral indices, time series analysis, and object-based change detection, enabling more accurate and efficient change assessment.

3.1.2. Classification Algorithms

The repetitive practice of using a multi-sensor image system to capture information provides valuable data for managing land-based resources. Remote sensing also offers the standardized data collection procedure, data integration, and analysis within a geographic information system [21,22]. Remote sensing satellites have proven valuable in employing various classification techniques to map LULC changes within watersheds such as supervised, unsupervised classification, and object-based image analysis (Table 1).

Table 1. Classification algorithms for mapping LULC classification.

Algorithm/Techniques	Sensor Used	Performance Range	References
Supervised Classification			
Support Vector Machine (SVM)	Landsat OLI, ETM+, TM, Terra ASTER, Hyperion Hyperspectral imagery and Quickbird	88–98%	[23–25]
Random Forest (RF)	Synthetic Aperture Radar (SAR) Sentinel 2 MSI, Landsat 8, SPOT, RapiEye, LiDAR	88–95%	[20,26,27]
Convolutional Neural Network (CNN)	Aerial photograph	91–98%	[28]
Classification and Regression Tree (CART)	Sentinel 2 and Landsat OLI, LiDAR	85–90%	[29,30]
Deep Neural Network (DNN)	Landsat TM and OLI, Sentinel 2	92–95%	[31]
Decision Tree (DT)	Landsat TM and ETM+, Sentinel 2	85–90%	[20,32]
Spectral Angle Mapper	Landsat 8, hyperspectral, RapidEye	89–90%	[24,33,34]
Recurrent Neural Network (RNN)	Very High Spatial Resolution (VHSR)	75–86%	[35]
Artificial Neural Network (ANN)	Landsat ETM+	70–85%	[32]
Maximum Likelihood (MLC)	Landsat TM	67–72%	[20,36,37]
Unsupervised Classification			
ISODATA	MODIS	54–69%	[38]
K-Nearest neighbor	Landsat TM and ETM+, Sentinel 2	87–91%	[39]
Object-Based Image Analysis			
Object-based image analysis (OBIA)	Lidar, Sentinel 2	87–91%	[40]

Unsupervised classification is a computer-automated process that groups pixels that are statistically similar into categories using a clustering algorithm, such as K-means [41] or ISODATA [38]. Unsupervised classification is valuable when the field data are lacking or knowledge about the study area is unavailable [20,42]. However, the spectral properties change over time with the image data, atmospheric condition, and the sun's angle at the time the image was captured; hence, detailed spectral knowledge of different features may be required. In contrast, supervised classification allows the analyst to select training samples for each land cover class and guide the computer to identify the spectral features of similar areas for each class [43]. This is particularly achievable using classification algorithms such as maximum likelihood (MLC), support vector machine (SVM), an artificial neural network (ANN) and random forest (RF), among others. Based on the finding (Figure 2), classification techniques such as MLC have been widely employed in studying LULC change in water resources. For example, the study by Ding et al. [44] derived the LULC change map from the Landsat TM using the MLC algorithm for classification and achieved an overall accuracy of 88% and a Kappa coefficient of 0.85%. Similarly, the study by Tadesse et al. [45] derived the LULC change from Landsat TM, ETM+ and OLI using MLC and achieved overall accuracies of 87%, 89%, and 93%, respectively, and Kappa coefficients of 0.83, 0.83, and 0.88, respectively. The MLC considers spectral variation within each category and the overlap covering the different classes. However, this method is time-consuming since it requires more pixels in each training dataset to specify each class [20] The MLC technique also produces lower classification accuracy results when compared to other classification techniques such as ANN, SVM, and RF [20,30,46].

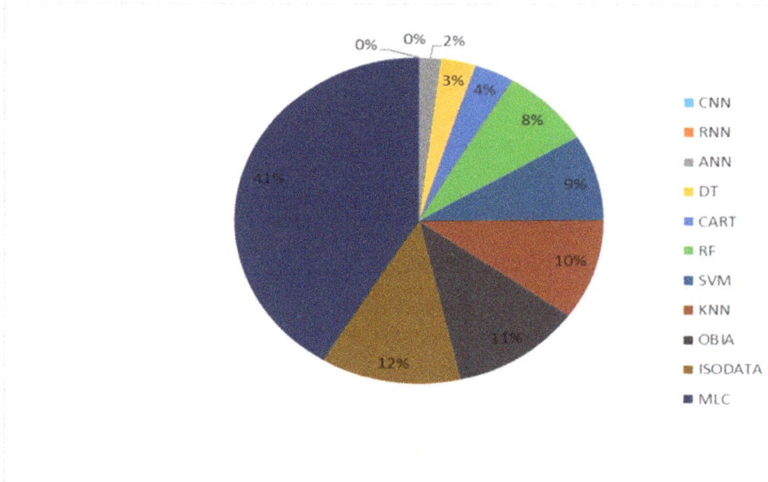

Figure 2. Classification algorithms (%) used to detect and map LULC using multi-sensors.

Thus, ANN, SVM, and RF machine learning (ML) algorithms are more robust and effective for classification techniques [20,47,48]. Supervised classification requires labelled training data to learn and understand the patterns associated with each category. Once the training data are labelled, these ML techniques can build models based on the labelled examples and use this knowledge to categorize new, unseen text accurately and efficiently, which is a challenge with MLC. However, the challenges associated with these methods are that they require extensive training (data hungry), require human supervision (expertise), and are computationally extensive when producing accurate LULC change maps [49]. Although individual classifiers achieve better accuracy results, they often fail to predict true classes with high accuracy. Therefore, studies have combined different classifiers to boost performance and reduce the classification error [50–52]. However, the ensemble fails on new data when individual classifiers are too complex for the training data present or their training error becomes too large quickly [49,53]. The problem associated with the ensemble is finding the right balance between the individual models' complexity and their fit to the data. Thereby, performing more ensemble iterations can reduce the error of the combined classifier on the training data.

Obtaining high-quality and sufficient reference datasets, as is required in most of these machine learning algorithms, is still an enormous task in most of the sub-Saharan African countries [54]. Acquiring reference data through field surveys is still challenging due to the inaccessibility of some areas, costs associated with travel and the time investment needed. To counter this limitation, other studies have adopted the migration of reference data from a specific time (year) to another time (target year) in order to address the lack of accurate and reliable current data. Likewise, good training data combined with the high spatial resolution of an image, together with an ensemble ML, often provide a high classification accuracy [55]. For instance, the study by Zhou [40] used Light Detection and

Ranging (LiDAR) to map LULC change using object-based image analysis classification and achieved an overall accuracy of 91% and a Kappa coefficient of 0.87%; additional examples are provided in Table 1.

Deep Learning (DL) algorithms have the capacity to extract automatic and hierarchical features from a large dataset, which makes them able to work with remote sensing data [31,56]. CNNs are one DL algorithm that are used to solve problems associated with spatial data. CNNs outshine other algorithms when aiming to capture spatial features such LULC patterns, textures, and shapes. They can also handle both multispectral and hyperspectral data. They can automatically learn and extract meaningful representations from the image data, leading to a high classification accuracy [28]. CNN transfer learning enables the learned features to be leveraged and reduces the need for extensive labelled training data. Meanwhile, RNNs are best at processing temporal and spatial dependencies in sequential remote sensing data, such as time series and spatial data in a sequential structure. RNNs are also able to flexibly work with hyperspectral imagery because they can handle data in which the number of spectral bands may vary across different samples [35]. Understanding the different spectral bands can be valuable for the classification of LULC. This provides opportunities to classify complex contextual images and improve the classification accuracy. However, these techniques have never been used to assess and monitor the impacts of LULC on water resources in semi-arid tropical environments. The major drawback of using these algorithms is the requirement for extensive datasets during the training process, particularly when dealing with large amounts of data. The computation involved in training and testing can be costly. RNNs can struggle to capture long-term dependencies effectively with remote sensing data, especially when analyzing time series with a large time lag. CNNs are prone to overfitting, where the model becomes excessively specialized to the training data and fails to generalize to unseen data.

3.1.3. Spectral Classification

Band Based Classification

Band-based classification is performed using the individual spectral bands of the remote sensing image [57]. This method relies solely on the spectral information captured by the different bands of the remote sensing image. They cannot effectively enhance specific features of interest, as the classification is based on the raw spectral values of each band. The results of band-based classification may be more straightforward to interpret since they directly correspond to the individual spectral bands [58]. This method can be more sensitive to atmospheric effects, such as haze or aerosols, which may impact the accuracy of the classification [57,58]. It is relatively straightforward and simple to implement, as it involves using each band independently for classification.

Index Based Classification

Classification using different spectral indices in remote sensing is a common approach used to extract valuable information from satellite imagery. Spectral indices combine specific spectral bands to highlight various land cover characteristics, such as vegetation health, water content, and soil properties. Index-based classification and specific spectral indices, which are combinations of different spectral bands, are used to perform the classification [59]. Instead of using individual bands, index-based classification uses spectral indices that combine bands in specific ways to highlight certain features of interest. By using spectral indices, this method can enhance specific features, such as vegetation, water bodies, or soil, making it more effective in certain applications [59]. Index-based classification can be more robust against atmospheric effects since the spectral indices can mitigate some of the atmospheric influences. Implementing index-based classification may require more pre-processing steps to calculate the spectral indices, adding some complexity to the process. The commonly used indices are the normalized vegetation index (NDVI), the normalized difference water index (NDWI), the normalized difference built-up index (NDBI), and the soil adjusted vegetation index (SAVI). Integrating spectral indices with other data sources, such as ancillary data and ground truth information, can enhance the accuracy and reliability of LULC change classification results.

3.1.4. Change Detection

Change detection is the process of identifying and quantifying areas of change in the LULC, or other environmental variables over time. Change detection provides a means by which to monitor and understand the dynamics of environmental changes [60]. It helps identify the magnitude, location, and patterns of changes, enabling researchers and decision makers to gain insights into ecosystem dynamics, urban growth, deforestation, agricultural expansion, and other changes affecting the environment [27]. It facilitates a better understanding of the dynamic nature of the Earth's surface and enables informed decision making for sustainable development and environmental protection. The accuracy of change detection relies on factors such data quality, spatial and temporal resolution, methodology and reference, or ground truth data [61]. The accuracy of the quality input data used for change detection plays a significant role. High-quality, well-calibrated data with minimal noise and distortions contribute to better levels of accuracy. Possessing reliable reference data or ground truth information is crucial for evaluating the change detection accuracy. Ground truth data, obtained via field surveys or high-accuracy sources and remote sensing technology, are used to verify the detected changes, and assess the method's performance [62,63].

Techniques such as pixel-by-pixel differencing, image ratioing, and image thresholding are used to identify and highlight differences between images. Time series analysis involves studying data collected over a period of time to detect patterns or trends. Statistical methods, such as regression analysis, seasonal decomposition, and moving averages, are employed to reveal changes in temporal data [64]. Remote sensing technologies, including satellite and aerial imagery, play a crucial role in monitoring LULC. Geographic Information Systems (GIS) aid in spatially analyzing and visualizing the detected changes. With the advent of machine learning and artificial intelligence, change detection has seen significant advancements. Supervised and unsupervised learning algorithms, such as SVM, RF, and CNNs, can be trained to automatically detect changes in various data types [27,65]. Light Detection and Ranging (LiDAR) technology utilizes laser pulses to measure distances to the Earth's surface and generate high-resolution 3D maps [29]. It is especially useful in detecting changes in topography, vegetation, and infrastructure. Radar-based change detection employs microwaves to penetrate clouds and vegetation, making it an all-weather and day-and-night imaging technique. It is useful for monitoring land subsidence, urban growth, and natural disasters. Data mining techniques can be applied to large datasets to discover hidden patterns or anomalies that are indicative of changes. This technology is widely used in fraud detection, network intrusion detection, and more. Time-of-flight cameras use light signals to measure distances, enabling real-time 3D imaging and change detection applications, such as object tracking and gesture recognition [66]. The challenges associated with the technique include spatial and temporal resolution, spectral heterogeneity, spectral similarity, radiometric variation, scale and context, computational resources and processing, and data availability.

3.1.5. Web-Based Platforms and Open Data Initiatives

The emergence of web-based platforms, such as Google Earth Engine (GEE) and Sentinel Hub, has facilitated easy access to remote sensing data and analysis tools [67]. Open data initiatives promoted by space agencies and governments have further promoted the sharing of remote sensing datasets, enabling broader participation in LULC change monitoring. These advancements in remote sensing technology and techniques have significantly improved the assessment and monitoring of LULC changes. They provide a more comprehensive understanding of the dynamics and impacts of human activities on the Earth's surface. Continued progress in remote sensing, along with ongoing research and development, will further enhance our ability to monitor and manage LULC changes effectively.

3.2. Water Quality and Quantity

Water quantity refers to the various aspects that contribute to the overall measurement and understanding of water availability and supply. Some common components that can be measured using remote sensing technology include precipitation, evapotranspiration, runoff and streamflow, groundwater, reservoir and lake monitoring, and soil moisture. Remote sensing can estimate precipitation patterns and distribution by measuring cloud properties, rainfall rates, and storm characteristics using sensors such as radar or microwave radiometers [22,68]. Remote sensing can quantify the amount of water lost from the land surface through evaporation and plant transpiration. It involves estimating energy fluxes and vegetation indices using optical or thermal sensors in order to assess evapotranspiration rates [69,70]. It can also help to estimate runoff and streamflow by monitoring changes in water levels and river discharge using altimeters, radar sensors, or optical imagery [71,72]. This enables the water movement in river networks to be assessed. More insights into groundwater resources can be obtained by monitoring changes in the land surface elevation using satellite-based radar interferometry (InSAR) or gravity data from the Gravity Recovery and Climate Experiment (GRACE) mission [48,73]. The soil moisture content can be estimated by measuring the microwave radiation emitted or reflected by the Earth's surface [68,74]. This information helps to assess water availability in the root zone and supports agricultural water management. In reservoir and lake monitoring, remote sensing enables the monitoring of water levels, surface area, and volume changes in reservoirs and lakes using radar or optical imagery [75]. This information is crucial for water supply management and flood control.

Remote sensing data may have limitations regarding its spatial and temporal resolution, sensitivity to atmospheric conditions, calibration and validation, complex terrain and land cover heterogeneity, and limited data accessibility for specific parameters. Therefore, it is important to consider these limitations when using remote sensing for water quantity assessments. Integrating remote sensing with other data sources, utilizing complementary techniques, and incorporating appropriate modelling approaches can help to mitigate these limitations and improve the accuracy and reliability of water quantity assessments. In addition, by leveraging remote sensing data, scientists and water resource managers can assess and monitor these components of water quantity on various scales, providing valuable insights into water availability, distribution, and movement.

Water quality refers to the chemical, physical, biological, and radiological characteristics of water that determine its suitability for various uses and its impact on the environment and human health [28]. The water quality parameters that can be assessed using remote sensing include the chlorophyll-a concentration, water turbidity, dissolved organic matter (DOM), water temperature, total suspended solids (TSS), water pH, and harmful algal blooms (HABs) [76]. For further details, the study by Gholizadeh et al. [77] elaborates more on the water quality parameters and limitations of remote sensing for assessing water quality. However, most studies monitoring water quantity have focused more on water balance (evapotranspiration) and monitoring water (reservoir and lakes), with little attention paid to runoff, ground water recharge and soil moisture. Water quality parameters such as pH, TSS, temperature, and DO are commonly evaluated in water quality assessments [78]. Meanwhile, HABs and the chlorophyll-a concentration have received little attention.

3.3. Impacts of LULC Changes on Water Resources

Rapid population growth, socioeconomic factors and a lack of natural resource conservation policies are major contributing factors to LULC changes worldwide. According to Dwarakish and Ganasri [43], slope, the distance from the river, soil erosion, altitude, and built-up areas are significant contributing factors to LULC changes. Nonetheless, they are not considered in most studies when assessing and monitoring the impact of LULC on water resources. Changes in LULC can have a wide-ranging impact on various aspects of the environment. These LULC changes can alter landscape patterns, hydrological processes (surface flow), physical factors (stream morphology and temperature increase), biology (biodiversity and ecosystems) and water quality and quantity (nutrients and pollution increase). Semi-arid tropical climates are characterized by erratic climate change. The semi-arid tropical climatic conditions need to be frequently monitored for effective watershed or catchment management, and sustainable water resources. LULC changes, such as agriculture, urbanization, and mining, are likely to improve livelihoods, contribute towards local and national economic development, facilitate food security and advance biofuel energy, making them a priority for the development of countries. Yet, these are the most substantial factors causing negative environmental modification. Agriculture is directly associated with the removal of natural vegetation, increased soil erosion [79], algal bloom [80], increased greenhouse gas emissions and nutrient imbalances. Natural vegetation, such as riparian vegetation, can act as an important habitat for a variety of species and can trap sediments and pollutants in water. Therefore, alterations in the natural vegetation affect the ecosystem services provided by the streams, which are important for reducing flood streams, managing runoff, and preventing erosion [64,79,81]. Ultimately, these changes affect the water storage provided by aquifers, ecological processes, functions and services, and hydrological factors. This will have implications for hydrological factors, thus affecting water supply and availability [26]. This will lead to changes in the drainage network, which is important for surface runoff and drainage patterns. Ecosystem services play an important role in watershed and water management. However, they are often overlooked in water resources management.

Urbanization and mining are associated with impervious surfaces, urban heat islands, the increased susceptibility of areas to floods, the loss of drainage systems and changing hydrological systems [82,83]. These artificial surfaces often result in increased runoff, and further generate a path for the transportation of pollutants into water bodies [84]; this reduces the infiltration into and storage capacity of water in shallow aquifers and increases the chances of severe floods occurring [85]. Urban heat islands can be aggravated by a warming climate, particularly during heat waves, which change the water balance (evapotranspiration) and hydrological factors of the catchment. These changes affect the timing and magnitude of evaporation loss and the water yield, which govern the soil moisture content and the flow patterns of hydrological regimes. At the end, increased streamflow and precipitation occur, and the frequency of large floods and larger sedimentation increases [85]. To better manage LULC and water resources effectively, it is important to assess hydrological components by using advanced tools, with which it is very important to attain sustainable water resources at a catchment scale. Land cover changes are located within the spatial representation of landscape. Modeling LULC patterns offers a better understanding of past and future LULC and its related implications and guides future land use policies. Taking these studies in semi-arid tropical areas into account will ensure the sustainability of the catchment via the planning and management of the watersheds.

Consequently, these LULC changes have a serious impact on a wide range of ecological processes and result in several global environmental problems, such as land degradation, desertification, biodiversity loss, habitat loss, and species transfer. These impacts affect the water supply, irrigation, fishing, and power generation, reduce food production and land productivity, and decrease many countries' Gross Domestic Products (GDP). Contaminated surface water results in health risks for humans, increases the financial cost of purification and human consumption, and affects economic development. The impact of LULC on water resources is caused by a lack of proper management strategies and land use planning surrounding water resources. A lack of awareness regarding water pollution, especially in Sub-Saharan countries, further compromises water quality. The enforcement of policies and regulations regarding the discharge of pollutants from agricultural sectors, urbanization, waste-water treatments, and industries needs to be strengthened. This will promote sustainable water quality, watershed management and improve global economies.

3.4. The Role Played by Remote Sensing Platforms in Assessing and Monitoring LULC Changes and Their Impacts on Water Resources

Seasonal and Long-Term Monitoring

The effects of LULC changes on water quality differ spatially and temporarily due to climatic conditions (e.g., temperature and rainfall), and topography (e.g., slope and landscape patterns). The hydrological factors (e.g., flow) also play an important role in the movement and transportation of the pollutants into rivers and the degree to which they disrupt the ecosystems. Remote sensing platforms, with different image acquisitions (Table 2), offer an opportunity for the seasonal and long-term monitoring of the effects of LULC changes on water resources. Long-term monitoring offers the important data needed to measure the changes in natural water resources over time and predict trends; this is in order to implement, plan, monitor and manage water quality. The tabulated information below is helpful in designing an assessment evaluating the effects of LULC changes on water resources and may be used in the selection of appropriate sensors.

Table 2. Available sensors that can be applied for assessing and monitoring the effects of LULC changes on water resources.

Sensor/Platform	Resolution (m)	Spectral Bands	Swath Width (km)	Revisit Time (days)	Acquisition Cost
AVHHR	1100	5	2900	1	Free
IKONOS	4	5	11	1–2	High
ASTER	15, 30, 90	144	60	16	Free
GRACE				10	Free
Hyperspectral	<1		>100		Very high
Landsat ETM+	30	8	185	16	Free
Landsat TM	30	7	185	16	Free
Landsat OLI	30	11	185	16	Free
LIDAR	0.45	5	1–2		Very high
MODIS	500, 1000	7	2330	1	Free
MERIS	300	15	1150	3	Free
Radar	0.3, 0.56	2			Very high
Rapid Eye	5	5	77	5.5	High
Sentinel 1 SAR	5, 5 $\times$ 20, 20 $\times$ 40	4	20, 80, 250, 400	6–12	Free
Sentinel 2 MSI	10, 20, 60	13	290	5	Free
SPOT	10, 20	4	120	26	High
Quickbird	2.4	5	16.5	1–3.5	High
Worldview	<1	8	16.4	1–3.7	Very high

The long-term monitoring of the effects of LULC changes on water resources has used aerial photographs due to their long period of existence. For example, the study by Schilling et al. [86] used aerial photographs to map LULC, and the results indicated that the change in grassland to row crop increased nitrate levels up from 8.0 to 11.6 mg L in two Squaw creek subbasins in the USA for a period of ten years. Aerial photographs offer a high spatial resolution and are very good for analyzing ground surface events and detecting different LULC [87]. However, they are limited to a smaller scale compared to satellite data, and the images lack repeatable acquisition. They require qualified and experienced personnel to interpret the image [88], and interpretating the images is costly and time-consuming.

Landsat ((Thematic Mapper (TM), Multispectral Scanner System (MSS), Enhanced Thematic Mapper (ETM)) multispectral sensors have been widely used to conduct long-term monitoring of the impacts of LULC changes on water quality at regional and local scales. The study by Kibena [89] used Landsat TM to map the effects of LULC change on water quality from the year 1995 to 2012. They observed that the LULC, namely grassland, forest and bare land, had been converted into settlement and agricultural land, which had increased pollution in Lake Chevero in Zimbabwe. The results showed that the total phosphorus (TP) increased from 130 to 376 kg/day, the total nitrogen (TN) increased from 290 to 494 kg/day, the DO increased from 0.1 to 6.8 mg/L, the chemical oxygen demand (COD) increased from 11 to 569 mg/L, the biochemical oxygen demand (BOD) increased from 5 to 341 mg/L, Phosphate Phosphorus (PO_4-P) increased from 0.01 to 4.45 mg/L, Ammonia Nitrogen (NH_3-N) increased from 0.001 to 6.800 mg/L and the electrical conductivity (EC) increased from 38 to 642 mg/L. Similarly, Zhang et al. [90] further observed that grassland, forest, water, and bare land were converted into farmland and constructed land from 1990 to 2016, which increased the level of chemical oxygen demand, manganese variant (COD_{MN}) from 0.92 to 1.09 mg/L, BOD from 0.63 to 0.85 mg/L, TP from 0.006 to 0.007 mg/L, and TN from 0.12 to 0.20 mg/L. The temporal range of the Landsat images provided the researchers with the ability to predict the effects of LULC on water quality; hence, there are still challenges related to the use of the Thematic Mapper (TM) and Multispectral Scanner (MSS), in that they are no longer operating, and with the Enhanced Thematic Mapper Plus (ETM+) images, which are persistent with the Scan Line Corrector (SLC). The malfunctioning of the sensor's SLC leads to a data loss of approximately 22% of the normal scene area [91]. Therefore, these medium-resolution sensors (i.e., Landsat TM and MSS) fail to deliver real-time images; in addition, the loss of data for the operational ETM+ sensor has resulted in considerable challenges regarding the estimation of the impacts of LULC change on water quality.

In 1986, the French government launched the Satellite Pour L'Observation de la Terra (SPOT), and it was the first earth resource satellite to have a pointable optic with high resolution, which increases the high opportunity of the imaging areas [92]. The SPOT sensor has the ability to obtain information every day at any time due to the frequency revisit time, and can map LULC change, ranging from the regional scale to global scale [93]. Plessis et al. [94] used Landsat TM and SPOT for classification and successfully predicted the future concentration of the water quality for the years 2015, 2020, 2030 and 2050, which increased with changes in the LULC. However, SPOT imagery is costly, which often hinders the adoption of its products in many studies.

Moderate-Resolution Imaging Spectroradiometer (MODIS) instruments use NASA Aqua and Terra satellites, thus providing nearly daily repeated coverage of the Earth's surface with 36 spectral bands and a swath width of approximately 2330 km [87]. MODIS plays a significant role in mapping LULC change and dynamics at a coarse spatial resolution [88]. The sensor is freely available and presents an opportunity for the long-term and seasonal monitoring of LULC changes at a large scale due to its long period of existence and its revisit time [93]. For example, the study by Juma [95] used MODIS to map LULC change and demonstrated an increase in agriculture and residential growth due to population growth from 1990 to 2008. The results showed that Nitrate Nitrogen (NO_3-N) increased from 10 μg^{-1} to 98 μgL^{-1}, PO_4-P increased from 4 μgL^{-1} to 57 μgmL^{-1} and chlorophyll also increased due to poor practices in agriculture, which resulted in the proliferation of an alien invasive water hyacinth species. However, the limitation of using the MODIS sensor is the difficulty involved in linking the coarse spatial resolution with field data, and the difficulties involved in monitoring small areas using the sensor.

On the other hand, the Advanced Very High-Resolution Radiometer (AVHRR) is freely available and has a high probability of obtaining a cloud-free view of the land surface compared to multispectral sensors (e.g., Landsat) [96], and is very useful for long-term monitoring. The coarse spatial resolution of AVHRR can cover large areas and fails to distinguish the earth's features. This makes it difficult to detect or view detailed information about features. However, AVHRR has not been used at a national or global scale because of the difficulties involved in linking coarse spatial resolution data and field measurement [97]. Still, not all LULC changes affect water quality, as demonstrated by the study by Kaushal [98]. The authors found that the transformation of row crop cover to perennial grassland decreased the amount of NO_3-N. Also, Khare et al. [99] indicated that between 1974 and 2007, residential areas increased from 10% to 21%, while agriculture decreased from 36% to 19%, and forest decreased from 13% to 8%. This conversion decreased the TN from 2 mg/L to 1.5 mg/L. In addition, the expansion of residential areas may cause an increase in fecal coliform in water bodies since they pave the way for the entry of pollutants into rivers. On the contrary, the expansion of forest cover decreases the number of pollutants entering into rivers, although it increases dissolved oxygen, which is important for the lives of fish and macroinvertebrates [100]. Therefore, many studies agree that the expansion of forest and grassland cover is important an mitigation strategy for improving water quality [101,102].

The seasonal monitoring of the river catchment plays a substantial role in evaluating the temporal variations in river pollution due to LULC changes for effective land use management and watershed management. This is important for developing proper management strategies for water resources. The seasonal monitoring of the quality of pollution can further indicate the current, ongoing, and emerging problems. Most studies, for instance those by Tahiru et al. [7], Namugize et al. [14] and Beck et al. [71], have observed that built-up areas are positively correlated with most of the water quality parameters; this is due to increased surface runoff caused by impervious surfaces, which create paths for the transportation of contaminants including TP, TN, NH_3-N. Thus, policies pertaining to land use planning, especially in urban areas, should be implemented and enforced in order to sustain water resources. Forests and grasslands have a relationship with DO, and cultivated land/agriculture are positively associated with TN, TP, turbidity, and pH due to various farming practices, such as the use of pesticides, herbicides, and fertilizers. As such, most concentrations of these contaminants are higher during the dry season than in the wet season. However, DO is higher in the wet season than the dry season, while TP and TN are higher in both seasons (dry and wet). The high quantities of contaminants observed in the dry season may be driven by high water discharge and low water retention [71].

For instance, Pullanikkatil et al. [103] seasonally monitored the impacts of LULC on the Likangala river catchment in Malawi and found that turbidity increased in the wet season, with 190.5 NTU downstream in the area dominated by wetland and settlement. Meanwhile, the pH increased during the dry season, with 880.83 mg/L, and decreased by 218.72 mg/L in the wet season downstream, which was dominated by wetland and settlements where people nearby practice fishing. The EC varied from 4 to 466 $\mu s\ cm^{-1}$ during the wet season and between 40 and 3520 $\mu s\ cm^{-1}$ in the dry season in the upstream and downstream, respectively. The TDS varies from 20 to 1760 mg^{-1} in the dry season and from 2 to 233 mg^{-1} during the wet season. Similarly, the study by Zhang [101] seasonally monitored the effect of land use on water quality and observed that water quality parameters such NH_3-N and CODmn were higher in the dry season (7.29 mg/L and 7.8 mg/L) and lower in the wet season (0.048 mg/L and 2.2 mg/L) in built-up areas (dominated by urban areas). Meanwhile, the DO was lower in the wet season (3 mg/L) in the built-up area and higher in the dry season (12.9 mg/L) in the forest–grassland area. However, the TP was higher both seasons, with (0.28 mg/L) in the wet season and (0.29 mg/L) in the dry season. The study concluded that the high levels of NH_3-N and COD_{mn} were influenced by urban areas, which increased the surface runoff of contaminants into the water bodies. Meanwhile, studies such as those by Kaushal et al. [98] and Rothenberger et al. [97] found that NO_3-N and NH_4-N, respectively, are higher in the wet season than in the dry season. The high concentration of pollutants in the wet season may be attributed to an increased run-off that washes off the soil, releasing a large amount of sediment, nutrients, and pesticides into surface water. An in-depth understanding of LULC change and seasonal factors can help to implement effective catchment management strategies for the protection of these water resources.

Progress has been noted in detecting, mapping, and monitoring LULC change and its effects on water quality using remotely sensed data over the years (Figure 3a). The use of satellite remote sensing in mapping and modelling the impact of LULC change on water quality has recently attracted increased attention, as evidenced in the number of publications between the years 2001 and 2021 (Figure 1). According to our analysis, extensive research has been conducted mostly using multispectral sensors; these include a range of Landsat sensors (TM, MSS, ETM+), and a few studies have used Sentinels, SPOT and MODIS (Figure 3b).

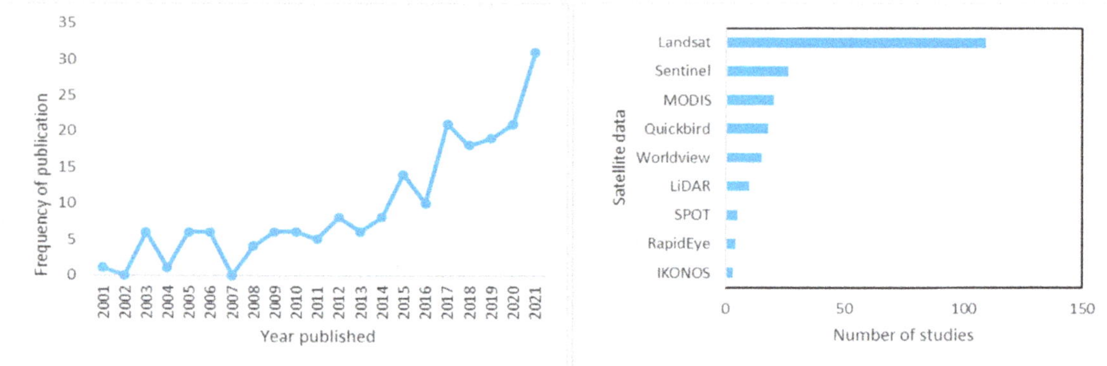

(**a**) (**b**)

Figure 3. (**a**) Progress of remote sensing publications, and (**b**) sensors that were used in monitoring the effect of land use and land cover change on water quantity and quality.

The advancement of this research could be attributed to the significant increase in earth observation technologies, their relatively low cost and their time efficiency when managing large areas [9]. For instance, remote sensing has the ability to perform the spatiotemporal monitoring of LULC change, water quality, and natural resources, which is also important for assessing water quality and quantity in terms of river pollution. However, Landsat TM, MSS and ETM+ have been the most used sensors [104,105]. This could be attributed to the fact that Landsat is the longest mission (since 1972) that has been supplying remotely sensed data for a wide variety of applications without charges. However, the Landsat sensors have been useful for assessing and monitoring the impacts of LULC change on water quality. It cannot be denied that medium- and low-spatial-resolutions can limit the detection and mapping of LULC and water quality when the area affected is smaller than the pixel size [91].

Advances in earth observation technology with improved image acquisition characteristics have progressively expanded our ability to distinguish features of the earth. Platforms such as Sentinel 1 synthetic aperture radar (SAR) provide an opportunity to integrate optical and radar data to improve the mapping capacity on cloudy days [50]. This is necessary for monitoring areas such as semi-arid environments, which receive most of their rainfall seasonally (wet season). Therefore, Sentinel 1 uses multiple sensors and sensing periods to accurately map heterogeneous LULC [51]. Schulz et al. [51] and Hu et al. [106] combined an ensemble Sentinel 1 and 2 Multispectral Instrument (MSI) to map LULC at a local to regional scale and achieved improved results compared to using only the optical sensor exclusively. The water quality parameters that were assessed using sentinel 1 were TN, TSS, COD and TP. Munthali et al. [66] and Chen et al. [78] used Sentinel 2 and normalized the vegetation index to monitor the TSS concentration at various buffer scales and showed that a 300 m scale most effectively explained the variation in TSS concentrations (R^2) of 0.83, $p < 0.001$). Sentinel (SAR and MSI) sensors are freely available and have successfully monitored LULC and water quality separately. However, they have not been used to their full capacity in assessing and monitoring the relationship between LULC and water resources [77]. Other studies have used the LULC prototype supplied by the European space agency (ESA) for instance, Copernicus global land cover (100 m × 100 m pixel size) [107] and Globcover (300 m × 300 m pixel size) [32]. Although the LULC prototypes provide useful data, they are need to be updated frequently for effective management since the LULC changes over time. The development of hyperspectral sensors presents a unique opportunity for the extraction of the LULC [108], biological and physicochemical parameters of water quality [77,109].

Specifically, hyperspectral remote sensing provides several narrow and quasi contiguous bands that enhance discrimination among different land uses and land covers. Thus, the narrow band of hyperspectral sensors provides an opportunity to generate spectral reflectance curves for each pixel, which are unique in order to differentiate different classes of LULC, including water and pollutants [17,110]. Hyperspectral sensors have been reported in the literature to capture unique spectral signatures of water quality indicators, such as salinity, chlorophyll content (chla), turbidity, TSS and colored dissolved organic matter (CDOM) [21]. This may help us to understand and quantify the relationships among the spatial, structural, biological, and chemical processes occurring in the natural water ecosystem. The narrow bands of the sensors have allowed researchers to develop and implement water quality indices that have been effective in estimating water quality, thus including the Maximum Chlorophyll Index (MCI), Normalized Difference Turbidity Index (NDTI), and Green Normalized Difference Vegetation Index (GNDVI). The study by Elhag et al. [111] used water quality parameter indices to estimate water quality, including MCI, NDTI and GNDVI. They achieved an outstanding coefficient of correlation (R) result of 0.96 with MCI, an R of 0.94 with NDTI, and an R of 0.94 with GNDVI. Although water quality indices are effective for monitoring water, they have been widely used in semi-arid tropical environments. The lack of robust and reliable water and quantity data needed to parameterize models remains a challenge.

The optimal spectral and spatial resolution remains a major challenge to the remote sensing community since not all water quality parameters can be detected and monitored using remotely sensed data because they are not optically active. Hyperspectral image analysis has not been fully explored due to its high cost and complex pre-processing procedure [109]. Regardless of their outstanding performance, only a few studies have attempted to use hyperspectral sensors to assess and monitor the relationship between LULC and its associated impacts on water resources. Advanced modern technologies such as Unnamed Aerial Vehicles (UAVs), commonly known as drones, have emerged as a potential alternative for mapping and monitoring LULC and water resources at a local scale [112,113]. They are flexible, affordable, and offer a very high spatial resolution. However, drones are limited to small areas and many of the affordable drones exclusively cover the true color (RGB) section, which does not offer sufficient data for extensive application in areas such as characterizing water quality [113,114]. These have never been fully explored in understanding the relationship between LULC and water quality. Based on the findings of this study, the growing interest in assessing and mapping the relationship between LULC and its impact on water resources in semi-arid areas has focused on using the Landsat image platform. Meanwhile, advanced earth observation platforms such as sentinels, UAVs, and hyperspectral technology have not been fully explored in assessing and monitoring the relationship between LULC changes and their impact on water quality and quantity. The use of advanced earth observation could be viable in monitoring the impacts of LULC on water resources in semi-arid tropical environments.

3.5. Algorithms Used for Quantifying the Effects of LULC Changes on Water Resources (Quality and Quantity)

Different algorithms are available for estimating the effects of LULC change on water quality using remotely sensed data. The techniques use different statistical modelling approaches that provide relatively accurate results and are easier to understand when compared to hydrological water quality modelling approaches. Most machine learning (ML) algorithms developed for remotely sensing the effect of LULC changes on water quality can be categorized as parametric or non-parametric. The most frequently used methods by researchers for modelling the relationship between LULC change and water quality parameters are the parametric machine learning algorithm (PMLA), which assumes the linear relationship between variables. Parametric algorithms, such as linear [115], multilinear [7,116] and Stepwise multilinear (SML) regression [117], have shown satisfactory performance. However, PMLA, such as linear and multilinear regression, have the limitations of multicollinearity and the overfitting of large data. Although the SML regression algorithm attempts to reduce the collinearity problems, it also eliminates variables that are ecologically and statistically important [117]. Seilheimer [118] used the linear mixed-effect model (LMEM) algorithm approach, which enables the robust simultaneous evaluation of the association and environmental gradient. Meanwhile, it accounts for the repeated measures embedded in the data structure and indicates a better prediction performance when compared to multiple linear regression. The LMEM assumes that observation with the cluster is always positively correlated and that some individuals competing in the cluster for the scarce resources are negatively correlated. Thereby, it ignores a small negative correlation, resulting in a deflated type-1 error, and an invalid standard error and confidence interval in regression analysis [119].

Researchers have introduced more advanced non-parametric machine learning algorithms (NPMLA), including Partial Least Square Regression (PLSR) and multivariate statistics, which have been reported to be robust and efficient when aiming to overcome the problems of overfitting and multicollinearity with high accuracy. The algorithms include Principal Component Analysis (PCA) [120,121], Discriminant analysis (DA) [122], Cluster analysis (CA), Redundancy Analysis (RDA) [123] and Hierarchical cluster analysis [122] (Figure 4). The NPMLA works well with large volumes of data. Singh et al. [124] reported that multivariate statistical techniques (CA, DA, and PCA) are important for evaluating and interpreting large and complex datasets and obtaining better information regarding water quality and the design of monitoring. Besides the robustness of the NPMLA, there are limitations associated with some of the machine learning (ML) algorithms.

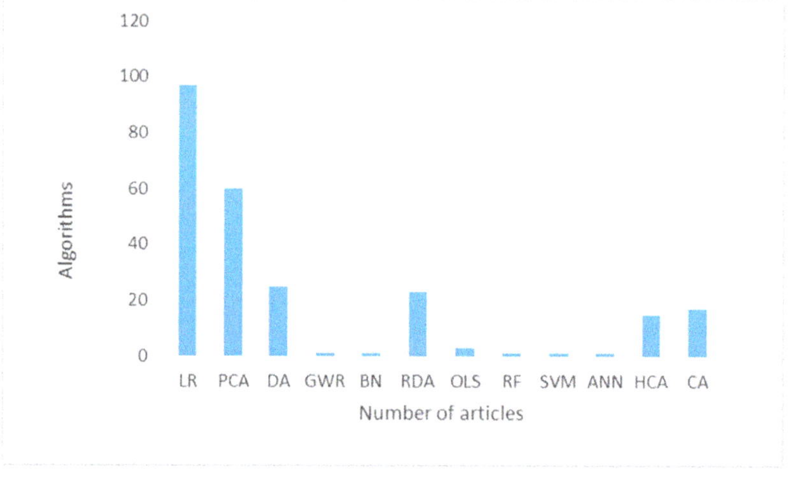

Figure 4. Algorithms used to assess and monitor the relationship between impacts of LULC and water quality.

The limitation of these multivariate statistical methods shows an existing relationship between water quality and LULC pattern, which may neglect some of the important spatial characteristics and hide local variation [101,125]. ML algorithms, including Ordinary Least Squares (OLS) and Geographically Weighted Regression (GWR), were introduced to address spatial non-stationary issues and to examine spatial autocorrelation, which is neglected by multivariate algorithms [122,125]. OLS is important for developing a relationship between independent LULC and water quality variables at a large scale by selecting the most significant variables in the regression [18]. The coefficient from the OLS model provides the most influential parameter of different water quality parameters.

However, OLS requires a large dataset to obtain reliable results [126]. The GWR considers the spatial variation and determines the relationship by incorporating the coordinates of location into the regression equation. The GWR model uses a single land use indicator as the independent variable and excludes the selection model because of the high potential of multicollinearity among different land-use variables; this would result in an invalid GWR model when variables experiencing multicollinearity are selected [127]. However, GWR statistical analysis is limited to small scales. Therefore, the implementation of models that cover large areas is required. Bayesian Network (BN) is an effective method for spatiotemporal analysis as it enables the interaction of variables in space and time [18], and is suitable for handling missing data prediction [18,128]. The model is essential for evaluating the complicated LULC change in water quality at various scales. However, the model is too complicated, too difficult to automate, requires specific software to run the model and qualified statistical expertise, and is sensitive to probabilities [129].

On the other hand, Artificial Intelligence (AI) has gained considerable attention because of its potential to leverage big data and solve the problems faced in traditional techniques (mathematical models). AI refers to the simulation of human intelligence in machines that are programmed to think, reason, learn, and perform tasks that typically require human intelligence [130]. The goal of AI is to create machines or systems that can mimic human cognitive functions, such as problem solving, pattern recognition, language understanding, and decision making [131]. Moreover, AI encompasses a broad range of techniques, algorithms, and methodologies to enable machines to perform intelligent tasks. AI computing technologies are on the edge of becoming the prevalent alternative to conventional data processing techniques [131]. Some of the key subfields of AI that are used in remote sensing include ML and DL. ML developed for earth observation data can support the challenges of spatial and temporal realm adaptations, hyperspectral data, the integration of multisource information and large-volume data analyses [132,133]. Advances in ML technology have created a unique opportunity for the development of accurate large-scale prediction and prescriptive models [134]. DL is used to improve results due to its accuracy in classification and prediction when trained with extremely big data; in addition, it can extract features from raw data. DL helps to capture the potential relationship between environmental variables for remote sensing retrieval, fusion, downscaling and superiority in multiscale and multilevel feature extraction. However, it is not clear how to best use ML or DL for expanding the range of increasingly accessible satellite data for LULC change research, particularly under environmental and socioeconomic impacts.

3.6. Multi-Spatial Scale Relationship between LULC Changes and Water Quality and Quantity

Water quality monitoring often relies on the conventional methods of conducting in situ measures using a handheld multiparameter instrument and laboratory analysis. However, the accuracy and precision of collected in situ data may be questionable due to human error in the field and laboratory [77]. The use of in situ data provides an accurate measure; therefore, integrating them with remote sensing data provides accurate measurements for the cost- and time-effective management of water pollution. Therefore, remote sensing has been widely used to assess and monitor the environmental effects had by LULC change on water quality at both the local and global scales [89,93,135,136]. However, when determining the relationship between LULC change and water quality, it is important to consider scale (spatial and temporal). Scale plays an important role in reflecting the different impacts of LULC changes on water quality. Therefore, in order to better manage the impacts of LULC changes on water resources, it is essential to consider streams as a complex ecosystem that operate at different spatial and temporal scales [136]. Three types of spatial scale, namely the buffer, sub-catchment, and catchment scale, have been used to estimate the impacts of LULC changes on water quality. However, there is no consensus regarding which of the abovementioned spatial scales explains a better relationship between LULC changes and their impacts on water quality and quantity.

Buffer zones are strongly influenced by water from the upslope, which is divided into three categories: surface flow and shallow subsurface flow [137,138]. When choosing a buffer scale, one should consider the structure and the function of the catchment, since there are two types of buffer scale; these include a circular and riparian buffer [135]. Riparian buffers are determined by the soil, vegetation and hydrology characteristics of the buffer and the interaction with the upslope and downslope. Meanwhile, circular buffers are effective for the diversity of water bodies such as lakes, streams, and dams in the lakeshore areas [139]. Most studies use a riparian buffer rather than a circular buffer to measure the influence of LULC change on water quality and quantity. For example, the study by [139] observed that many impacts came from all land uses, such as constructed land (CL), wetland (WL), original forest (OF), artificial forest (AF), and original land (OL), which reflected TN at 2 km; meanwhile, WL and CL continued to affect TN at the 4 km buffer. Moreover, Song et al. [135] found that urban areas influenced TN in all buffers from 500 m to 1000 m. Similarly, Li et al. [140] further indicated that a 300 m buffer is the strongest for the land use type to affect COD. Riparian buffers are storage areas that can be both the source and the sink of pollutants if no degradation process exists in the buffer zone. However, the drawback of using riparian buffers is that there is no uniform way of defining the width of a riparian zone [84]. Overcoming this problem requires a wider use of a riparian buffer to maximize its effectiveness in improving the water quality. The other problem with using a riparian buffer is the inability to address all water-related problems, since they are only effective in buffer areas that are not degraded. In the case of a degraded buffer zone, the scale may fail to reflect some of the impacts of land use on water quality and quantity.

Meanwhile, Gyawali et al. [141] indicated that the sub-catchment scale is more effective in reflecting the impacts of land use on water than the buffer and the whole catchment. Their results indicated that agriculture influenced dissolved oxygen (DO), and that urban and water bodies influenced dissolved solids (DS), biological oxygen dissolve (BOD), and temperature at the sub-catchment scale. Wan et al. [96] further revealed that LULC showed varied impacts of the same LULC category over different sub-catchments. However, other LULCs that are not nearby sub-catchments may have an influence on or contribute to impacts on the water quality, because pollutants from LULC in the upper catchment may be transported downstream. The scale of the stream reach might be improper as the pollutants are diluted by the flow or absorbed by plants [38,142]. Other studies propound that the whole catchment plays an important role in influencing the impacts of LULC on the water quality [12,136,143]. However, it is difficult to sample larger catchments at an appropriate spatial and temporal resolution. In addition, the spatial scale between LULC change and water quality differs spatially, and the characteristics of the stream, human disturbance and data accuracy all have different degrees of influence in multi-scale studies [135]. Therefore, for the better management of pollution, the application of different spatial scales may provide an effective method by which to understand the relationship between LULC and its impacts on water quality and quantity. Gyawali et al. [141] revealed that all LULCs, i.e., agriculture, forest, urban, and water bodies, affect temperature, DO, BOD, solid sediments (SS), DS, Fecal Coliform (FC) and EC at all three spatial scales. Similarly, Tanaka et al. [143] also confirmed that water quality indicators have a different response to LULC patterns when evaluated at different spatial scales.

4. Discussion

The findings based on the search keywords indicate that the use of remote sensing in assessing and monitoring land use and land cover (LULC) changes has been significant. Remote sensing techniques have undergone significant advancements, particularly with the introduction of advanced algorithms such as machine learning (ML) and deep learning (DL), which offer increased opportunities to explore and effectively manage environmental issues. Numerous studies [33,39,66,70] have focused on accurately mapping and detecting LULC changes using remote sensing. However, the utilization of machine learning and deep learning classifiers remains relatively limited compared to the traditional use of methods like maximum likelihood classification (MLC). Over the decades, the assessment and monitoring of LULC changes have extensively employed geographic information systems (GIS) and remote sensing. Earlier methods involved the use of aerial photographs and field observations for the interpretation of LULC to produce maps. With time, there have been notable advancements in LULC classification techniques, leading to improved mapping accuracy facilitated by advanced satellite imagery. Accurate LULC maps play a crucial role in informing decision-making processes for the planning and management of natural land resources.

4.1. Challenges in Remote Sensing the Effects of LULC Changes on Water Resources

Identifying the non-point source of pollution is still a challenging task due to an ongoing discussion and varying thoughts regarding the scale. Some studies argue that sub-catchment influences can be used [141], others suggest that the whole catchment would be the most optimum to use [116], and others have supported that riparian buffer [101] play an important role in influencing the water quality. Therefore, separating the impacts of LULC on water resources remains problematic, due to the extensive time scale over which impacts from LULC spread through the hydrologic system. The confounding effects of climate and weather, as well as large-scale observation field studies, often lack control, thus making it difficult to assign the temporal changes to causal mechanisms [144]. Thus far, most of the remote sensing techniques have been applied to LULC changes using a statistical model that links the relationship to in situ water quality parameters. However, the statistical models do not consider physically based hydrologic models, which are important in representing hydrological processes using spatially distributed data such as climate parameters (e.g., precipitation, temperature, and evapotranspiration), vegetation and soil moisture, and slope distribution [145]. The data contain crucial information about surface water flow and the integration of surface and ground water and can be used to describe the land surface topography characteristics [43]. Hydrological process models, such as the soil and water resources tool (SWAT), have been presented by many reviewers as an effective management tool for watershed models, simulating the stream flow better than other models [75,145,146]. The integration of these models with remote sensing data will provide effective management for monitoring the impacts of land use on water resources in semi-arid tropical environments. The reviews by Dwarakish and Ganasri [43], as well as Dong et al. [147], elaborate more on integrating remotely sensed data and hydrological models in LULC and water resources.

Efficient and accurate LULC using remote sensing therefore requires a high spatial detailed image for the classification method [125]. However, it is important to note that factors such as image resolution, radiometric conditions, and atmospheric effects can impact the effectiveness and accuracy of classification algorithms in remote sensing. Radiometric and atmospheric corrections play a crucial role in mitigating these challenges. However, performing accurate corrections can be complex, particularly due to the dynamic and spatially varying nature of atmospheric conditions. To address this issue, it is essential for users to have access to suitable atmospheric correction software that aligns with the specific requirements of their study and the chosen approach. Selecting the correct or appropriate atmospheric correction method is crucial in order to ensure reliable and accurate results in remote sensing analyses. By carefully considering the data needs and research objectives, users can make informed decisions and choose the most suitable atmospheric correction software for their specific study [108]. The accurate assessment of LULC change is crucial, particularly when it comes to capturing small classes such as roads and built-up areas, as they can have significant impacts on water quality. Achieving precise results for these classes requires the use of sensors with a very high spatial resolution, typically less than 5 m. While sensors like LiDAR, WorldView-2, and Quickbird offer the required level of detail for accurate LULC mapping, their utilization in studies has been limited. The main challenges associated with these sensors are their cost and low temporal resolution. Acquiring data from these sensors for long-term studies can be expensive, and their restricted availability may limit their use to smaller study areas. To conduct the effective long-term monitoring of LULC change, it is necessary to analyze time series of remotely sensed imagery. However, obtaining and analyzing consistent time series data from multiple sensors pose challenges. It is essential to ensure that the acquired images are captured under similar environmental conditions, such as the same time of the year, sun angle, and spectral bands, to minimize errors and maintain data consistency. Overcoming these challenges in acquiring and analyzing time series and multi-sensor data is crucial for improving our understanding of LULC change and its impact on water resources. Efforts should be made to address these limitations and develop methodologies that enable the accurate and consistent analysis of LULC change over time [60,148]. To minimize errors, the accuracy of a fraction of a pixel must be attained, meaning that variations in solar illumination atmospheric scattering and absorption and detector performance must be normalized, i.e., the radiometric properties of each image must adjust to those of the reference image [60,148].

One drawback of minimizing the issue of radiometric calibration is that any errors identified in the classification maps of individual dates will also be present in the final change detection map. To validate the results, it is important to compare the classified maps with ground-truth data. Researchers often utilize a confusion matrix as the preferred method for validation, which includes metrics such as overall accuracy, user accuracy, producer accuracy, and the Kappa coefficient. However, the Kappa coefficient has been criticized for its limitations in accurately assessing the results. It fails to convert the sample confusion matrix into an estimated population matrix, which can affect the reliability of the assessment [149,150]. To perform this hardware, software and qualified personnel are required for processing and analyzing the dataset. This hinders the application of remote sensing in monitoring the effects of LULC on water quality. Moreover, open-source software solutions, such as R-software, QuantumGIS or GRASS [151], are freely available to manipulate remote sensing products. However, these open-source software lack clear documentation and steep learning curves that hinder their adoption and use. Therefore, offering training across the discipline could potentially increase the adoption of these software [152]. The lack of reliability in ground-truth data has also inhibited the progress of classification in remote sensing.

Different algorithms have different strengths and require different input parameters [18]. Many factors, such as the spatial resolution of the remotely sensed data, the scale of the study area, the availability of software, the capacity of the analyst skills and knowledge, affect the modelling approach [149]. The developed model from remote sensing data requires adequate calibration and validation using in situ measurement and can only be used in the absence of clouds [149]. ML algorithms rely on a large number of training samples, which are difficult to obtain in the real world. However, remote sensing big data may provide significant solutions to the lack of data, although they often cause computational challenges, for instance, the need for scalable data storage, dynamic workflow management, and flexible computing resource provisioning [149].

The Google Earth Engine (GEE) [153], which is a cloud-based semi-automated platform that offers basic calculation functions for both raster and vector data, can successfully handle remote sensing big data on the cloud [153,154]. These advancements offer new possibilities for integrating and combining techniques to assess and monitor the connection between LULC change and water resources. However, the Google Earth Engine (GEE) platform has its limitations. It does not support the execution of deep learning algorithms due to computational constraints and the unavailability of such algorithms on the platform. Consequently, users can only gather the data on the platform and perform deep learning algorithms outside of the GEE platform. Additionally, using the platform requires a robust internet connection, which can be challenging in developing countries and may result in limited adoption [155,156]. GEE can cause an error with large computation complexity because of the memory limitation [146]. Remote sensing data are often multimodal, which requires the development of a novel ML model to extract joint features from the heterogeneous spectral, spatial, and temporal information. The integration of hydrologists, statisticians and expert remote sensing analysts is limited, leading to satellite remote sensing data regularly being underutilized and undervalued [152]. The limited sharing of data, particularly in Sub-Saharan Africa, significantly hampers our capacity to examine land function changes beyond land use and land cover alterations. Understanding the spatial variability in the land's ability to offer unintended services and identifying the factors influencing it are crucial for regional policy and spatial planning. Despite these challenges, this review emphasizes the need to shift towards adopting satellite data applications in assessing and monitoring the connection between LULC changes and water resources. This shift involves leveraging multiple data sources and employing advanced data processing techniques to enhance our comprehension of these complex systems.

4.2. Progress and Future Direction on Remote Sensing of LULC Changes on Water Resources

Progress has been made regarding the utility of remote sensing in semi-arid tropical environments, particularly in long-term monitoring, with a limited number of studies on seasonal monitoring. There is still a gap in the real-time use of modern earth observation techniques, such as Sentinels, which are freely available. Sentinels, with improved spectral resolution and revisit time (5 days), bring new opportunities for the biweekly and seasonal monitoring of the effect of LULC on water quantity and quality. The assessment and monitoring of LULC changes and their impacts on water resources using hyperspectral, drones and balloons has not attracted much attention. Only a few studies have attempted nonparametric machine learning algorithms [149,154]. Numerous researchers have found that LULC change correlates with water quantity and quality; still, there is no clear understanding of how LULC changes affect water quantity and quality. The quantification of the relationship between LULC change and water resources is a complex system. Understanding the relationship between LULC and water resource dynamics cannot be solely based on a single factor. Therefore, factors such as hydrology, spatial patterns and ecosystem services are often overlooked when assessing and monitoring the impact of LULC changes on water resources. Research efforts need to be promoted to evaluate the relationship between LULC and spatial patterns, ecosystem services and hydrological processes, particularly in semi-arid tropical environments that are susceptible to climate variability. This research could be crucial to supporting LULC planning, effective watershed management, and making informed decisions regarding water resource management in order to ensure the sustainable management of landscape composition and configuration, ecosystem services and hydrological factors.

Multispectral sensors such as Sentinel 2 and Landsat images tend to be limited by clouds and a relatively coarser spatial and temporal resolution [112,113]. Drone images are not affected by clouds because they are flown at a lower altitude and can be used to collect data over some inaccessible and remote areas [157]. As the fourth industrial revolution is progressing, the adoption of drones is advised in future studies as an innovative source of near real-time spatial data for mapping and monitoring the relationship between LULC and water resources. Thus far, no studies have been conducted using drones and comparing satellite sensors and drones in monitoring the impacts of LULC change on water quantity and quality in semi-arid tropical environments. Balloons equipped with a digital camera also have capabilities with regard to filling the gap between satellites and aircraft on the earth observation platform. Balloons filled with helium gas can be flown at lower altitudes than airplanes when detecting relatively small objects (small rivers, roads) with minimal expense [158]. The limitations associated with balloons are that they need to be appropriately used to avoid the geometrical distortion of images [158]. Geometrically distorted images provide false locations for the objects detected. Due to minimal expense, there is a need to also test this platform.

More research is needed to find the best variable and prediction model that can be integrated with a free multispectral dataset [108]. Different algorithms have weaknesses and strengths. Thus far, using an ensemble of different individual sensors and classifiers has been the most effective approach regarding the advance of LULC mapping. The use of more advanced non-parametric algorithms in estimating the effects of LULC on water quantity and quality has also been underutilized, despite their higher predictive accuracy, when compared to parametric algorithms, even when using broadband multispectral sensors. The application of the GWR and BN statistical model is robust; therefore, the use of this model should be adopted in future studies since the algorithms alone could not determine the physical processes of water resources. Integrating remotely sensed data with physical-based hydrologic models is recommended in future studies for effective management and decision making in watershed management. More research is needed to develop advanced models using multi-source data, and improved algorithms and applications are vital and required. The water quality parameter indices also showed effectiveness in estimating water quality; therefore, integrating them with ML algorithms in future might improve the models. There is also a need to assess the spatial pattern since it provides an understanding of the spatial processes underlying the distribution. The use of AI capabilities enables ML approaches to draw the complex nonlinear relationship between the land surfaces variable and water quality parameters [47,49]. AI has shown much promise for a wide range of physical environmental problems, from identifying critical situations, to aiding human interpretation, to discovering new relationships in large datasets [49]. This provides new opportunities with which to speed up the analysis process of large datasets, improve models and successfully transform the exploitation of environmental data in the future. In this regard, the use of multispectral sensor and parametric analysis in characterizing the impacts of LULC on surface water is required if the sustainable utilization of water management is to be achieved when aiming to address the rapidly growing population and its water needs. This information holds great importance for water managers, catchment managers, and land planners as it enables them to tailor their land and water management strategies according to the spatial variability and seasonal changes in the impacts of LULC on water quality and quantity. By understanding these relationships, it becomes feasible to implement local to regional framework policies that promote the sustainable utilization of land and water resources, thus facilitating effective land and water management practices.

5. Conclusions

The objective of this study was to conduct a systematic review to assess and monitor the progress of remote sensing applications in mapping land use and land cover (LULC) changes and their impacts on surface water quality and quantity. The literature indicates that the use of remote sensing in this field has gained significant attention in recent years. However, most studies have used multispectral Landsat images, while the potential of other sensors with an improved revisit time, a medium spatial resolution, and enhanced radiometric capabilities (e.g., Sentinel 1 and 2) has received less attention. These sensors offer opportunities for seasonal monitoring. Additionally, the integration of multisource remote sensing data, such as Sentinel 1, Worldview, radar, Quickbird, and LiDAR, has expanded tour ability to obtain, although their limitations need to be assessed. Combining data from different sensors enhances our understanding of the impact of LULC dynamics on water resources. Platforms like drones and helium-filled balloons enable the near-real-time acquisition of fine-resolution data, which can enhance the accuracy of LULC mapping via improved training and validation processes. The use of advanced classification algorithms, namely ML (e.g., SVM, ANN, RF, BN) and DL algorithms (e.g., RNN, CNN, DNN), which improve the accuracy and efficiency of land cover classification, is recommended. Moreover, the underutilization of advanced machine learning and deep learning algorithms in tropical semi-arid regions and globally suggests the potential for adopting these approaches in order to accurately estimate the effects of LULC changes on water resources. The use of artificial intelligence (AI) technology can enhance the data processing capabilities of big data analysis, thereby improving the quality of outcomes in future studies. The GEE platform, known for its capacity to handle remote sensing big data, could be a valuable and time-efficient tool for mapping LULC changes. To effectively manage water pollution resulting from different LULC patterns and improve water quality at local and regional scales, it is crucial to fully embrace these innovative technologies and methodologies. These advancements have enhanced our ability to assess and monitor the effects of land use changes on water resources, enabling more effective planning and management strategies. The progress made in remote sensing applications contributes to a better understanding of the complex interactions between LULC and water resources, thus supporting sustainable water management and environmental conservation efforts.

Author Contributions: Conceptualization, T.D., K.K.A. and M.R.R.; methodology, M.J.M.; formal analysis, M.J.M.; writing—original draft preparation, M.J.M.; writing—review and editing T.D. and B.T.M. All authors have read and agreed to the published version of the manuscript.

Funding: This research was funded by South African Water Research Commission, grant number C2019/2020-00166 and National Research Foundation (NRF) grant: PMDS2207123918.

Data Availability Statement: Data sharing is not applicable as new data were not analyzed.

Acknowledgments: Authors would like to thank the South African Water Research Commission (WRC) Project Number: C2019/2020-00166 and National Research Foundation (NRF) grant: PMDS22071239185 for funding this work. We also thank various scientists who have contributed towards this work.

Conflicts of Interest: The authors declare no conflict of interest.

References

1. Chemura, A.; Rwasoka, D.; Mutanga, O.; Dube, T.; Mushore, T. The impact of land-use/land cover changes on water balance of the heterogeneous Buzi sub-catchment, Zimbabwe. *Remote Sens. Appl. Soc. Environ.* **2020**, *18*, 100292. [CrossRef]
2. Leenaars, J.G.B.; Kempen, B.; van Oostrum, A.J.M.; Batjes, N.H. *Africa Soil Profiles Database: A Compilation of Georeferenced and Standardized Legacy Soil Profile Data for Sub-Saharan Africa*; Arrouays, D., McKenzie, N., Hempel, J., de Forges, A.R., McBratney, A.B., Eds.; Taylor & Francis Group: London, UK, 2014; pp. 51–57.
3. Huang, J.; Zhan, J.; Yan, H.; Wu, F.; Deng, X. Evaluation of the impacts of land use on water quality: A case study in the Chaohu lake basin. *Sci. World J.* **2013**, *2013*, 329187. [CrossRef] [PubMed]
4. Rapholo, M.T.; Makia, L.D. Are smallholder farmers' perceptions of climate variability supported by climatological evidence? Case study of a semi-arid region in South Africa. *Int. J. Clim. Chang. Strat. Manag.* **2020**, *12*, 571–585. [CrossRef]
5. Gan, T.; Ito, M.; Hülsmann, S.; Qin, X.; Lu, X.; Liong, S.; Rutschman, P.; Disse, M.; Koivusalo, H. Possible climate change/variability and human impacts, vulnerability of drought-prone regions, water resources and capacity building for Africa. *Hydrol. Sci. J.* **2016**, *61*, 1209–1226. [CrossRef]
6. Kusangaya, S.; Warburton, M.L.; van Garderen, E.A.; Jewitt, G.P.W. Impacts of climate change on water resources in southern Africa: A review. *Phys. Chem. Earth Parts A/B/C* **2014**, *67–69*, 47–54. [CrossRef]
7. Tahiru, A.A.; Doke, D.A.; Baatuuwie, B.N. Effect of land use and land cover changes on water quality in the Nawuni Catchment of the White Volta Basin, Northern Region, Ghana. *Appl. Water Sci.* **2020**, *10*, 198. [CrossRef]
8. Bilgin, A. An assessment of water quality in the Coruh Basin (Turkey) using multivariate statistical techniques. *Environ. Monit. Assess.* **2015**, *187*, 721. [CrossRef]
9. Fiquepron, J.; Garcia, S.; Stenger, A. Land use impact on water quality: Valuing forest services in terms of the water supply sector. *J. Environ. Manag.* **2013**, *126*, 113–121. [CrossRef]
10. Lambin, E.F.; Geist, H.J.; Lepers, E. Dynamics of Land-Use and Land-Cover Change in Tropical Regions. *Annu. Rev. Environ. Resour.* **2003**, *28*, 205–241. [CrossRef]
11. Liping, C.; Yujun, S.; Saeed, S. Monitoring and predicting land use and land cover changes using remote sensing and GIS techniques—A case study of a hilly area, Jiangle, China. *PLoS ONE* **2018**, *13*, e0200493. [CrossRef]
12. Ahearn, D.S.; Sheibley, R.W.; Dahlgren, R.A.; Anderson, M.; Johnson, J.; Tate, K.W. Land use and land cover influence on water quality in the last free-flowing river draining the western Sierra Nevada, California. *J. Hydrol.* **2005**, *313*, 234–247. [CrossRef]
13. Wu, F.; Mu, Y.; Chang, H.; Zhao, X.; Giesy, J.P.; Wu, K.B. Predicting Water Quality Criteria for Protecting Aquatic Life from Physicochemical Properties of Metals or Metalloids. *Environ. Sci. Technol.* **2013**, *47*, 446–453. [CrossRef] [PubMed]
14. Namugize, J.N.; Jewitt, G.; Graham, M. Effects of land use and land cover changes on water quality in the uMngeni river catchment, South Africa. *Phys. Chem. Earth Parts A/B/C* **2018**, *105*, 247–264. [CrossRef]
15. Rather, I.A.; Dar, A.Q. Assessing the impact of land use and land cover dynamics on water quality of Dal Lake, NW Himalaya, India. *Appl. Water Sci.* **2020**, *10*, 219. [CrossRef]
16. Bufebo, B.; Elias, E. Land Use/Land Cover Change and Its Driving Forces in Shenkolla Watershed, South Central Ethiopia. *Sci. World J.* **2021**, *2021*, 9470918. [CrossRef] [PubMed]

17. Giri, S.; Qiu, Z. Understanding the relationship of land uses and water quality in Twenty First Century: A review. *J. Environ. Manag.* **2016**, *173*, 41–48. [CrossRef] [PubMed]
18. Ullah, K.A.; Jiang, J.; Wang, P. Land use impacts on surface water quality by statistical approaches. *Glob. J. Environ. Sci. Manag.* **2018**, *4*, 231–250. [CrossRef]
19. Ozbay, G.; Fan, C.; Yang, Z. Relationship between Land Use and Water Quality and its Assessment Using Hyperspectral Remote Sensing in Mid-Atlantic Estuaries. In *Water Quality*; InTech: Houston, TX, USA, 2017. [CrossRef]
20. Otukei, J.R.; Blaschke, T. Land cover change assessment using decision trees, support vector machines and maximum likelihood classification algorithms. *Int. J. Appl. Earth Obs. Geoinf.* **2010**, *12* (Suppl. S1), S27–S31. [CrossRef]
21. Mwita, E.; Menz, G.; Misana, S.; Becker, M.; Kisanga, D.; Boehme, B. Mapping small wetlands of Kenya and Tanzania using remote sensing techniques. *Int. J. Appl. Earth Obs. Geoinf.* **2012**, *21*, 173–183. [CrossRef]
22. Weng, Q. Thermal infrared remote sensing for urban climate and environmental studies: Methods, applications, and trends. *ISPRS J. Photogramm. Remote Sens.* **2009**, *64*, 335–344. [CrossRef]
23. Kadavi, P.R.; Lee, C.-W. Land cover classification analysis of volcanic island in Aleutian Arc using an artificial neural network (ANN) and a support vector machine (SVM) from Landsat imagery. *Geosci. J.* **2018**, *22*, 653–665. [CrossRef]
24. Jayakumar, S. Support Vector Machine and Spectral Angle Mapper Classifications of High Resolution Hyper Spectral Aerial Image. *Korean J. Remote Sens.* **2009**, *25*, 233–242. Available online: https://www.researchgate.net/publication/263440372 (accessed on 23 April 2021).
25. Zhou, M.; Li, C.R.; Ma, L.; Guan, H.C. Land cover classification from full-waveform lidar data based on support vector machines. *Int. Arch. Photogramm. Remote Sens. Spatial Inf. Sci.-ISPRS Arch.* **2016**, *XLI-B3*, 447–452. [CrossRef]
26. Ghorbanian, A.; Kakooei, M.; Amani, M.; Mahdavi, S.; Mohammadzadeh, A.; Hasanlou, M. Improved land cover map of Iran using Sentinel imagery within Google Earth Engine and a novel automatic workflow for land cover classification using migrated training samples. *ISPRS J. Photogramm. Remote Sens.* **2020**, *167*, 276–288. [CrossRef]
27. Kulithalai Shiyam Sundar, P.; Deka, P.C. Spatio-Temporal Classiication and Prediction of Land Use and Land Cover Change for the Vembanad Lake System, Kerala-a Machine Learning Approach. *Environ. Sci. Pollut. Res.* **2021**, *29*, 86220–86236. [CrossRef]
28. Scott, G.J.; England, M.R.; Starms, W.A.; Marcum, R.A.; Davis, C.H. Training Deep Convolutional Neural Networks for Land–Cover Classification of High-Resolution Imagery. *IEEE Geosci. Remote Sens. Lett.* **2017**, *14*, 549–553. [CrossRef]
29. Chehata, N.; Guo, L.; Mallet, C. Airborne lidar feature selection for urban classification using random forests. *Laser Scanning* **2009**, *XXXVIII*, 207–212. Available online: https://hal.science/hal-02384719 (accessed on 27 March 2022).
30. Talukdar, S.; Singha, P.; Mahato, S.; Shahfahad; Pal, S.; Liou, Y.-A.; Rahman, A. Land-Use Land-Cover Classification by Machine Learning Classifiers for Satellite Observations—A Review. *Remote Sens.* **2020**, *12*, 1135. [CrossRef]
31. Vali, A.; Comai, S.; Matteucci, M. Deep Learning for Land Use and Land Cover Classification based on Hyperspectral and Multispectral Earth Observation Data: A Review. *Remote Sens.* **2020**, *12*, 2495. [CrossRef]
32. Pal, M.; Mather, P.M. An assessment of the effectiveness of decision tree methods for land cover classification. *Remote Sens. Environ.* **2003**, *86*, 554–565. [CrossRef]
33. Makinde, E.O.; Salami, A.T.; Olaleye, J.B.; Okewusi, O.C. Object Based and Pixel Based Classification Using Rapideye Satellite Imager of ETI-OSA, Lagos, Nigeria. *Geoinform. FCE CTU* **2016**, *15*, 59–70. [CrossRef]
34. Christovam, L.E.; Pessoa, G.G.; Shimabukuro, M.H.; Galo, M.L.B.T. Land use and land cover classification using hyperspectral imagery: Evaluating the performance of spectral angle mapper, support vector machine and random forest. *Int. Arch. Photogramm. Remote Sens. Spat. Inf. Sci.-ISPRS Arch.* **2019**, *XLII-2/W13*, 1841–1847. [CrossRef]
35. Ienco, D.; Gaetano, R.; Dupaquier, C.; Maurel, P. Land Cover Classification via Multitemporal Spatial Data by Deep Recurrent Neural Networks. *IEEE Geosci. Remote Sens. Lett.* **2017**, *14*, 1685–1689. [CrossRef]
36. KTanji, K.K.; Kielen, N.C. *Agricultural Drainage Water Management in Arid and Semi-Arid Areas*; FAO Irrigation and drainage paper 61; FAO: Rome, Italy, 2002.
37. Munthali, M.G.; Davis, N.; Adeola, A.M.; Botai, J.O.; Kamwi, J.M.; Chisale, H.L.W.; Orimoogunje, O.O.I. Local Perception of Drivers of Land-Use and Land-Cover Change Dynamics across Dedza District, Central Malawi Region. *Sustainability* **2019**, *11*, 832. [CrossRef]
38. Kim, K.-H.; Pauleit, S. Landscape character, biodiversity and land use planning: The case of Kwangju City Region, South Korea. *Land Use Policy* **2007**, *24*, 264–274. [CrossRef]
39. Haapanen, R.; Ek, A.R.; Bauer, M.E.; Finley, A.O. Delineation of forest/nonforest land use classes using nearest neighbor methods. *Remote Sens. Environ.* **2004**, *89*, 265–271. [CrossRef]
40. Zhou, W. An Object-Based Approach for Urban Land Cover Classification: Integrating LiDAR Height and Intensity Data. *IEEE Geosci. Remote Sens. Lett.* **2013**, *10*, 928–931. [CrossRef]
41. Chakraborty, S.; Paul, D.; Das, S.; Xu, J. Entropy Weighted Power k-Means Clustering. In Proceedings of the 23rd International Conference on Artificial Intelligence and Statistics (AISTATS) 2020, Palermo, Italy, 26–28 August 2020.
42. Bruzzone, L.; Prieto, D. Unsupervised retraining of a maximum likelihood classifier for the analysis of multitemporal remote sensing images. *IEEE Trans. Geosci. Remote Sens.* **2001**, *39*, 456–460. [CrossRef]
43. Dwarakish, G.; Ganasri, B. Impact of land use change on hydrological systems: A review of current modeling approaches. *Cogent Geosci.* **2015**, *1*, 1115691. [CrossRef]

44. Ding, J.; Jiang, Y.; Fu, L.; Liu, Q.; Peng, Q.; Kang, M. Impacts of Land Use on Surface Water Quality in a Subtropical River Basin: A Case Study of the Dongjiang River Basin, Southeastern China. *Water* **2015**, *7*, 4427–4445. [CrossRef]
45. Tadesse, L.; Suryabhagavan, K.; Sridhar, G.; Legesse, G. Land use and land cover changes and Soil erosion in Yezat Watershed, North Western Ethiopia. *Int. Soil Water Conserv. Res.* **2017**, *5*, 85–94. [CrossRef]
46. Arabameri, A.; Lee, S.; Tiefenbacher, J.P.; Ngo, P.T.T. Novel ensemble of MCDM-artificial intelligence techniques for groundwater-potential mapping in arid and semi-arid regions (Iran). *Remote Sens.* **2020**, *12*, 490. [CrossRef]
47. Peng, T.; Zhang, C.; Zhou, J.; Nazir, M.S. Negative correlation learning-based RELM ensemble model integrated with OVMD for multi-step ahead wind speed forecasting. *Renew. Energy* **2020**, *156*, 804–819. [CrossRef]
48. Banerjee, P.; Singh, V.S.; Chatttopadhyay, K.; Chandra, P.C.; Singh, B. Artificial neural network model as a potential alternative for groundwater salinity forecasting. *J. Hydrol.* **2012**, *398*, 212–220. [CrossRef]
49. Yuan, H.; Yang, G.; Li, C.; Wang, Y.; Liu, J.; Yu, H.; Feng, H.; Xu, B.; Zhao, X.; Yang, X. Retrieving Soybean Leaf Area Index from Unmanned Aerial Vehicle Hyperspectral Remote Sensing: Analysis of RF, ANN, and SVM Regression Models. *Remote Sens.* **2017**, *9*, 309. [CrossRef]
50. Bullock, E.L.; Woodcock, C.E.; Souza, C., Jr.; Olofsson, P. Satellite-based estimates reveal widespread forest degradation in the Amazon. *Glob. Chang. Biol.* **2020**, *26*, 2956–2969. [CrossRef]
51. Schulz, D.; Yin, H.; Tischbein, B.; Verleysdonk, S.; Adamou, R.; Kumar, N. Land use mapping using Sentinel-1 and Sentinel-2 time series in a heterogeneous landscape in Niger, Sahel. *ISPRS J. Photogramm. Remote Sens.* **2021**, *178*, 97–111. [CrossRef]
52. Shen, Z.; Hou, X.; Li, W.; Aini, G.; Chen, L.; Gong, Y. Impact of landscape pattern at multiple spatial scales on water quality: A case study in a typical urbanised watershed in China. *Ecol. Indic.* **2015**, *48*, 417–427. [CrossRef]
53. Tulyakov, S.; Jeni, L.A.; Cohn, J.F.; Sebe, N. Viewpoint-Consistent 3D Face Alignment. *IEEE Trans. Pattern Anal. Mach. Intell.* **2018**, *40*, 2250–2264. [CrossRef]
54. Zhang, J.; Zhang, Y. Remote sensing research issues of the National Land Use Change Program of China. *ISPRS J. Photogramm. Remote Sens.* **2007**, *62*, 461–472. [CrossRef]
55. Nofallah, S.; Mehta, S.; Mercan, E.; Knezevich, S.; May, C.J.; Weaver, D.; Witten, D.; Elmore, J.G.; Shapiro, L. Machine learning techniques for mitoses classification. *Comput. Med. Imaging Graph.* **2021**, *87*, 101832. [CrossRef]
56. Witten, J.; Witten, Z. Deep learning regression model for antimicrobial peptide design. *BioRxiv* **2019**, 692681. [CrossRef]
57. Crossley, S.A.; Cobb, T.; McNamara, D.S. Comparing count-based and band-based indices of word frequency: Implications for active vocabulary research and pedagogical applications. *System* **2013**, *41*, 965–981. [CrossRef]
58. Kwan, C.; Ayhan, B.; Budavari, B.; Lu, Y.; Perez, D.; Li, J.; Bernabe, S.; Plaza, A. Deep Learning for Land Cover Classification Using Only a Few Bands. *Remote Sens.* **2020**, *12*, 2000. [CrossRef]
59. Dahigamuwa, T.; Yu, Q.; Gunaratne, M. Feasibility Study of Land Cover Classification Based on Normalized Difference Vegetation Index for Landslide Risk Assessment. *Geosciences* **2016**, *6*, 45. [CrossRef]
60. Verbesselt, J.; Hyndman, R.; Newnham, G.; Culvenor, D. Detecting trend and seasonal changes in satellite image time series. *Remote Sens. Environ.* **2010**, *114*, 106–115. [CrossRef]
61. Hassan, Z.; Shabbir, R.; Ahmad, S.S.; Malik, A.H.; Aziz, N.; Butt, A.; Erum, S. Dynamics of land use and land cover change (LULCC) using geospatial techniques: A case study of Islamabad Pakistan. *Springerplus* **2016**, *5*, 812. [CrossRef] [PubMed]
62. Fatima, N.; Javed, A. Assessment of Land Use Land Cover Change Detection Using Geospatial Techniques in Southeast Rajasthan. *J. Geosci. Environ. Prot.* **2021**, *09*, 299–319. [CrossRef]
63. Tewabe, D.; Fentahun, T. Assessing land use and land cover change detection using remote sensing in the Lake Tana Basin, Northwest Ethiopia. *Cogent Environ. Sci.* **2020**, *6*, 1778998. [CrossRef]
64. Griffiths, P.; Kuemmerle, T.; Kennedy, R.E.; Abrudan, I.V.; Knorn, J.; Hostert, P. Using annual time-series of Landsat images to assess the effects of forest restitution in post-socialist Romania. *Remote Sens. Environ.* **2012**, *118*, 199–214. [CrossRef]
65. Campos-Taberner, M.; García-Haro, F.J.; Martínez, B.; Izquierdo-Verdiguier, E.; Atzberger, C.; Camps-Valls, G.; Gilabert, M.A. Understanding deep learning in land use classification based on Sentinel-2 time series. *Sci. Rep.* **2020**, *10*, 17188. [CrossRef]
66. Munthali, M.G.; Botai, J.O.; Davis, N.; Adeola, A.M. Multi-Temporal Analysis of Land Use and Land Cover Change Detection for Dedza District of Malawi Using Geospatial Techniques. 2019. Available online: http://www.ripublication.com (accessed on 15 August 2021).
67. Kumar, L.; Mutanga, O. Google Earth Engine Applications Since Inception: Usage, Trends, and Potential. *Remote Sens.* **2018**, *10*, 1509. [CrossRef]
68. Das, N.N.; Entekhabi, D.; Dunbar, R.S.; Chaubell, M.J.; Colliander, A.; Yueh, S.; Jagdhuber, T.; Chen, F.; Crow, W.; O'Neill, P.E.; et al. The SMAP and Copernicus Sentinel 1A/B microwave active-passive high resolution surface soil moisture product. *Remote Sens. Environ.* **2019**, *233*, 111380. [CrossRef]
69. Becker, R.; Koppa, A.; Schulz, S.; Usman, M.; der Beek, T.A.; Schüth, C. Spatially distributed model calibration of a highly managed hydrological system using remote sensing-derived ET data. *J. Hydrol.* **2019**, *577*, 123944. [CrossRef]
70. Saddique, N.; Mahmood, T.; Bernhofer, C. Quantifying the impacts of land use/land cover change on the water balance in the afforested River Basin, Pakistan. *Environ. Earth Sci.* **2020**, *79*, 448. [CrossRef]
71. Beck, D.A.; Johnson, G.R.; Spolek, G.A. Amending greenroof soil with biochar to affect runoff water quantity and quality. *Environ. Pollut.* **2011**, *159*, 2111–2118. [CrossRef]

72. Guzha, A.; Rufino, M.; Okoth, S.; Jacobs, S.; Nóbrega, R. Impacts of land use and land cover change on surface runoff, discharge and low flows: Evidence from East Africa. *J. Hydrol. Reg. Stud.* **2018**, *15*, 49–67. [CrossRef]
73. Becht, A.; Bürger, C.; Kostic, B.; Appel, E.; Dietrich, P. High-resolution aquifer characterization using seismic cross-hole tomography: An evaluation experiment in a gravel delta. *J. Hydrol.* **2007**, *336*, 171–185. [CrossRef]
74. Ahlmer, A.-K.; Cavalli, M.; Hansson, K.; Koutsouris, A.J.; Crema, S.; Kalantari, Z. Soil moisture remote-sensing applications for identification of flood-prone areas along transport infrastructure. *Environ. Earth Sci.* **2018**, *77*, 533. [CrossRef]
75. Tanksali, A.; Soraganvi, V.S. Assessment of impacts of land use/land cover changes upstream of a dam in a semi-arid watershed using QSWAT. *Model. Earth Syst. Environ.* **2021**, *7*, 2391–2406. [CrossRef]
76. Lam, Q.D.; Schmalz, B.; Fohrer, N. Assessing the spatial and temporal variations of water quality in lowland areas, Northern Germany. *J. Hydrol.* **2012**, *438–439*, 137–147. [CrossRef]
77. Gholizadeh, M.H.; Melesse, A.M.; Reddi, L. A Comprehensive Review on Water Quality Parameters Estimation Using Remote Sensing Techniques. *Sensors* **2016**, *16*, 1298. [CrossRef] [PubMed]
78. Chen, W.; Wang, J.; Cao, X.; Ran, H.; Teng, D.; Chen, J.; He, X.; Zheng, X. Possibility of using multiscale normalized difference vegetation index data for the assessment of total suspended solids (TSS) concentrations in surface water: A specific case of scale issues in remote sensing. *Environ. Res.* **2021**, *194*, 110636. [CrossRef] [PubMed]
79. Ribaudo, M.; Johansson, R. Water quality: Impacts of agriculture. In *Water Quality: Physical, Chemical and Biological Characteristics*; Nova Science Publishers: Hauppauge, NY, USA, 2011; pp. 201–209.
80. Glasgow, H.B.; Burkholder, J.M.; Reed, R.E.; Lewitus, A.J.; Kleinman, J.E. Real-time remote monitoring of water quality: A review of current applications, and advancements in sensor, telemetry, and computing technologies. *J. Exp. Mar. Biol. Ecol.* **2004**, *300*, 409–448. [CrossRef]
81. Chotpantarat, S.; Boonkaewwan, S. Impacts of land-use changes on watershed discharge and water quality in a large intensive agricultural area in Thailand. *Hydrol. Sci. J.* **2018**, *63*, 1386–1407. [CrossRef]
82. Yohannes, H.; Elias, E. Contamination of Rivers and Water Reservoirs in and Around Addis Ababa City and Actions to Combat It. *Environ. Pollut. Clim. Chang.* **2017**, *1*, 8. [CrossRef]
83. Cai, Y.; Zhang, H.; Zheng, P.; Pan, W. Quantifying the Impact of Land use/Land Cover Changes on the Urban Heat Island: A Case Study of the Natural Wetlands Distribution Area of Fuzhou City, China. *Wetlands* **2016**, *36*, 285–298. [CrossRef]
84. Giri, S.; Qiu, Z.; Zhang, Z. Assessing the impacts of land use on downstream water quality using a hydrologically sensitive area concept. *J. Environ. Manag.* **2018**, *213*, 309–319. [CrossRef]
85. Kebede, W.; Tefera, M.; Habitamu, T.; Alemayehu, T. Impact of Land Cover Change on Water Quality and Stream Flow in Lake Hawassa Watershed of Ethiopia. *Agric. Sci.* **2014**, *5*, 647–659. [CrossRef]
86. Schilling, K.E.; Hubbard, T.; Luzier, J.; Spooner, J. *Walnut Creek Watershed Restoration and Water Quality Monitoring Project: Final Report*; Iowa Department of Natural Resources: Iowa City, IA, USA, 2006. Available online: https://iro.uiowa.edu/esploro/outputs/9984111963002771/filesAndLinks?institution=01IOWA_INST&index=null (accessed on 10 May 2021).
87. Guo, M.; Li, J.; Sheng, C.; Xu, J.; Wu, L. A Review of Wetland Remote Sensing. *Sensors* **2017**, *17*, 777. [CrossRef]
88. Gallant, A.L. The Challenges of Remote Monitoring of Wetlands. *Remote Sens.* **2015**, *7*, 10938–10950. [CrossRef]
89. Kibena, J.; Nhapi, I.; Gumindoga, W. Assessing the relationship between water quality parameters and changes in landuse patterns in the Upper Manyame River, Zimbabwe. *Phys. Chem. Earth Parts A/B/C* **2014**, *67–69*, 153–163. [CrossRef]
90. Zhang, Z.; Zhang, F.; Du, J.; Chen, D.; Zhang, W. Impacts of land use at multiple buffer scales on seasonal water quality in a reticular river network area. *PLoS ONE* **2021**, *16*, e0244606. [CrossRef]
91. Dube, T.; Mutanga, O. Evaluating the utility of the medium-spatial resolution Landsat 8 multispectral sensor in quantifying aboveground biomass in uMgeni catchment, South Africa. *ISPRS J. Photogramm. Remote Sens.* **2015**, *101*, 36–46. [CrossRef]
92. Ozesmi, S.L.; Bauer, M.E. Satellite remote sensing of wetlands. *Wetl. Ecol. Manag.* **2002**, *10*, 381–402. [CrossRef]
93. Xie, Y.; Sha, Z.; Yu, M. Remote sensing imagery in vegetation mapping: A review. *J. Plant Ecol.* **2008**, *1*, 9–23. [CrossRef]
94. du Plessis, A.; Harmse, T.; Ahmed, F. Quantifying and Predicting the Water Quality Associated with Land Cover Change: A Case Study of the Blesbok Spruit Catchment, South Africa. *Water* **2014**, *6*, 2946–2968. [CrossRef]
95. Juma, D.W.; Wang, H.; Li, F. Impacts of population growth and economic development on water quality of a lake: Case study of Lake Victoria Kenya water. *Environ. Sci. Pollut. Res.* **2014**, *21*, 5737–5746. [CrossRef] [PubMed]
96. Van der Meer, F.; Schmidt, K.S.; Bakker, W.; Bijker, W. *Environmental Modelling with GIs and Remote Sensing*; CRC Press: Boca Raton, FL, USA, 2002.
97. Rothenberger, M.B.; Burkholder, J.M.; Brownie, C. Long-Term Effects of Changing Land Use Practices on Surface Water Quality in a Coastal River and Lagoonal Estuary. *Environ. Manag.* **2009**, *44*, 505–523. [CrossRef] [PubMed]
98. Kaushal, S.S.; Duan, S.; Doody, T.R.; Haq, S.; Smith, R.M.; Johnson, T.A.N.; Newcomb, K.D.; Gorman, J.; Bowman, N.; Mayer, P.M.; et al. Human-accelerated weathering increases salinization, major ions, and alkalinization in fresh water across land use. *Appl. Geochem.* **2017**, *83*, 121–135. [CrossRef]
99. Khare, Y.P.; Martinez, C.J.; Toor, G.S. Water Quality and Land Use Changes in the Alafia and Hillsborough River Watersheds, Florida, USA. *JAWRA J. Am. Water Resour. Assoc.* **2012**, *48*, 1276–1293. [CrossRef]
100. Liao, B.; Lin, H.; Langevin, S.; Gao, W.; Leppard, G. Effects of temperature and dissolved oxygen on sludge properties and their role in bioflocculation and settling. *Water Res.* **2011**, *45*, 509–520. [CrossRef] [PubMed]

101. de Mello, K.; Valente, R.A.; Randhir, T.O.; dos Santos, A.C.A.; Vettorazzi, C.A. Effects of land use and land cover on water quality of low-order streams in Southeastern Brazil: Watershed versus riparian zone. *Catena* **2018**, *167*, 130–138. [CrossRef]
102. Price, K. Effects of watershed topography, soils, land use, and climate on baseflow hydrology in humid regions: A review. *Prog. Phys. Geogr.* **2011**, *35*, 465–492. [CrossRef]
103. Pullanikkatil, D.; Palamuleni, L.; Ruhiiga, T. Impact of land use on water quality in the Likangala catchment, southern Malawi. *Afr. J. Aquat. Sci.* **2015**, *40*, 277–286. [CrossRef]
104. Wijesiri, B.; Deilami, K.; Goonetilleke, A. Evaluating the relationship between temporal changes in land use and resulting water quality. *Environ. Pollut.* **2018**, *234*, 480–486. [CrossRef] [PubMed]
105. Yu, S.; Xu, Z.; Wu, W.; Zuo, D. Effect of land use on the seasonal variation of streamwater quality in the Wei River basin, China. *Proc. Int. Assoc. Hydrol. Sci.* **2015**, *368*, 454–459. [CrossRef]
106. Hu, B.; Xu, Y.; Huang, X.; Cheng, Q.; Ding, Q.; Bai, L.; Li, Y. Improving Urban Land Cover Classification with Combined Use of Sentinel-2 and Sentinel-1 Imagery. *ISPRS Int. J. Geo-Inf.* **2021**, *10*, 533. [CrossRef]
107. Teixeira, Z.; Teixeira, H.; Marques, J.C. Systematic processes of land use/land cover change to identify relevant driving forces: Implications on water quality. *Sci. Total Environ.* **2014**, *470–471*, 1320–1335. [CrossRef]
108. Sagan, V.; Peterson, K.T.; Maimaitijiang, M.; Sidike, P.; Sloan, J.; Greeling, B.A.; Maalouf, S.; Adams, C. Monitoring inland water quality using remote sensing: Potential and limitations of spectral indices, bio-optical simulations, machine learning, and cloud computing. *Earth-Sci. Rev.* **2020**, *205*, 103187. [CrossRef]
109. Flores-Anderson, A.I.; Griffin, R.; Dix, M.; Romero-Oliva, C.S.; Ochaeta, G.; Skinner-Alvarado, J.; Moran, M.V.R.; Hernandez, B.; Cherrington, E.; Page, B.; et al. Hyperspectral Satellite Remote Sensing of Water Quality in Lake Atitlán, Guatemala. *Front. Environ. Sci.* **2020**, *8*, 7. [CrossRef]
110. Schilling, K.E.; Spooner, J. Effects of Watershed-Scale Land Use Change on Stream Nitrate Concentrations. *J. Environ. Qual.* **2006**, *35*, 2132–2145. [CrossRef] [PubMed]
111. Elhag, M.; Gitas, I.; Othman, A.; Bahrawi, J.; Gikas, P. Assessment of Water Quality Parameters Using Temporal Remote Sensing Spectral Reflectance in Arid Environments, Saudi Arabia. *Water* **2019**, *11*, 556. [CrossRef]
112. Syifa, M.; Park, S.J.; Lee, C.W. Detection of the Pine Wilt Disease Tree Candidates for Drone Remote Sensing Using Artificial Intelligence Techniques. *Engineering* **2020**, *6*, 919–926. [CrossRef]
113. Singh, P.; Gupta, A.; Singh, M. Hydrological inferences from watershed analysis for water resource management using remote sensing and GIS techniques. *Egypt. J. Remote Sens. Space Sci.* **2014**, *17*, 111–121. [CrossRef]
114. Sibanda, M.; Mutanga, O.; Chimonyo, V.G.P.; Clulow, A.D.; Shoko, C.; Mazvimavi, D.; Dube, T.; Mabhaudhi, T. Application of Drone Technologies in Surface Water Resources Monitoring and Assessment: A Systematic Review of Progress, Challenges, and Opportunities in the Global South. *Drones* **2021**, *5*, 84. [CrossRef]
115. Verma, P.; Singh, P.; Srivastava, S.K. Impact of land use change dynamics on sustainability of groundwater resources using earth observation data. *Environ. Dev. Sustain.* **2020**, *22*, 5185–5198. [CrossRef]
116. Sliva, L.; Williams, D.D. Buffer Zone versus Whole Catchment Approaches to Studying Land Use Impact on River Water Quality. *Water Res.* **2001**, *35*, 3462–3472. [CrossRef]
117. Chen, Y.; Shuai, J.; Zhang, Z.; Shi, P.; Tao, F. Simulating the impact of watershed management for surface water quality protection: A case study on reducing inorganic nitrogen load at a watershed scale. *Ecol. Eng.* **2014**, *62*, 61–70. [CrossRef]
118. Seilheimer, T.S.; Zimmerman, P.L.; Stueve, K.M.; Perry, C.H. Landscape-scale modeling of water quality in Lake Superior and Lake Michigan watersheds: How useful are forest-based indicators? *J. Great Lakes Res.* **2013**, *39*, 211–223. [CrossRef]
119. Nielsen, N.M.; Smink, W.A.C.; Fox, J.-P. Small and negative correlations among clustered observations: Limitations of the linear mixed effects model. *Behaviormetrika* **2021**, *48*, 51–77. [CrossRef]
120. Lee, J.Y.; Yang, J.S.; Kim, D.K.; Han, M.Y. Relationship between land use and water quality in a small watershed in South Korea. *Water Sci. Technol.* **2010**, *62*, 2607–2615. [CrossRef] [PubMed]
121. Chen, Q.; Zhang, Y.; Hallikainen, M. Water quality monitoring using remote sensing in support of the EU water framework directive (WFD): A case study in the Gulf of Finland. *Environ. Monit. Assess.* **2007**, *124*, 157–166. [CrossRef] [PubMed]
122. Liu, J.; Shen, Z.; Chen, L. Assessing how spatial variations of land use pattern affect water quality across a typical urbanized watershed in Beijing, China. *Landsc. Urban Plan.* **2018**, *176*, 51–63. [CrossRef]
123. Kändler, M.; Blechinger, K.; Seidler, C.; Pavlů, V.; Šanda, M.; Dostál, T.; Krása, J.; Vitvar, T.; Štich, M. Impact of land use on water quality in the upper Nisa catchment in the Czech Republic and in Germany. *Sci. Total Environ.* **2017**, *586*, 1316–1325. [CrossRef]
124. Singh, K.P.; Malik, A.; Mohan, D.; Sinha, S. Multivariate statistical techniques for the evaluation of spatial and temporal variations in water quality of Gomti River (India)—A case study. *Water Res.* **2004**, *38*, 3980–3992. [CrossRef]
125. Brown, S.; Versace, V.L.; Laurenson, L.; Ierodiaconou, D.; Fawcett, J.; Salzman, S. Assessment of Spatiotemporal Varying Relationships Between Rainfall, Land Cover and Surface Water Area Using Geographically Weighted Regression. *Environ. Model. Assess.* **2011**, *17*, 241–254. [CrossRef]
126. Shrestha, P.M. *Comparison of Ordinary Least Square Regression, Spatial Autoregression, and Geographically Weighted Regression for Modeling Forest Structural Attributes Using a Geographical Information System (GIS)/Remote Sensing (RS) Approach*; University of Calgary: Calgary, AB, Canada, 2006.
127. Tu, Y.; Chen, B.; Lang, W.; Chen, T.; Li, M.; Zhang, T.; Xu, B. Uncovering the Nature of Urban Land Use Composition Using Multi-Source Open Big Data with Ensemble Learning. *Remote Sens.* **2021**, *13*, 4241. [CrossRef]

128. Wan, R.; Cai, S.; Li, H.; Yang, G.; Li, Z.; Nie, X. Inferring land use and land cover impact on stream water quality using a Bayesian hierarchical modeling approach in the Xitiaoxi River Watershed, China. *J. Environ. Manag.* **2014**, *133*, 1–11. [CrossRef]
129. Wikle, C.K. Hierarchical bayesian models for predicting the spread of ecological processes. *Ecology* **2003**, *84*, 1382–1394. [CrossRef]
130. Boukabara, S.-A.; Krasnopolsky, V.; Stewart, J.Q.; Maddy, E.S.; Shahroudi, N.; Hoffman, R.N. Leveraging Modern Artificial Intelligence for Remote Sensing and NWP: Benefits and Challenges. *Bull. Am. Meteorol. Soc.* **2019**, *100*, ES473–ES491. [CrossRef]
131. Ashapure, A.; Jung, J.; Chang, A.; Oh, S.; Maeda, M.; Landivar, J. A comparative study of RGB and multispectral sensor-based cotton canopy cover modelling using multi-temporal UAS data. *Remote Sens.* **2019**, *11*, 2757. [CrossRef]
132. Wang, X.; Xie, H. A Review on Applications of Remote Sensing and Geographic Information Systems (GIS) in Water Resources and Flood Risk Management. *Water* **2018**, *10*, 608. [CrossRef]
133. Bhattarai, A.; Dhakal, S.; Gautam, Y.; Bhattarai, R. Prediction of Nitrate and Phosphorus Concentrations Using Machine Learning Algorithms in Watersheds with Different Landuse. *Water* **2021**, *13*, 3096. [CrossRef]
134. Jung, J.; Maeda, M.; Chang, A.; Bhandari, M.; Ashapure, A.; Landivar-Bowles, J. The potential of remote sensing and artificial intelligence as tools to improve the resilience of agriculture production systems. *Curr. Opin. Biotechnol.* **2021**, *70*, 15–22. [CrossRef]
135. Song, Y.; Song, X.; Shao, G.; Hu, T. Effects of Land Use on Stream Water Quality in the Rapidly Urbanized Areas: A Multiscale Analysis. *Water* **2020**, *12*, 1123. [CrossRef]
136. de Mello, K.; Taniwaki, R.H.; de Paula, F.R.; Valente, R.A.; Randhir, T.O.; Macedo, D.R.; Leal, C.G.; Rodrigues, C.B.; Hughes, R.M. Multiscale land use impacts on water quality: Assessment, planning, and future perspectives in Brazil. *J. Environ. Manag.* **2020**, *270*, 110879. [CrossRef]
137. Sidle, R.C.; Ziegler, A.D.; Vogler, J.B. Contemporary changes in open water surface area of Lake Inle, Myanmar. *Sustain. Sci.* **2007**, *2*, 55–65. [CrossRef]
138. Tetzlaff, D.; Seibert, J.; Soulsby, C. Inter-catchment comparison to assess the influence of topography and soils on catchment transit times in a geomorphic province; the Cairngorm mountains, Scotland. *Hydrol. Process. Int. J.* **2009**, *23*, 1874–1886. [CrossRef]
139. Su, Z.H.; Lin, C.; Ma, R.H.; Luo, J.H.; Liang, Q.O. Effect of land use change on lake water quality in different buffer zones. *Appl. Ecol. Environ. Res.* **2015**, *13*, 639–653. [CrossRef]
140. Li, S.; Gu, S.; Tan, X.; Zhang, Q. Water quality in the upper Han River basin, China: The impacts of land use/land cover in riparian buffer zone. *J. Hazard. Mater.* **2009**, *165*, 317–324. [CrossRef]
141. Gyawali, S.; Techato, K.; Monprapusson, S. Assessing the Impact of Land Use on Water Quality across Multiple Spatial Scales in U-tapao River Basin, Thailand. *J. Inst. Sci. Technol.* **2015**, *20*, 54–60. [CrossRef]
142. Ceballos, G.; Davidson, A.; List, R.; Pacheco, J.; Manzano-Fischer, P.; Santos-Barrera, G.; Cruzado, J. Rapid Decline of a Grassland System and Its Ecological and Conservation Implications. *PLoS ONE* **2010**, *5*, e8562. [CrossRef] [PubMed]
143. Tanaka, M.O.; de Souza, A.L.T.; Moschini, L.E.; de Oliveira, A.K. Influence of watershed land use and riparian characteristics on biological indicators of stream water quality in southeastern Brazil. *Agric. Ecosyst. Environ.* **2016**, *216*, 333–339. [CrossRef]
144. Stonestrom, D.A.; Scanlon, B.R.; Zhang, L. Introduction to special section on Impacts of Land Use Change on Water Resources. *Water Resour. Res.* **2009**, *45*. [CrossRef]
145. Thakur, J.K.; Singh, S.K.; Ekanthalu, V.S. Integrating remote sensing, geographic information systems and global positioning system techniques with hydrological modeling. *Appl. Water Sci.* **2017**, *7*, 1595–1608. [CrossRef]
146. Jordan, T.E.; Weller, D.E.; Pelc, C.E. Effects of Local Watershed Land Use on Water Quality in Mid-Atlantic Coastal Bays and Subestuaries of the Chesapeake Bay. *Estuaries Coasts* **2018**, *41*, 38–53. [CrossRef]
147. Dong, C. Remote sensing, hydrological modeling and in situ observations in snow cover research: A review. *J. Hydrol.* **2018**, *561*, 573–583. [CrossRef]
148. Wulder, M.A.; White, J.C.; Goward, S.N.; Masek, J.G.; Irons, J.R.; Herold, M.; Cohen, W.B.; Loveland, T.R.; Woodcock, C.E. Landsat continuity: Issues and opportunities for land cover monitoring. *Remote Sens. Environ.* **2008**, *112*, 955–969. [CrossRef]
149. Naboureh, A.; Bian, J.; Lei, G.; Li, A. A review of land use/land cover change mapping in the China-Central Asia-West Asia economic corridor countries. *Big Earth Data* **2021**, *5*, 237–257. [CrossRef]
150. Foody, G.M. Explaining the unsuitability of the kappa coefficient in the assessment and comparison of the accuracy of thematic maps obtained by image classification. *Remote Sens. Environ.* **2020**, *239*, 111630. [CrossRef]
151. Neteler, M.; Bowman, M.H.; Landa, M.; Metz, M. GRASS GIS: A multi-purpose open source GIS. *Environ. Model. Softw.* **2012**, *31*, 124–130. [CrossRef]
152. Thamaga, K.H.; Dube, T. Remote sensing of invasive water hyacinth (Eichhornia crassipes): A review on applications and challenges. *Remote Sens. Appl. Soc. Environ.* **2018**, *10*, 36–46. [CrossRef]
153. Gorelick, N.; Hancher, M.; Dixon, M.; Ilyushchenko, S.; Thau, D.; Moore, R. Google Earth Engine: Planetary-scale geospatial analysis for everyone. *Remote Sens. Environ.* **2017**, *202*, 18–27. [CrossRef]
154. Eskandari, S.; Jaafari, M.R.; Oliva, P.; Ghorbanzadeh, O.; Blaschke, T. Mapping Land Cover and Tree Canopy Cover in Zagros Forests of Iran: Application of Sentinel-2, Google Earth, and Field Data. *Remote Sens.* **2020**, *12*, 1912. [CrossRef]
155. Gxokwe, S.; Dube, T.; Mazvimavi, D. Leveraging Google Earth Engine platform to characterize and map small seasonal wetlands in the semi-arid environments of South Africa. *Sci. Total Environ.* **2022**, *803*, 150139. [CrossRef]
156. Mudereri, B.T.; Abdel-Rahman, E.M.; Dube, T.; Niassy, S.; Khan, Z.; Tonnang, H.E.; Landmann, T. A two-step approach for detecting Striga in a complex agroecological system using Sentinel-2 data. *Sci. Total Environ.* **2021**, *762*, 143151. [CrossRef]

157. Al-Najjar, H.A.H.; Kalantar, B.; Pradhan, B.; Saeidi, V.; Halin, A.A.; Ueda, N.; Mansor, S. Land Cover Classification from fused DSM and UAV Images Using Convolutional Neural Networks. *Remote Sens.* **2019**, *11*, 1461. [CrossRef]
158. Kako, S.; Isobe, A.; Magome, S. Low altitude remote-sensing method to monitor marine and beach litter of various colors using a balloon equipped with a digital camera. *Mar. Pollut. Bull.* **2012**, *64*, 1156–1162. [CrossRef]

remote sensing MDPI

Article

Machine Learning in the Analysis of Multispectral Reads in Maize Canopies Responding to Increased Temperatures and Water Deficit

Josip Spišić [1], Domagoj Šimić [2], Josip Balen [1], Antun Jambrović [2] and Vlatko Galić [2,*]

[1] Faculty of Electrical Engineering, Computer Science and Information Technology, Josip Juraj Strossmayer University of Osijek, HR31000 Osijek, Croatia; josip.spisic@ferit.hr (J.S.); josip.balen@ferit.hr (J.B.)
[2] Agricultural Institute Osijek, HR31000 Osijek, Croatia; domagoj.simic@poljinos.hr (D.Š.); antun.jambrovic@poljinos.hr (A.J.)
* Correspondence: vlatko.galic@poljinos.hr

Abstract: Real-time monitoring of crop responses to environmental deviations represents a new avenue for applications of remote and proximal sensing. Combining the high-throughput devices with novel machine learning (ML) approaches shows promise in the monitoring of agricultural production. The 3×2 multispectral arrays with responses at 610 and 680 nm (red), 730 and 760 nm (red-edge) and 810 and 860 nm (infrared) spectra were used to assess the occurrence of leaf rolling (LR) in 545 experimental maize plots measured four times for calibration dataset (n = 2180) and 145 plots measured once for external validation. Multispectral reads were used to calculate 15 simple normalized vegetation indices. Four ML algorithms were assessed: single and multilayer perceptron (SLP and MLP), convolutional neural network (CNN) and support vector machines (SVM) in three validation procedures, which were stratified cross-validation, random subset validation and validation with external dataset. Leaf rolling occurrence caused visible changes in spectral responses and calculated vegetation indexes. All algorithms showed good performance metrics in stratified cross-validation (accuracy >80%). SLP was the least efficient in predictions with external datasets, while MLP, CNN and SVM showed comparable performance. Combining ML with multispectral sensing shows promise in transition towards agriculture based on data-driven decisions especially considering the novel Internet of Things (IoT) avenues.

Keywords: machine learning; maize; stress; heat; classification; validation; python; IoT

1. Introduction

Human population growth has led to increasing food requirements and resource depletion, intensifying the use of modern technologies in agriculture over the last few decades. Thus, major achievements in sensing technologies, wireless communication and artificial intelligence have been made by research efforts in agriculture globally [1,2]. In the context of climate change, real-time monitoring of drought is of primary interest which yielded a forked remote sensing approach, deploying satellite imagery to spot drought occurrence [3–7], or utilizing the recent developments in affordable sensor solutions, generating data in a more (unmanned aerial vehicles—UAV) or less (pole, machine or tower mounted) remote manner [8–12]. In conventional agricultural production, the only objective data collected on plant side (grain yield) are collected when the plant is already dead, so the real-time monitoring practices represent a paradigmatic shift for most farmers around the world.

Studies show that there is a growing occurrence of heatwave days accompanied by longer spans of drought [13]. These deleterious climate changes come with increased sensitivity of maize and soybean to heat and drought, despite ever strong breeding efforts for tolerance to abiotic stress [14]. In maize, there is a large number of morpho-physiological adjustments in response to drought stress [15–17]. However, many of these changes are irreversible, such as low fertilization rate, increased susceptibility to diseases or depleted stands [18], and once they are expressed, the economic losses are unavoidable. There are also some changes that are different among maize cultivars [19] and they are reversible in nature providing an excellent signal of the current plant water status.

One such trait is transverse rolling of leaf blades (Figure S1), usually caused by hydronastic changes. Many plant species use this mechanism as a drought avoidance strategy [20]. It represents a mechanism of adaptation in plants to control stress mostly by the means of auto-stress (cell-wall tension increase/decrease), reduction in the light interception and transpiration, thus preventing dehydration and overheating [21]. Not all genotypes express leaf rolling at the same conditions [22,23], nor the monotonic increase in leaf curvature shows the same maximum over different genotypes [24]. While in some genotypes it represents an avoidance strategy for stress conditions, in others it marks a tipping point for the physiological damage [25–27], thus implying the need to dynamically monitor for this trait in real time, for breeding and management purposes. So far, the use of different methods of leaf rolling quantification was reported in the literature, most of them grading responses in scales from 1 (no leaf rolling) to 5 (completely rolled leaves, dead or lax) [28–30], or 1 (no leaf rolling) to 9 (completely rolled leaf blades) [31]. All of the mentioned methods imply phenotyping at (i) drought stress treatment, or (ii) at the time of solar noon, when the strongest leaf rolling is expected to occur [24]. However, all of the conventional methods also imply a need for human screening limiting the ability to capture trait dynamics, accompanied with the increased error in rolling grading between scorers. In order to use phenotypic indicators, such as leaf rolling in some unfavorable conditions, in decision support systems, the possibility for their remote monitoring is of critical importance.

Leaf rolling causes several easily detectable changes in plant level. One such change is reduction in leaf area index (LAI), rendering the changes easily detectable by simple hemispherical photography [31]. However, there are also more subtle changes in spectral derivatives of leaf-rolled plants exposed to drought treatment, mostly caused by the accumulation/translocation of biochemicals and decrease in photosynthetic activity [32], allowing the drought detection using hyperspectral data along with several derived normalized vegetation indexes such as normalized difference vegetation index (NDVI) [33]. Normalized difference vegetation index and other normalized vegetation indices (VI) represent a useful and sensitive tool in vegetation monitoring converting the raw sensor reads to useful normalized and repeatable results [34,35]. Hyper/multispectral monitoring is already a proven method of vegetation monitoring [36,37] coming more and more to focus of researchers addressing high-throughput phenotyping and precision agriculture [38–40]. Moreover, different types of stress can be detected by combining ML with spectral monitoring [41–43]. There are many approaches for applied regression analysis of the remote sensing data being used [44,45]; however, such models are inefficient when accounting for nonlinearity of targets in multi-dimensional hyperplanes. On the other hand, these data properties are efficiently handled by modern machine learning (ML) algorithms, extracting numerical features from the data while retaining the information from the original dataset. Moreover, there is a growing body of evidence of the superiority in performance of ML algorithms in remote sensing data analysis for various agricultural applications such as vegetation classification [46], biomass and soil moisture analysis [47], crop stress phenotyping [38], precision farming [48,49] and many others [50]. Furthermore, these methods also have an ecological and humanitarian depth showing promise in helping to adhere to the Sustainable Development Goals [51] presented by the United Nations [52].

In this research, we attempted to use a low-cost sensor capturing spectral responses around several critical plant reflectance wavelengths and to apply machine learning to detect changes in plant morphology for the envisioned use in precision agriculture and plant breeding. Specifically, objectives were to determine the usefulness of a simple multispectral sensor to monitor leaf rolling in maize as a sign of stress, and to assess different machine learning models readily applied in the classification of labeled data.

2. Materials and Methods

2.1. Field Experiments

The field experiments were carried out at the experimental station of the Agricultural Institute Osijek (AIO) in Osijek Croatia (45°32′N 18°44′E). Fields are subject to barley–soybean–maize rotation and are, following the soil analysis, fertilized and maintained with non-limiting amounts of fertilizers following respective local best practices and regulations. For the purpose of development of new cultivars AIO organizes several levels of field trials with multiple cultivars, separated between early, early to medium, medium to late and late maturity breeding programs. Maturity was categorized by FAO system [53] with reference genotypes used as checks to categorize hybrids to groups 1 (early) to 7 (late). Besides breeding programs, AIO organizes demonstrational trials with 3 to 75 hybrids representing the pallet of latest breeding efforts. AIO is a certified seed producer, marketing maize hybrids in Southeast and Central Europe and the Middle East. Thus, there is a considerable diversity of hybrids present at breeding trials of every maturity group, aiming at adaptation to different agro-ecological scenarios. For the purpose of this study, we chose seven trials with different numbers of hybrids from the mentioned maturity groups, namely, single irrigated demonstrational trial (DTir) and two rainfed demonstrational trials (DTrf and SDTrf). Soil type in DTrf and DTir was anthropogenized eutric cambisol, while at SDTrf, soil was sandy loam. In the gradient of trial qualities, SDTrf thus represented a low-water availability trial, while DTir was not water limited during the screening. The DTir trial was irrigated twice with 40 mm/m^2 per irrigation, on 20th of June and 2nd of July. Irrigation was carried out by gear-driven full circle sprinklers. Trials of early (ET), early to medium (EMT), medium to late (MLT) and late maturity (LT) breeding programs were represented by randomized complete block trials with 25 hybrids in four replicates on anthropogenized eutric cambisol (ET, MLT and LT) and sandy loam (EMT). All trials were sown in a north–south orientation except SDTrf and EMT which were sown in an east–west orientation. Details on experimental design are shown in Table 1.

Table 1. Details about experimental design.

Experiment Design	No. Hybrids	Replicates	Plot Size (m^2)	FAO Relative Maturity	Planting Date	Anthesis Interval (Days)	Measurement Time
DTir	61	n.a.	50	180–720	7th April 2021	2.7.–20.7.	06:30–08:30 & 12:30–14:30
DTrf	64	n.a.	50	180–720	16th April 2021	29.6.–17.7.	
SDTrf	20	n.a.	50	350–680	22nd April 2021	5.7.–16.7.	
ET	25	4	8.4	210–290	8th April 2021	29.6.–7.7.	
EMT	25	4	8.4	350–490	22nd April 2021	3.7.–15.7.	
MLT	25	4	8.4	500–590	9th April 2021	8.7.–19.7.	
LT	25	4	8.4	>600	9th April 2021	9.7.–19.7.	

2.2. Measurements and Agroecological Conditions

Measurements were carried out with AMS (ams-OSRAM AG, Austria) AS7263 sensor unit with six spectral bands (3 × 2 photo diode array) responsive to wavelengths in red and near-infrared spectra (610, 680, 730, 760, 810 and 860 nm) with 20 nm full width at half maximum. Sensor consists of plastic housing, a lens and photodiode array with aperture of 0.75 mm and 20.5° viewing angle. The sensor was connected to Arduino Uno prototyping board and the data were logged based on a programmed button-interrupt to SD card. Each interrupt consisted of 10 consecutive measurements within 2000 ms and their average was logged with timestamp. The wiring was mounted to a 3D printer printed mount and set up on a 2.2 m telescopic tripod, and the 10 Ah power bank was used to power the device (Figure 1c). The sensor was set 2 m from ground at 90° to capture leaves in 0.6 m^2 of theoretical field width (to ground). The 2 m height was chosen as only 16 out of 545 plots showed height lower than 2 m (shortest hybrid in SDTrf was 1.77 m), so the sensor captured the leaves intersecting the field of view. Measurements were carried out during the morning and around solar noon (Figure 1a), when the weakest and the strongest leaf rolling is expected to occur (Table 1) [31]. The tripod with mounted sensor wiring was carried between the plots and set 1.5 m within rows for measurement (Figure 1b). The exact data for plant height were collected for three experiments DTir, DTrf and SDTrf with 3 m long ruler, and no connection was observed between multispectral reads and plant height (correlations < 0.3). The mean height was 214.9 cm, with standard deviation of 6.05 cm.

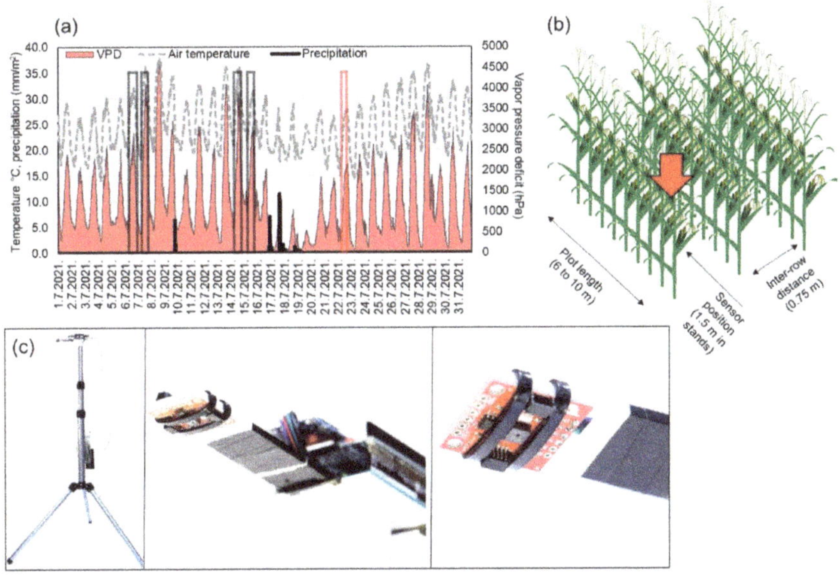

Figure 1. Schematic representation of measurement process. (**a**) shows diurnal temperatures (°C), vapor pressure deficit (hPa) and daily precipitation (mm/m^2) for July 2021. Gray boxes show measurement times, and the red box shows external validation set measurement time. (**b**) shows the position of tripod mounted sensor in plant stands, and (**c**) is a tripod mounted sensor wiring used to carry out the measurements. In (**c**), left image represents the tripod-mounted unit, in the middle is a close-up of wiring and a printed mount, while the right image shows AMS AS7263 sensor unit on a breakout board.

All 545 plots (Table 1) were assessed for four times with a sensor and labeled by a maize breeder for leaf rolling, two times during the morning and two times in the solar noon (Figure 1a) on sunny days, yielding 2180 labeled measurements and means of sensor reads. Additionally, DTir, DTrf and SDRtf were assessed on 22nd of July, to obtain an external validation set for testing the robustness of the modelling approach.

Timing of measurements (Figure 1a) was chosen to capture the window of highest maize susceptibility to drought [54], heat [55] and the combination of these two stressors. This window covers growth stages from floral transition to early grain filling. During the experiments, different hybrids transitioned between developmental stages; however, the aim of this study was not to analyze genotypic responses but rather the ability of a simple multispectral sensor to capture leaf rolling occurrence. Initial grades of leaf rolling (samples available as Supplementary Figure S1) were taken following methodology described in Bolaños and Edmeades [30] on scale 1 (green erect leaf blades) to 5 (rolled, lax or dead). However, the experimental design limited appropriate account for all factors affecting the rolling occurrence. Furthermore, as the leaf morphology affects the rolling maximum [24], the lower grades (higher than 1) also indicate leaf rolling, whose occurrence was of primary interest of this study. Reads were thus binary labeled with 0 for no leaf rolling (below 15% of the plants within the plot showing leaf rolling) and 1 (more than 15% of the plants within the plot showing leaf rolling). The 15% threshold represents the tolerable amount of secondary plants that are usually more susceptible to leaf rolling. The images of leaf-rolled plants in the field are available as Figure S1.

2.3. Data Analysis and Model Assesment

Six raw sensor reads were used to calculate 15 unique, simple vegetation indexes (VI). The indexes were calculated as absolute values of the quotient between differences of the subtracted and added values of each two pairs of wavelengths ($wl_{a...f}$):

$$VI_i = \left| \frac{wl_b - wl_a}{wl_a + wl_b} \right| \quad (1)$$

The raw sensor reads and VIs were scaled, centered and log-transformed and the principal component analysis (PCA) was carried out in R [56]. The raw sensor reads and the VI values (21 original features) were tested for differences between LR+ and LR− by the means of a Welch two-sample *t*-test. Prior to tests, the data were visually assessed for normality of distribution densities. The raw data and VIs were read into Python environment and four machine learning models were constructed. First model was the single layer perceptron (SLP) with single fully connected layer with 128 nodes with rectified linear unit (ReLu) activation function. The dense neural network layer was flattened and passed through softmax function to obtain predictions. For multilayer perceptron (MLP), another hidden layer was added with 32 fully connected nodes prior to flattening. Convolutional neural network (CNN) was setup with single 1D convolution layer of length 64, followed by two hidden fully connected layers with 48 and 24 nodes before flattening and passing to the softmax function (Figure 2). These three models were setup in TensorFlow library with training in 100 epochs with batch size of 16. Additionally, a support vector machine (SVM) model was built with scikit-learn module svm with linear kernel and penalty of the error term (C) set to 1. Three procedures of model validation were followed using calibration (n = 2180) and external (n = 145) labeled datasets:

1. Stratified 5-fold cross-validation with 85% (1853) of the 2180 records;
2. Validation based on a 15% random subset (327) of the 2180 records with random seed number 109;
3. Validation with an external validation set consisting of 145 separate records.

To ensure reproducibility of the results, the same subsets were used to validate each model.

The model performance was assessed by model accuracy and its standard deviation across folds in cross-validation, and by measuring accuracy, precision and recall in random-subset validation and validation with external dataset. The breeder's classifications at solar noon (explained above) to LR− and LR+ were used as ground truth in model evaluation metrics. The performance indicators were calculated as:

$$Accuracy = \frac{TN + TP}{n},\ Precision = \frac{TP}{TP + FP},\ Recall = \frac{TP}{TP + FN} \quad (2)$$

where TN is number of true negatives, TP is number of true positives, FP is a number of false positives, FN is number of false negatives and n is a number of relevant samples. Furthermore, F1 score was calculated as

$$F1 = 2 * \frac{Precision * Recall}{Precision + Recall} \quad (3)$$

F1 score represents a harmonic mean between the ability of a model to classify true positives among all positively labeled examples (precision) and the fraction of examples classified as positives among all positive plots (recall).

The CPU time was assessed on an Intel® i7 9750H 6-core, 12-thread processor with 12 MB internal cache memory. Full notebook with Python code is available from the corresponding author upon request.

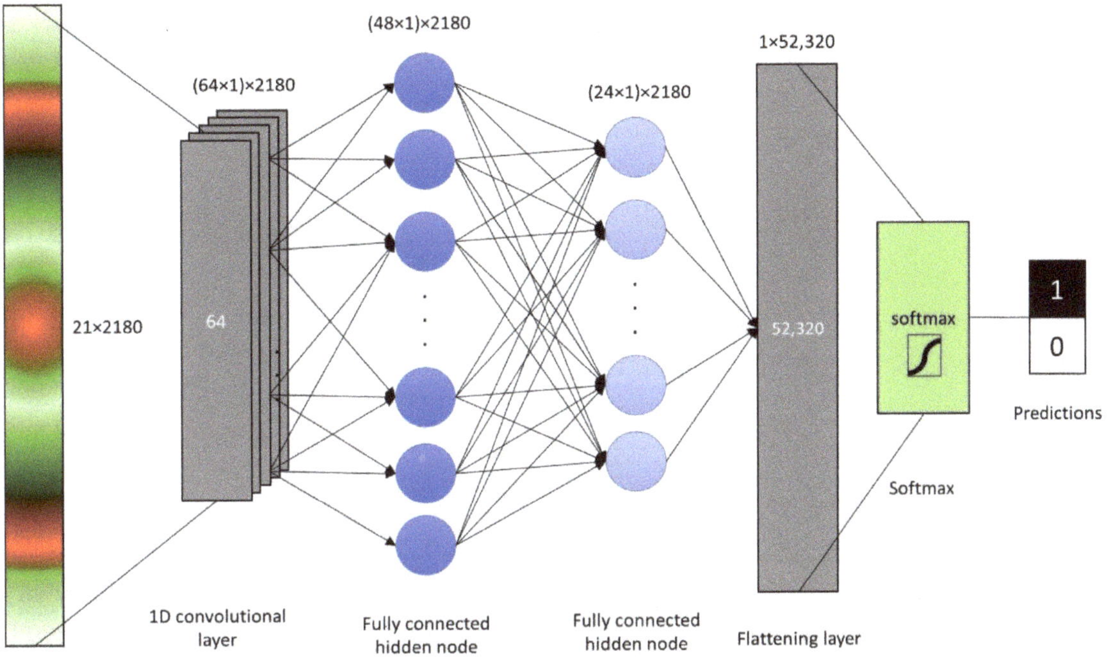

Figure 2. Schematic representation of the convolutional neural network (CNN) with single convolution layer, 2 hidden fully connected nodes, flattening layer and a softmax function.

3. Results

3.1. Changes in Multispectral Sensor Reads in Leaf Rolling Conditions

All measurements in the calibration set were carried out during the extremely dry conditions (Figure S2) and high temperature and VPD (Figure 1a). Conditions changed to very wet when the external validation set was assessed, leading to a low number (7) of plots showing leaf rolling. The analysis of original features showed recognizable patterns of increase/decrease in both calibration (n = 2180) and external (n = 145) datasets showing no leaf rolling (LR−) and leaf rolling (LR+, Table 2). High standard deviations of raw wavelengths indicate the changes in light quality. However, the deviations decreased in VI values reflecting the normalization of the data. Interestingly, all VI values decreased in the LR+ in both datasets except VI680610 which increased slightly in LR+. According to the two-sample t-test, all differences between LR− and LR+ were significant in the calibration set in both original features and Vis. In the external set, significant differences were observed only in reads at 610 and 680 nm among original features. In VIs, a lack of significant differences was observed among all indexes with reads at 730 nm as denominator, and 860 as numerator.

Principal component analysis (Figure 3) showed diverse and substantial correlations between original variables and their projections (PCs), seen as red arrows, e.g., eigenvectors. Full list of loading weights is available online as Table S1. First three principal components explained 74.0% of total variability in the dataset, separate PCs explaining 43.2, 19.0 and 11.8%, respectively. Principal component analysis confirmed patterns from Table 2, rendering two partially overlapping, but separable groups of wavelength changes in leaf rolling conditions in three latent variables. Despite the overlap in part of the responses in lower-dimensional hyperplane (3 PCs), the ability of PCA to capture 74% of variance between groups in only three components with spread in eigenvectors indicates high information density in a small number of underlying features.

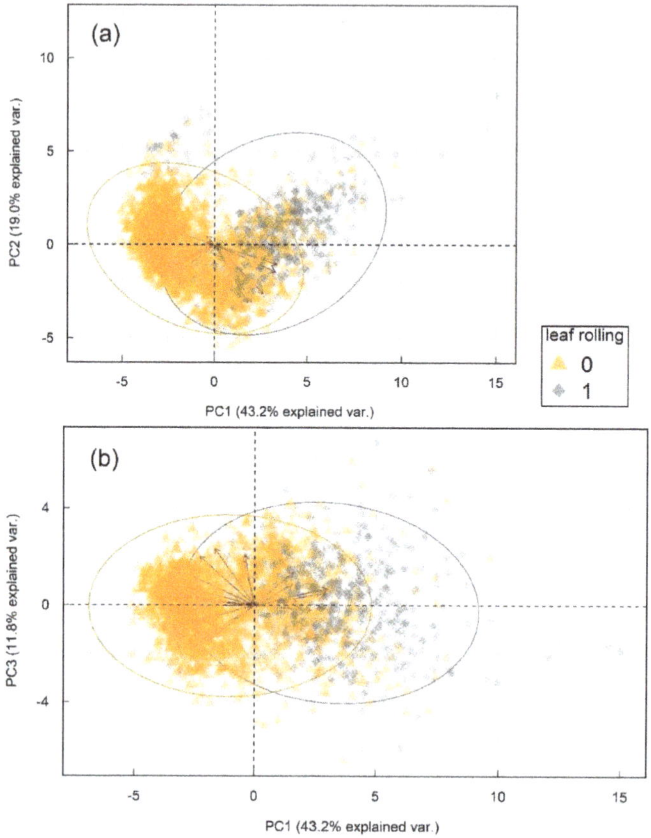

Figure 3. Principal component analysis of the multispectral sensor reads and the corresponding normalized difference vegetation indexes. (**a**) shows principal components 1 and 2, and principal components 1 and 3 are shown in (**b**). Arrows represent eigenvectors of the variables (Table S1) and ellipses represent 95% confidence intervals for each group based on Gaussian distribution.

Table 2. Raw sensor output at six wavelengths and 15 values of normalized difference vegetation indexes (VI) calculated from unique combinations of six wavelengths expressed as a mean ± standard deviation for maize plots showing leaf rolling (LR+) and no leaf rolling (LR−) in calibration (n = 2180) and external (n = 144) datasets measured by a multispectral sensor. Column p denotes significance according to two-sample t-test at values of α < 0.05 (*), <0.01 (**) and <0.001 (***). p values of differences >0.05 are denoted as non-significant (n.s.).

Feature	Calibration Set (n = 2180)		p	External Set (n = 144)		p
	LR− (0)	LR+ (1)		LR− (0)	LR+ (1)	
n	1631	549		139	7	
610	3309 ± 1974	5769 ± 1971	***	4131 ± 1239	8047 ± 2398	**
680	3207 ± 2010	6459 ± 2151	***	3886 ± 991	7220 ± 2118	**
730	7916 ± 4211	10,721 ± 3514	***	12,097 ± 3171	12,927 ± 4985	n.s.

Table 2. *Cont.*

Feature	Calibration Set (n = 2180) LR− (0)	LR+ (1)	p	External Set (n = 144) LR− (0)	LR+ (1)	p
n	**1631**	**549**		**139**	**7**	
760	12,809 ± 5873	15,143 ± 4967	***	19,996 ± 5037	18,570 ± 5429	n.s.
810	14,895 ± 6610	16,900 ± 5552	***	22,564 ± 5671	19,548 ± 5309	n.s.
860	14,831 ± 6418	17,208 ± 5139	***	22,762 ± 6197	20,578 ± 4996	n.s.
VI680610	0.082 ± 0.069	0.109 ± 0.082	***	0.101 ± 0.081	0.147 ± 0.107	n.s.
VI730610	0.417 ± 0.107	0.299 ± 0.094	***	0.489 ± 0.091	0.218 ± 0.075	***
VI760610	0.6 ± 0.107	0.445 ± 0.102	***	0.651 ± 0.095	0.395 ± 0.048	***
VI810610	0.647 ± 0.096	0.488 ± 0.097	***	0.683 ± 0.093	0.418 ± 0.065	***
VI860610	0.645 ± 0.101	0.497 ± 0.099	***	0.68 ± 0.109	0.44 ± 0.095	***
VI730680	0.431 ± 0.13	0.249 ± 0.126	***	0.508 ± 0.089	0.253 ± 0.215	*
VI760680	0.611 ± 0.116	0.397 ± 0.132	***	0.67 ± 0.062	0.426 ± 0.15	**
VI810680	0.656 ± 0.11	0.44 ± 0.139	***	0.699 ± 0.072	0.45 ± 0.138	**
VI860680	0.657 ± 0.104	0.452 ± 0.121	***	0.699 ± 0.078	0.474 ± 0.125	**
VI760730	0.25 ± 0.079	0.173 ± 0.059	***	0.245 ± 0.08	0.193 ± 0.075	n.s.
VI810730	0.32 ± 0.099	0.225 ± 0.088	***	0.299 ± 0.113	0.219 ± 0.102	n.s.
VI860730	0.318 ± 0.128	0.236 ± 0.104	***	0.301 ± 0.134	0.246 ± 0.141	n.s.
VI810760	0.088 ± 0.06	0.069 ± 0.049	***	0.083 ± 0.057	0.041 ± 0.022	***
VI860760	0.102 ± 0.073	0.086 ± 0.063	***	0.093 ± 0.074	0.087 ± 0.052	n.s.
VI860810	0.057 ± 0.043	0.053 ± 0.037	*	0.05 ± 0.037	0.047 ± 0.032	n.s.

Calculation of different normalized vegetation indices represents a convenient mean of auto-normalization of the raw sensor reads. The 15 normalized vegetation indices (termed VI) assessed in this study (Table 2) showed significant variability between leaf rolling and plots without leaf rolling. Interestingly, between wavelengths from the red spectra (610 and 680 nm), lower difference was observed between LR− and LR+ (Figure 4). The same pattern was observed in VIs assessing wavelengths > 700 nm. The largest differences between LR− and LR+ reads were observed in VIs combining wavelengths > 760 nm and <700 nm.

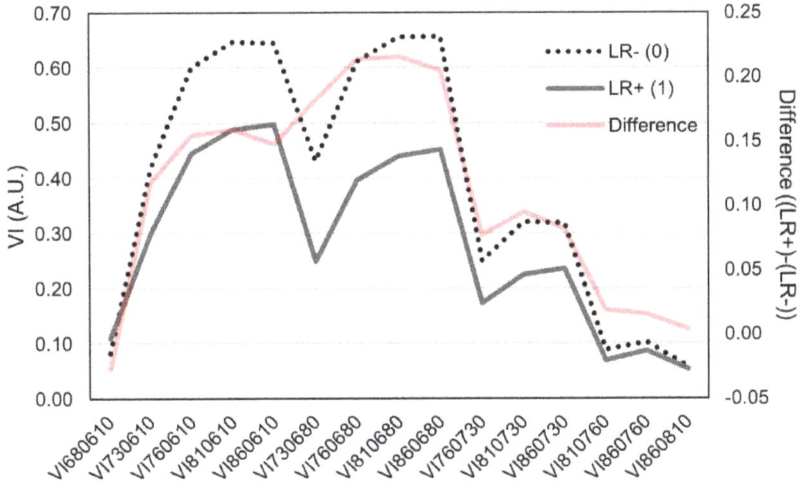

Figure 4. Plot of different normalized difference vegetation indices (VI) in arbitrary units (A.U.) between plots showing leaf rolling (LR+) and plots without leaf rolling (LR−) and their respective differences. Standard deviations of the VIs are.

3.2. Assesment Different of Machine Learning Algorithms for Prediction of Leaf Rolling

In stratified 5-fold cross-validation, considerable variability was detected between the performance of different models. Highest prediction accuracy with lowest standard deviation was observed for SLP, followed by CNN, SVM and MLP, respectively (Table 3). Highest precision in cross-validation was observed for SVM, accompanied by second-highest standard deviation between folds. Recall was the highest for SLP, followed by CNN. The highest F1 was observed for CNN. The compute times increased with model complexity in the order SLP < MLP < CNN < SVM. However, when attempting to generalize the results of the calibrated models with the random 15% subset, the prediction accuracies changed. The SLP was shown to be the least accurate model, however, with high recall. According to the precision metrics, the model aimed at target many times (many false positives) which was followed by many hits. Such results indicate overfitting in the model architecture possibly caused by many nodes (128) and only a single layer. Briefly, the model was able to extract features linked to leaf rolling in a stratified set, but when attempting to generalize an unrelated dataset, the performance metrics dramatically decreased.

Table 3. Results of two validation procedures for single and multilayer perceptron (SLP and MLP), convolutional neural network (CNN) and support vector machines (SVM) with calibration dataset (n = 2180). Number in brackets for stratified 5-fold cross-validation represents standard deviation of accuracy across folds.

Model	Validation Procedure	No. Observations	Accuracy	Precision	Recall	F1	CPU Time (s)
SLP	5-fold cross-validation	1853/2180	83.76 (2.29)	69.67 (8.26)	66.52 (5.72)	65.3 (3.02)	38.6
MLP			81.27 (1.75)	64.42 (4.52)	63.91 (10.44)	61.13 (5.50)	49.8
CNN			83.49 (2.53)	68.88 (6.14)	64.13 (8.10)	64.55 (4.82)	94.0
SVM			82.68 (1.14)	70.85 (8.01)	53.91 (7.44)	60.52 (3.50)	469.0
SLP	Random subset	327/2180	58.41	39.46	98.88	56.41	10.0
MLP			86.54	74.19	77.53	75.82	12.6
CNN			87.77	75.26	82.02	78.49	24.0
SVM			88.69	84.21	71.91	77.58	136.0

Support vector machines model showed highest accuracy, but according to the high precision and lowest recall, it was the most conservative model, yielding a low number of false positives. High accuracy and generalization ability of this model is in accordance with results of PCA (Figure 3) and ability of simple dimension reduction (L2 norm in SVC linear kernel) to facilitate efficient feature extraction.

Contrarily, MLP and CNN showed fewer conservative values of precision with second highest accuracy (CNN) and 9.04% lower precision compared to SVM. Multilayer perceptron and CNN showed good generalization ability due to the added layers that mitigated the overfitting found for SLP. The best overall performance was captured by SVM and CNN, with higher accuracy and precision in SVM and higher recall in CNN. This was also confirmed by the highest values of harmonic mean accuracy of the model (F1) which was the highest for CNN, followed by SVM. Overall, compute times for the random subset training and predictions were significantly shorter (10 s to 136 s) compared to the cross-validation procedure.

The performance from validation procedures was mostly in alignment with the test using an external validation dataset. Due to the abundant rain between measurements of calibration and external datasets (Figure 1), there were only seven plots with detectable leaf rolling in the external dataset (Table 2), four of which were detected in SDTrf with sandy loam (not shown). However, this represented an appropriate test of the model robustness, due to the changes in many aspects of plant vitality. Single-layer perceptron maintained poor generalization ability, although with all seven LR+ plots properly classified (Figure 5a).

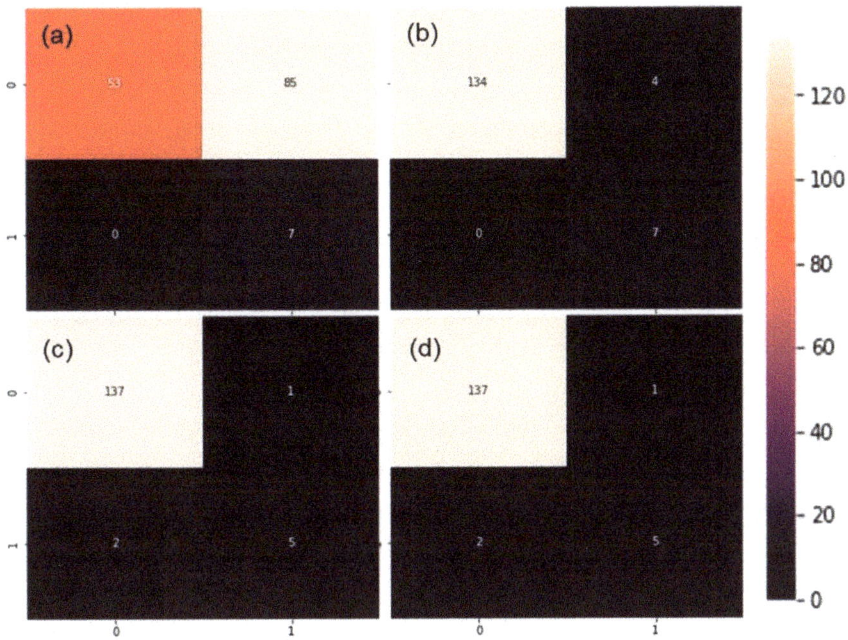

Figure 5. Performance of single and multilayer perceptron (**a**,**b**), convolutional neural network (**c**) and support vector machine (**d**) calibrated models with external dataset. Quadrants from left to right, top to bottom represent true negatives, false positives, false negatives and true positives.

As in the validation with the random subset, this validation was also followed by a high number of false positives reducing the F1 value to only 14.1 percent (Figure 6) caused by very low precision (7.6%). High number of good classifications (7/7) followed by a relatively conservative number of false positives and no true negatives in MLP (Figure 5b) rewarded the highest F1 score of 77.7% in the external set validation (Figure 6). Marginally lower F1 scores were obtained for CNN and SVM (both 76.9%), which were caused by the inability of the models to correctly classify all seven LR+ plots.

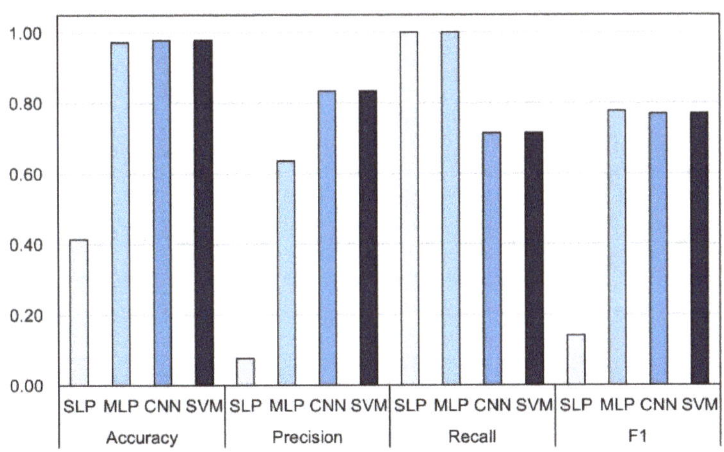

Figure 6. Performance metrics of four tested models: single and multilayer perceptron (SLP and MLP), convolutional neural network (CNN) and support vector machines (SVM) in external dataset.

4. Discussion

Phenotyping for drought responses in real time represents a new frontier for crop breeding and precision agriculture [57]; however, the advancements in this field are limited by the high costs of measurement equipment such as unmanned aerial vehicles and hyperspectral cameras. This implies the need for new low-cost proximal or remote sensing solutions, efficiently assessing plant physiological status in real time. Hot and dry conditions during the phenotyping procedure of our study (Figure 1a and Figure S2) in a large number of maize hybrids allowed us to robustly assess a large number of experimental plots with a prototype of multispectral proximal sensing node intended for use in the Internet of Things (IoT) applications. The sensor was used to assess leaf rolling, a trait that was shown to be involved in adaptation to drought and heat conditions [22]. There are various methods for the assessment of leaf rolling [23,29,31,58] and it is well known that different hybrids show different levels of leaf folding, especially given the varying water availability and temperature changes. However, the aim of our work was to assess if the leaf rolling occurrence, despite the varying levels of phenotypic expression, could be spotted or predicted based only on basic reads of spectral responses in red and near-infrared parts of electromagnetic spectra.

At the onset of leaf rolling, many physiological changes take place, such as reduction in photosynthetic activity [59] and changes in metabolic genetic regulatory mechanisms [60]. The reduction in photosynthetic activity should be mostly visible at wavelengths between 710 and 740 nm capturing fluorescence overlap of both photosystem I and II, and 685 nm representing a peak of fluorescence of photosystem II [61] which is within the spectral peaks captured by 680 and 730 nm diodes (according to 20 nm full width at half-maximum) in our study. This was corroborated by the reduction in VIs assessing 680 and 730 nm wavelengths (Table 2, Figure 4).

Stress adaptation, such as the reduction in photochemical activity, also involves translocation of the biochemicals [32]. Additionally, in responses such as leaf rolling, the previously unexposed plant parts become intercepted by sunlight, such as abaxial parts of leaves, having the different pigment mixtures compared to adaxial parts [62]. The reflectance between 700 nm and 980 nm, where the spectral responses of the brown pigments are located, along with chlorophyll fluorescence signals might provide the insight in plant biochemistry, and consequently, physiological status [63]. The exposure of plant abaxial surfaces to sunlight reveals red-brown pigments due to the water deficit [64], thus changing the leaf optical properties [65]. According to the results presented in Weber et al. [66], combined water stress and heat stress, as in our study (Figure 1 and Figure S2), are expected to produce the most visible response in leaf reflectance at wavelengths near the reflectance of brown pigments. This was also confirmed in our results, where the difference between LR− and LR+ increased in indices combining wavelengths > 700 nm and <700 nm (Figure 4). Among these indices is also the commonly used NDVI combining reflectance at approximately 680 and 770 nm [67]. Normalized vegetation indices are traditionally used to assess the vegetation cover from satellite imagery [34], but the advancement of analytic solutions allows their deployment for analysis of a wide range of quantitative and qualitative traits. However, one must note that the usage of multispectral sensor reads also bears the risk of reduced repeatability of the results due to the deviations of atmospheric and sensor effects [68], so the use of VIs is advised, such as Vis in this study.

Due to the rapidly changing climate [69], there is strong pressure on developing new proximal and remote, data-rich high-throughput plant phenotyping solutions, rectified by the lack of manpower and the increased demand for high-quality data [70]. The UAV and phenopole (phenotyping pole) solutions yield similar insight and information density of the reads, however, with more throughput in UAV solutions [71], but more temporal information with phenopoles, facilitating the monitoring of the trait onset dynamics. The sensor node used in our study aims to provide the low-cost solution to this problem, so that the increased density of the sensing nodes, providing the better sampling, could compensate for UAV's higher throughput, at a lower cost. Furthermore, given the application of our sensor in the envisioned IoT framework, the simultaneous, real-time data collection could provide higher information density compared to UAV, without the need for human intervention [72]. Combining such developments with machine learning methodology should converge to provide new layers of information in plant monitoring paradigm for transition to Agriculture 4.0 [73]. Modelling complex data with unknown hidden features can be efficiently carried out using ML methodology to explain considerable amounts of variance in agricultural production deviations [74]; however, the interpretability of the models is low and despite their high accuracy, they are unable to surrogate the science.

In our study, only marginal differences were observed between MLP, CNN showing good generalization ability due to the added layers that mitigated the overfitting found for SLP. In stratified cross-validation, SLP showed the highest F1 score, followed by the CNN. However, the severe drop in accuracy of SLP in un-stratified datasets can be viewed from the perspective of poor generalization abilities of the networks having a small number of layers with an excessive number of neurons [75], which is apparently the case with SLP presented in our study. Thus, the model was able to extract features linked to leaf rolling in a stratified set, but when attempting to generalize an unrelated dataset, the performance metrics dramatically decreased.

Study on plant *Bromus inermis* using hyperspectral indices and three ML algorithms, CNN, SVM and random forest, showed feasibility of drought classification in ML framework with the highest prediction accuracies observed for SVM [32]. In our study, SVM also showed the highest overall prediction accuracy in random subset validation. This is also corroborated by the results of PCA (Figure 3) and the ability of a simple dimension reduction technique such as Tikhonov regularization [76] (L2 norm in SVC linear kernel) to facilitate efficient feature extraction.

All models, except SLP, in our study performed well on the given dataset, and the observed differences were not discriminatory. Studies assessing vegetation classifications by the use of hyperspectral imagery with SVM, artificial neural networks and CNN by Hassan et al. [77,78], demonstrating very high classification accuracies reported similar conclusions. The added value in our research can be seen in the analysis of performance in tabular data using additional, unrelated validation dataset (Figures 5 and 6). It was shown that CNN and SVM yielded more conservative, similar performance metrics, while MLP showed the best overall performance, but with only a 1% increase in F1 score. On the other hand, the limitations of our study can be seen from two perspectives. As an early report, our dataset was only created in a single stage of plant development over a limited number of climatological scenarios [79], so further efforts with increased spatial and temporal resolution are needed. Additionally, the deep learning models represent a good way to cope with many types of data taking many conformations in multidimensional hyperplanes, however, with limited interpretability. Further assessment should thus include other ML models such as decision trees retaining more information on the effects of predictor variables.

The usability of ML was also demonstrated at many levels of agricultural production, by using multi/hyperspectral reads and imagery and climatological data, such as disease detection [41,48,80–83], nutrient deficiency assessment [84,85], stress detection [86–88] and in-season predictions of agronomic performance in maize [89,90], sorghum [91], soybean [92], wheat [93–95] and cocoa [96]. Corroborating these new avenues in agricultural sciences with the constant involvement of large companies in the development of new learning ML algorithms, and the optimization of the existing ones, Python open-source libraries Tensorflow [97] created and maintained by Google and Pytorch [98] created and maintained by Facebook, makes the future uses of ML incomprehensible.

5. Conclusions

This study demonstrated the ability of machine learning algorithms to use simple multispectral reads for efficient classification of maize leaf rolling. It was shown that there is variability between ML algorithms in terms of performance metrics, but also, computing times. There is a growing need for increased spatiotemporal resolution of plant monitoring with affordable remote sensing solutions, especially in the context of the Internet of Things (IoT). It was demonstrated that ML algorithms can efficiently extract information from multispectral reads and predict plant states such as leaf rolling. Since the envisioned use of the demonstrated sensor is an IoT framework, the inclination is towards less computationally intensive ML algorithms, without sacrificing performance. Thus, the use of MLP might represent the best overall option. Increasing the information density and using smart solutions for decision support in agriculture could facilitate the transition to the Agriculture 4.0 empowered by the nexus between food production and machine learning. Further research should test the framework in multiple topographies and water/nutrient availability scenarios to tackle the abilities of simple and affordable sensing solutions in the dissection of biological systems showing the highest order of complexity.

Supplementary Materials: The following supporting information can be downloaded at: https://www.mdpi.com/article/10.3390/rs14112596/s1, Figure S1. (a) Maize plot showing no leaf rolling (LR−) and a plot with leaf rolling (b); Figure S2. Precipitation during the months of June and July 2021 with 50-year average-based percentiles of dry and wet conditions for Osijek. Gray boxes show measurement times, and the red box shows external validation set measurement time; Figure S3. Learning rates of neural network algorithms from 85% random subset training procedure; Table S1: loading weights of the first three principal components

Author Contributions: Conceptualization, V.G. and J.S.; methodology, V.G. and D.Š.; software, V.G. and J.S.; validation, A.J., J.B. and D.Š.; formal analysis, J.S., J.B., D.Š., and V.G.; field experiments, A.J.; investigation, V.G. and J.S.; resources, A.J., J.B. and D.Š.; data curation, V.G.; writing—original draft preparation, J.S. and V.G.; writing—review and editing, J.B. and D.Š.; visualization, V.G.; funding acquisition, D.Š. All authors have read and agreed to the published version of the manuscript.

Funding: This work is supported by the project "IoT-field: An Ecosystem of Networked Devices and Services for IoT Solutions Applied in Agriculture" co-financed by the European Union from the European Regional Development Fund within the Operational Programme Competitiveness and Cohesion 2014–2020 of the Republic of Croatia.

Data Availability Statement: The data presented in this study are openly available in FigShare at https://doi.org/10.6084/m9.figshare.19904491.v1, reference number 19904491.v1.

Conflicts of Interest: The authors declare no conflict of interest.

References

1. King, A. The future of agriculture. *Nature* **2017**, *544*, S21–S23. [CrossRef] [PubMed]
2. Wang, D.; Cao, W.; Zhang, F.; Li, Z.; Xu, S.; Wu, X. A Review of Deep Learning in Multiscale Agricultural Sensing. *Remote Sens.* **2022**, *14*, 559. [CrossRef]
3. Polpanich, O.; Bhatpuria, D.; Fernanda, T.; Santos, S. *Leveraging Multi-Source Data and Digital Technology to Support the Monitoring of Localized Water Changes in the Mekong Region*; SEI: Oaks, PA, USA, 2022.
4. Trnka, M.; Hlavinka, P.; Možný, M.; Semerádová, D.; Štěpánek, P.; Balek, J.; Bartošová, L.; Zahradníček, P.; Bláhová, M.; Skalák, P.; et al. Czech Drought Monitor System for monitoring and forecasting agricultural drought and drought impacts. *Int. J. Climatol.* **2020**, *40*, 5941–5958. [CrossRef]
5. Johnson, L.F.; Trout, T.J. Satellite NDVI assisted monitoring of vegetable crop evapotranspiration in california's san Joaquin Valley. *Remote Sens.* **2012**, *4*, 439–455. [CrossRef]
6. Takeuchi, W.; Darmawan, S.; Shofiyati, R.; Khiem, M.V.; Oo, K.S.; Pimple, U.; Heng, S. Near-real time meteorological drought monitoring and early warning system for croplands in Asia. In Proceedings of the ACRS 2015: The 36th Asian Conference on Remote Sensing "Fostering Resilient Growth in Asia", Quezon City, Philippines, 19–23 October 2015; pp. 171–178.
7. Minamiguchi, N. The application of geospatial and disaster information for food insecurity and agricultural drought monitoring and assessment by the FAO GIEWS and Asia FIVIMS. *Work. Reducing Food Insecurity Assoc.* **2005**, *27*, 28.
8. Zhang, Y.; Han, W.; Niu, X.; Li, G. Maize crop coefficient estimated from UAV-measured multispectral vegetation indices. *Sensors* **2019**, *19*, 5250. [CrossRef]
9. Ramos-Giraldo, P.; Reberg-Horton, C.; Locke, A.M.; Mirsky, S.; Lobaton, E. Drought Stress Detection Using Low-Cost Computer Vision Systems and Machine Learning Techniques. *IT Prof.* **2020**, *22*, 27–29. [CrossRef]

10. dos Santos, R.A.; Mantovani, E.C.; Filgueiras, R.; Fernandes-Filho, E.I.; da Silva, A.C.B.; Venancio, L.P. Actual evapotranspiration and biomass of maize from a red-green-near-infrared (RGNIR) sensor on board an unmanned aerial vehicle (UAV). *Water* **2020**, *12*, 1–20.
11. Du, S.; Liu, L.; Liu, X.; Guo, J.; Hu, J.; Wang, S.; Zhang, Y. SIFSpec: Measuring solar-induced chlorophyll fluorescence observations for remote sensing of photosynthesis. *Sensors* **2019**, *19*, 3009. [CrossRef]
12. Zhang, Y.; Zhang, Q.; Liu, L.; Zhang, Y.; Wang, S.; Ju, W.; Zhou, G.; Zhou, L.; Tang, J.; Zhu, X.; et al. ChinaSpec: A Network for Long-Term Ground-Based Measurements of Solar-Induced Fluorescence in China. *J. Geophys. Res. Biogeosci.* **2021**, *126*, e2020JG006042. [CrossRef]
13. Lhotka, O.; Kyselý, J.; Farda, A. Climate change scenarios of heat waves in Central Europe and their uncertainties. *Theor. Appl. Climatol.* **2018**, *131*, 1043–1054. [CrossRef]
14. Lobell, D.B.; Roberts, M.J.; Schlenker, W.; Braun, N.; Little, B.B.; Rejesus, R.M.; Hammer, G.L. Greater sensitivity to drought accompanies maize yield increase in the U.S. Midwest. *Science* **2014**, *344*, 516–519. [CrossRef] [PubMed]
15. Sah, R.P.; Chakraborty, M.; Prasad, K.; Pandit, M.; Tudu, V.K.; Chakravarty, M.K.; Narayan, S.C.; Rana, M.; Moharana, D. Impact of water deficit stress in maize: Phenology and yield components. *Sci. Rep.* **2020**, *10*, 2944. [CrossRef] [PubMed]
16. Harrison, M.T.; Tardieu, F.; Dong, Z.; Messina, C.D.; Hammer, G.L. Characterizing drought stress and trait influence on maize yield under current and future conditions. *Glob. Chang. Biol.* **2014**, *20*, 867–878. [CrossRef]
17. Ribaut, J.; Betran, J.; Monneveux, P.; Setter, T. Drought Tolerance in Maize. In *Handbook of Maize: Its Biology*; Bennetzen, J.L., Hake, S., Eds.; Springer Science + Business Media, LLC: Berlin, Germany, 2009; ISBN 9780387794181.
18. Aslam, M.; Maqbool, M.A.; Cengiz, R. *Drought Stress in Maize (Zea mays L.)*; Springer: Berlin, Germany, 2015; ISBN 978-3-319-25440-1.
19. Masuka, B.; Araus, J.L.; Das, B.; Sonder, K.; Cairns, J.E. Phenotyping for Abiotic Stress Tolerance in Maize. *J. Integr. Plant Biol.* **2012**, *54*, 238–249. [CrossRef]
20. Beebe, S.E.; Rao, I.M.; Blair, M.W.; Acosta-Gallegos, J.A. Phenotyping common beans for adaptation to drought. *Front. Physiol.* **2013**, *4*, 35. [CrossRef]
21. Moulia, B. Leaves as shell structures: Double curvature, auto-stresses, and minimal mechanical energy constraints on leaf rolling in grasses. *J. Plant Growth Regul.* **2000**, *19*, 19–30. [CrossRef]
22. Monneveux, P.; Sanchez, C.; Tiessen, A. Future progress in drought tolerance in maize needs new secondary traits and cross combinations. *J. Agric. Sci.* **2008**, *146*, 287–300. [CrossRef]
23. Gao, L.; Yang, G.; Li, Y.; Fan, N.; Li, H.; Zhang, M.; Xu, R.; Zhang, M.; Zhao, A.; Ni, Z.; et al. Fine mapping and candidate gene analysis of a QTL associated with leaf rolling index on chromosome 4 of maize (*Zea mays* L.). *Theor. Appl. Genet.* **2019**, *132*, 3047–3062. [CrossRef]
24. Sirault, X.R.R.; Condon, A.G.; Wood, J.T.; Farquhar, G.D.; Rebetzke, G.J. "Rolled-upness": Phenotyping leaf rolling in cereals using computer vision and functional data analysis approaches. *Plant Methods* **2015**, *11*, 52. [CrossRef]
25. Fernandez, D.; Castrillo, M. Maize Leaf Roling Initiation. *Photosynthetica* **1999**, *37*, 493–497. [CrossRef]
26. Cal, A.J.; Sanciangco, M.; Rebolledo, M.C.; Luquet, D.; Torres, R.O.; McNally, K.L.; Henry, A. Leaf morphology, rather than plant water status, underlies genetic variation of rice leaf rolling under drought. *Plant Cell Environ.* **2019**, *42*, 1532–1544. [CrossRef] [PubMed]
27. Kenchanmane Raju, S.K.; Adkins, M.; Enersen, A.; Santana de Carvalho, D.; Studer, A.J.; Ganapathysubramanian, B.; Schnable, P.S.; Schnable, J.C. Leaf Angle eXtractor: A high-throughput image processing framework for leaf angle measurements in maize and sorghum. *Appl. Plant Sci.* **2020**, *8*, 1–9. [CrossRef] [PubMed]
28. O'Toole, J.C.; Cruz, R.T.; Singh, T.N. Leaf rolling and transpiration. *Plant Sci. Lett.* **1979**, *16*, 111–114. [CrossRef]
29. Premachandra, G.S.; Saneoka, H.; Fujita, K.; Ogata, S. Water Stress and Potassium Fertilization in Field Grown Maize (Zea mays L.): Effects on Leaf Water Relations and Leaf Rolling. *J. Agron. Crop Sci.* **1993**, *170*, 195–201. [CrossRef]
30. Bolaños, J.; Edmeades, G.O. The importance of the anthesis-silking interval in breeding for drought tolerance in tropical maize. *F. Crop. Res.* **1996**, *48*, 65–80. [CrossRef]
31. Baret, F.; Madec, S.; Irfan, K.; Lopez, J.; Comar, A.; Hemmerlé, M.; Dutartre, D.; Praud, S.; Tixier, M.H. Leaf-rolling in maize crops: From leaf scoring to canopy-level measurements for phenotyping. *J. Exp. Bot.* **2018**, *69*, 2705–2716. [CrossRef] [PubMed]
32. Dao, P.D.; He, Y.; Proctor, C. Plant drought impact detection using ultra-high spatial resolution hyperspectral images and machine learning. *Int. J. Appl. Earth Obs. Geoinf.* **2021**, *102*, 102364. [CrossRef]
33. Calderón, R.; Navas-Cortés, J.A.; Lucena, C.; Zarco-Tejada, P.J. High-resolution airborne hyperspectral and thermal imagery for early detection of Verticillium wilt of olive using fluorescence, temperature and narrow-band spectral indices. *Remote Sens. Environ.* **2013**, *139*, 231–245. [CrossRef]
34. Xue, J.; Su, B. Significant remote sensing vegetation indices: A review of developments and applications. *J. Sens.* **2017**, *2017*, 1353691. [CrossRef]
35. Dmitriev, P.A.; Kozlovsky, B.L.; Kupriushkin, D.P.; Lysenko, V.S.; Rajput, V.D.; Ignatova, M.A.; Tarik, E.P.; Kapralova, O.A.; Tokhtar, V.K.; Singh, A.K.; et al. Identification of species of the genus *Acer* L. using vegetation indices calculated from the hyperspectral images of leaves. *Remote Sens. Appl. Soc. Environ.* **2022**, *25*, 100679. [CrossRef]
36. Zarco-Tejada, P.J.; González-Dugo, V.; Berni, J.A.J. Fluorescence, temperature and narrow-band indices acquired from a UAV platform for water stress detection using a micro-hyperspectral imager and a thermal camera. *Remote Sens. Environ.* **2012**, *117*, 322–337. [CrossRef]

37. Mohammed, G.H.; Colombo, R.; Middleton, E.M.; Rascher, U.; van der Tol, C.; Nedbal, L.; Goulas, Y.; Pérez-Priego, O.; Damm, A.; Meroni, M.; et al. Remote sensing of solar-induced chlorophyll fluorescence (SIF) in vegetation: 50 years of progress. *Remote Sens. Environ.* **2019**, *231*, 111177. [CrossRef] [PubMed]
38. Virnodkar, S.S.; Pachghare, V.K.; Patil, V.C.; Jha, S.K. Remote Sensing and Machine Learning for Crop Water Stress Determination in Various Crops: A Critical Review. *Precis. Agric.* **2020**, *21*, 1121–1155.
39. Niazian, M.; Niedbała, G. Machine learning for plant breeding and biotechnology. *Agriculture* **2020**, *10*, 436. [CrossRef]
40. Singh, A.; Ganapathysubramanian, B.; Singh, A.K.; Sarkar, S. Machine Learning for High-Throughput Stress Phenotyping in Plants. *Trends Plant Sci.* **2016**, *21*, 110–124. [CrossRef]
41. Behmann, J.; Mahlein, A.K.; Rumpf, T.; Römer, C.; Plümer, L. A review of advanced machine learning methods for the detection of biotic stress in precision crop protection. *Precis. Agric.* **2015**, *16*, 239–260. [CrossRef]
42. Barradas, A.; Correia, P.M.P.; Silva, S.; Mariano, P.; Pires, M.C.; Matos, A.R.; da Silva, A.B.; Marques da Silva, J. Comparing machine learning methods for classifying plant drought stress from leaf reflectance spectra in arabidopsis thaliana. *Appl. Sci.* **2021**, *11*, 6392. [CrossRef]
43. Feng, X.; Zhan, Y.; Wang, Q.; Yang, X.; Yu, C.; Wang, H.; Tang, Z.Y.; Jiang, D.; Peng, C.; He, Y. Hyperspectral imaging combined with machine learning as a tool to obtain high-throughput plant salt-stress phenotyping. *Plant J.* **2020**, *101*, 1448–1461. [CrossRef]
44. Verrelst, J.; Camps-Valls, G.; Muñoz-Marí, J.; Rivera, J.P.; Veroustraete, F.; Clevers, J.G.P.W.; Moreno, J. Optical remote sensing and the retrieval of terrestrial vegetation bio-geophysical properties—A review. *ISPRS J. Photogramm. Remote Sens.* **2015**, *108*, 273–290. [CrossRef]
45. Zhang, L.; Peng, L.; Zhang, T.; Cao, S.; Peng, Z. Infrared small target detection via non-convex rank approximation minimization joint l2,1 norm. *Remote Sens.* **2018**, *10*, 1821. [CrossRef]
46. Maxwell, A.E.; Warner, T.A.; Fang, F. Implementation of machine-learning classification in remote sensing: An applied review. *Int. J. Remote Sens.* **2018**, *39*, 2784–2817. [CrossRef]
47. Ali, I.; Greifeneder, F.; Stamenkovic, J.; Neumann, M.; Notarnicola, C. Review of machine learning approaches for biomass and soil moisture retrievals from remote sensing data. *Remote Sens.* **2015**, *7*, 16398–16421. [CrossRef]
48. Zheng, C.; Abd-elrahman, A.; Whitaker, V. Remote sensing and machine learning in crop phenotyping and management, with an emphasis on applications in strawberry farming. *Remote Sens.* **2021**, *13*, 531. [CrossRef]
49. Zhao, R.; Li, Y.; Ma, M. Mapping paddy rice with satellite remote sensing: A review. *Sustainability* **2021**, *13*, 503. [CrossRef]
50. Kattenborn, T.; Leitloff, J.; Schiefer, F.; Hinz, S. Review on Convolutional Neural Networks (CNN) in vegetation remote sensing. *ISPRS J. Photogramm. Remote Sens.* **2021**, *173*, 24–49. [CrossRef]
51. Holloway, J.; Mengersen, K. Statistical machine learning methods and remote sensing for sustainable development goals: A review. *Remote Sens.* **2018**, *10*, 1365. [CrossRef]
52. Mitter, H.; Techen, A.K.; Sinabell, F.; Helming, K.; Schmid, E.; Bodirsky, B.L.; Holman, I.; Kok, K.; Lehtonen, H.; Leip, A.; et al. Shared Socio-economic Pathways for European agriculture and food systems: The Eur-Agri-SSPs. *Glob. Environ. Chang.* **2020**, *65*, 102159. [CrossRef]
53. Dwyer, L.M.; Stewart, D.W.; Carrigan, L.; Ma, B.L.; Neave, P.; Balchin, D. Guidelines for comparisons among different maize maturity. *Agron. J.* **1999**, *91*, 946–949. [CrossRef]
54. Araus, J.L.; Serret, M.D.; Edmeades, G.O. Phenotyping maize for adaptation to drought. *Front. Physiol.* **2012**, *3*, 305. [CrossRef]
55. Liu, X.; Wang, X.; Wang, X.; Gao, J.; Luo, N.; Meng, Q.; Wang, P. Dissecting the critical stage in the response of maize kernel set to individual and combined drought and heat stress around flowering. *Environ. Exp. Bot.* **2020**, *179*, 104213. [CrossRef]
56. R Core Team. *R: A Language and Environment for Statistical Computing*; R Foundation for Statistical Computing: Vienna, Austria, 2021.
57. Araus, J.L.; Cairns, J.E. Field high-throughput phenotyping: The new crop breeding frontier. *Trends Plant Sci.* **2014**, *19*, 52–61. [CrossRef] [PubMed]
58. Saruhan, N.; Saglam, A.; Kadioglu, A. Salicylic acid pretreatment induces drought tolerance and delays leaf rolling by inducing antioxidant systems in maize genotypes. *Acta Physiol. Plant.* **2012**, *34*, 97–106. [CrossRef]
59. Saglam, A.; Kadioglu, A.; Demiralay, M.; Terzi, R. Leaf rolling reduces photosynthetic loss in maize under severe drought. *Acta Bot. Croat.* **2014**, *73*, 315–332. [CrossRef]
60. Kadioglu, A.; Terzi, R.; Saruhan, N.; Saglam, A. Current advances in the investigation of leaf rolling caused by biotic and abiotic stress factors. *Plant Sci.* **2012**, *182*, 42–48. [CrossRef]
61. Kim, E.; Ahn, T.K.; Kumazaki, S. Changes in antenna sizes of photosystems during state transitions in granal and stroma-exposed thylakoid membrane of intact chloroplasts in arabidopsis mesophyll protoplasts. *Plant Cell Physiol.* **2015**, *56*, 759–768. [CrossRef]
62. Lu, Z.; Quiñones, M.A.; Zeiger, E. Abaxial and adaxial stomata from Pima cotton (*Gossypium barbadense* L.) differ in their pigment content and sensitivity to light quality. *Plant. Cell Environ.* **1993**, *16*, 851–858. [CrossRef]
63. Peñuelas, J.; Filella, L. Technical focus: Visible and near-infrared reflectance techniques for diagnosing plant physiological status. *Trends Plant Sci.* **1998**, *3*, 151–156. [CrossRef]
64. Moore, J.P.; Hearshaw, M.; Ravenscroft, N.; Lindsey, G.G.; Farrant, J.M.; Brandt, W.F. Desiccation-induced ultrastructural and biochemical changes in the leaves of the resurrection plant *Myrothamnus flabellifolia*. *Aust. J. Bot.* **2007**, *55*, 482–491. [CrossRef]
65. Hughes, N.M.; Vogelmann, T.C.; Smith, W.K. Optical effects of abaxial anthocyanin on absorption of red wavelengths by understorey species: Revisiting the back-scatter hypothesis. *J. Exp. Bot.* **2008**, *59*, 3435–3442. [CrossRef]

66. Weber, V.S.; Araus, J.L.; Cairns, J.E.; Sanchez, C.; Melchinger, A.E.; Orsini, E. Prediction of grain yield using reflectance spectra of canopy and leaves in maize plants grown under different water regimes. *F. Crop. Res.* **2012**, *128*, 82–90. [CrossRef]
67. Martin, D.E.; López, J.D.; Lan, Y. Laboratory evaluation of the GreenSeeker^{TM} hand-held optical sensor to variations in orientation and height above canopy. *Int. J. Agric. Biol. Eng.* **2012**, *5*, 43–47.
68. Huang, S.; Tang, L.; Hupy, J.P.; Wang, Y.; Shao, G. A commentary review on the use of normalized difference vegetation index (NDVI) in the era of popular remote sensing. *J. For. Res.* **2021**, *32*, 1–6. [CrossRef]
69. Dietz, K.J.; Zörb, C.; Geilfus, C.M. Drought and crop yield. *Plant Biol.* **2021**, *23*, 881–893. [CrossRef] [PubMed]
70. Reynolds, D.; Baret, F.; Welcker, C.; Bostrom, A.; Ball, J.; Cellini, F.; Lorence, A.; Chawade, A.; Khafif, M.; Noshita, K.; et al. What is cost-efficient phenotyping? Optimizing costs for different scenarios. *Plant Sci.* **2019**, *282*, 14–22. [CrossRef]
71. Gracia-Romero, A.; Kefauver, S.C.; Vergara-Díaz, O.; Hamadziripi, E.; Zaman-Allah, M.A.; Thierfelder, C.; Prassana, B.M.; Cairns, J.E.; Araus, J.L. Leaf versus whole-canopy remote sensing methodologies for crop monitoring under conservation agriculture: A case of study with maize in Zimbabwe. *Sci. Rep.* **2020**, *10*, 16008. [CrossRef]
72. Sun, D.; Robbins, K.; Morales, N.; Shu, Q.; Cen, H. Advances in optical phenotyping of cereal crops. *Trends Plant Sci.* **2022**, *27*, 191–208. [CrossRef]
73. Herrmann, I.; Berger, K. Remote and proximal assessment of plant traits. *Remote Sens.* **2021**, *13*, 1893. [CrossRef]
74. Lischeid, G.; Webber, H.; Sommer, M.; Nendel, C.; Ewert, F. Machine learning in crop yield modelling: A powerful tool, but no surrogate for science. *Agric. For. Meteorol.* **2022**, *312*, 108698. [CrossRef]
75. Hunter, D.; Yu, H.; Pukish, M.S.; Kolbusz, J.; Wilamowski, B.M. Selection of proper neural network sizes and architectures-A comparative study. *IEEE Trans. Ind. Inform.* **2012**, *8*, 228–240. [CrossRef]
76. Tikhonov, A.N.; Arsenin, V.Y. *Solutions of Ill-Posed Problems*; John Wiley & Sons: New York, NY, USA, 1977.
77. Hasan, H.; Shafri, H.Z.M.; Habshi, M. A Comparison between Support Vector Machine (SVM) and Convolutional Neural Network (CNN) Models for Hyperspectral Image Classification. *IOP Conf. Ser. Earth Environ. Sci.* **2019**, *357*, 012035. [CrossRef]
78. Hasan, M.; Ullah, S.; Khan, M.J.; Khurshid, K. Comparative analysis of SVM, ann and cnn for classifying vegetation species using hyperspectral thermal infrared data. *Int. Arch. Photogramm. Remote Sens. Spat. Inf. Sci.—ISPRS Arch.* **2019**, *42*, 1861–1868. [CrossRef]
79. Galic, V.; Franic, M.; Jambrovic, A.; Ledencan, T.; Brkic, A.; Zdunic, Z.; Simic, D. Genetic correlations between photosynthetic and yield performance in maize are different under two heat scenarios during flowering. *Front. Plant Sci.* **2019**, *10*, 566. [CrossRef] [PubMed]
80. Odilbekov, F.; Armonienė, R.; Henriksson, T.; Chawade, A. Proximal phenotyping and machine learning methods to identify septoria tritici blotch disease symptoms in wheat. *Front. Plant Sci.* **2018**, *9*, 1–11. [CrossRef]
81. Appeltans, S.; Pieters, J.G.; Mouazen, A.M. Detection of leek rust disease under field conditions using hyperspectral proximal sensing and machine learning. *Remote Sens.* **2021**, *13*, 1341. [CrossRef]
82. Ouhami, M.; Hafiane, A.; Es-Saady, Y.; El Hajji, M.; Canals, R. Computer Vision, IoT and Data Fusion for Crop Disease Detection Using Machine Learning: A Survey and Ongoing Research. *Remote Sens.* **2021**, *13*, 2486. [CrossRef]
83. Esposito, S.; Carputo, D.; Cardi, T.; Tripodi, P. Applications and Trends of Machine Learning in Genomics and Phenomics for Next-Generation Breeding. *Plants* **2020**, *9*, 34. [CrossRef]
84. Barbedo, J.G.A. Detection of nutrition deficiencies in plants using proximal images and machine learning: A review. *Comput. Electron. Agric.* **2019**, *162*, 482–492. [CrossRef]
85. Li, D.; Miao, Y.; Ransom, C.J.; Bean, G.M.; Kitchen, N.R.; Fernández, F.G.; Sawyer, J.E.; Camberato, J.J.; Carter, P.R.; Ferguson, R.B.; et al. Corn Nitrogen Nutrition Index Prediction Improved by Integrating Genetic, Environmental, and Management Factors with Active Canopy Sensing Using Machine Learning. *Remote Sens.* **2022**, *14*, 394. [CrossRef]
86. Zubler, A.V.; Yoon, J.Y. Proximal Methods for Plant Stress Detection Using Optical Sensors and Machine Learning. *Biosensors* **2020**, *10*, 193. [CrossRef]
87. Kumar, D.; Kushwaha, S.; Delvento, C.; Liatukas, Ž.; Chawade, A. Affordable Phenotyping of Winter Wheat under Field and Controlled Conditions for Drought Tolerance. *Agronomy* **2020**, *10*, 882. [CrossRef]
88. Deery, D.; Jimenez-Berni, J.; Jones, H.; Sirault, X.; Furbank, R. Proximal Remote Sensing Buggies and Potential Applications for Field-Based Phenotyping. *Agronomy* **2014**, *4*, 349–379. [CrossRef]
89. Mupangwa, W.; Chipindu, L.; Nyagumbo, I.; Mkuhlani, S.; Sisito, G. Evaluating machine learning algorithms for predicting maize yield under conservation agriculture in Eastern and Southern Africa. *SN Appl. Sci.* **2020**, *2*, 1–14. [CrossRef]
90. Yang, W.; Nigon, T.; Hao, Z.; Dias Paiao, G.; Fernández, F.G.; Mulla, D.; Yang, C. Estimation of corn yield based on hyperspectral imagery and convolutional neural network. *Comput. Electron. Agric.* **2021**, *184*, 106092. [CrossRef]
91. Habyarimana, E.; Baloch, F.S. Machine learning models based on remote and proximal sensing as potential methods for in-season biomass yields prediction in commercial sorghum fields. *PLoS ONE* **2021**, *16*, 1–23. [CrossRef]
92. Teodoro, P.E.; Teodoro, L.P.R.; Baio, F.H.R.; da Silva Junior, C.A.; Dos Santos, R.G.; Ramos, A.P.M.; Pinheiro, M.M.F.; Osco, L.P.; Gonçalves, W.N.; Carneiro, A.M.; et al. Predicting Days to Maturity, Plant Height, and Grain Yield in Soybean: A Machine and Deep Learning Approach Using Multispectral Data. *Remote Sens.* **2021**, *13*, 4632. [CrossRef]
93. Wang, X.; Huang, J.; Feng, Q.; Yin, D. Winter wheat yield prediction at county level and uncertainty analysis in main wheat-producing regions of China with deep learning approaches. *Remote Sens.* **2020**, *12*, 1744. [CrossRef]

94. Peichl, M.; Thober, S.; Samaniego, L.; Hansjürgens, B.; Marx, A. Machine-learning methods to assess the effects of a non-linear damage spectrum taking into account soil moisture on winter wheat yields in Germany. *Hydrol. Earth Syst. Sci.* **2021**, *25*, 6523–6545. [CrossRef]
95. Guberac, S.; Galić, V.; Rebekić, A.; Čupić, T.; Petrović, S. Optimising accuracy of performance predictions using available morphophysiological information in wheat breeding germplasm. *Ann. Appl. Biol.* **2021**, *178*, 367–376. [CrossRef]
96. Lamos-Díaz, H.; Puentes-Garzón, D.E.; Zarate-Caicedo, D.A. Comparison between Machine Learning Models for Yield Forecast in Cocoa Crops in Santander, Colombia. *Rev. Fac. Ing.* **2020**, *29*, e10853. [CrossRef]
97. Abadi, M.; Barham, P.; Chen, J.; Chen, Z.; Davis, A.; Dean, J.; Devin, M.; Ghemawat, S.; Irving, G.; Isard, M.; et al. TensorFlow: A system for large-scale machine learning. In Proceedings of the 12th USENIX Symposium on Operating Systems Design and Implementation (OSDI 16), Savannah, GA, USA, 2–4 November 2016.
98. Paszke, A.; Gross, S.; Massa, F.; Lerer, A.; Bradbury Google, J.; Chanan, G.; Killeen, T.; Lin, Z.; Gimelshein, N.; Antiga, L.; et al. PyTorch: An Imperative Style, High-Performance Deep Learning Library. *arXiv* **2019**, arXiv:1912.01703.

remote sensing

MDPI

Article

Estimating Community-Level Plant Functional Traits in a Species-Rich Alpine Meadow Using UAV Image Spectroscopy

Yi-Wei Zhang [1,2], Tiejun Wang [2], Yanpei Guo [1], Andrew Skidmore [2], Zhenhua Zhang [3], Rong Tang [1], Shanshan Song [1] and Zhiyao Tang [1,*]

1 Institute of Ecology, College of Urban and Environmental Sciences and Key Laboratory for Earth Surface Processes of the Ministry of Education, Peking University, Beijing 100871, China; zyw0624@pku.edu.cn (Y.-W.Z.); guoyanpei@pku.edu.cn (Y.G.); tangrong@stu.pku.edu.cn (R.T.); 2101112127@stu.pku.edu.cn (S.S.)

2 Faculty of Geo-Information Science and Earth Observation (ITC), University of Twente, P.O. Box 217, 7500 AE Enschede, The Netherlands; t.wang@utwente.nl (T.W.); a.k.skidmore@utwente.nl (A.S.)

3 Key Laboratory of Adaptation and Evolution of Plateau Biota and Qinghai Provincial Key Laboratory of Restoration Ecology of Cold Area, Northwest Institute of Plateau Biology, Chinese Academy of Sciences, Xining 810008, China; zhenhua@nwipb.cas.cn

* Correspondence: zytang@urban.pku.edu.cn; Tel.: +86-10-6275-4039

Abstract: Plant functional traits at the community level (plant community traits hereafter) are commonly used in trait-based ecology for the study of vegetation–environment relationships. Previous studies have shown that a variety of plant functional traits at the species or community level can be successfully retrieved by airborne or spaceborne imaging spectrometer in homogeneous, species-poor ecosystems. However, findings from these studies may not apply to heterogeneous, species-rich ecosystems. Here, we aim to determine whether unmanned aerial vehicle (UAV)-based hyperspectral imaging could adequately estimate plant community traits in a species-rich alpine meadow ecosystem on the Qinghai–Tibet Plateau. To achieve this, we compared the performance of four non-parametric regression models, i.e., partial least square regression (PLSR), the generic algorithm integrated with the PLSR (GA-PLSR), random forest (RF) and extreme gradient boosting (XGBoost) for the retrieval of 10 plant community traits using visible and near-infrared (450–950 nm) UAV hyperspectral imaging. Our results show that chlorophyll *a*, chlorophyll *b*, carotenoid content, starch content, specific leaf area and leaf thickness were estimated with good accuracies, with the highest R^2 values between 0.64 (nRMSE = 0.16) and 0.83 (nRMSE = 0.11). Meanwhile, the estimation accuracies for nitrogen content, phosphorus content, plant height and leaf dry matter content were relatively low, with the highest R^2 varying from 0.3 (nRMSE = 0.24) to 0.54 (nRMSE = 0.20). Among the four tested algorithms, the GA-PLSR produced the highest accuracy, followed by PLSR and XGBoost, and RF showed the poorest performance. Overall, our study demonstrates that UAV-based visible and near-infrared hyperspectral imaging has the potential to accurately estimate multiple plant community traits for the natural grassland ecosystem at a fine scale.

Keywords: community weighted means; natural grassland ecosystem; Qinghai–Tibet Plateau; remote sensing

1. Introduction

Plant functional traits may directly or indirectly affect plant fitness by influencing individuals' growth, reproduction and survival [1]. They reflect the morphological, physiological or phenological responses of species to the environment and serve as proxies for life strategies [2–4]. However, plant species primarily represent and survive as part of a plant community instead of as separate species or individuals. Therefore, the variation in traits among species cannot represent the characteristics of plant communities or vegetation, nor the ecological processes at the ecosystem level [5,6].

Plant communities are composed of sets of species with various abundances and maintain a dynamic balance through interactions between species [7]. Therefore, plant functional traits at the community level (plant community traits hereafter) include information not only on plant functional traits but also on species composition. Plant community traits offer a trait-based approach to address several key questions related to plant community assembling and productivity regulation [8,9] and will assist in exploring how vegetation responds to climate change [10].

The community-weighted mean (CWM) trait, which aggregates plant functional traits at the community level using the weighted-mean approach, is a commonly used indicator for the study of vegetation–environment relationships [7]. The CWM trait is typically calculated as the mean of plant functional trait values at the species level weighted by the relative abundance of taxa [5]. However, measuring CWM traits through field surveys is time-consuming and labour-intensive, while using trait data from published databases is hindered by differences in sampling and measurement criteria and usually ignores intraspecific variation of traits [11]. Moreover, functional traits derived from discrete sites could hardly reflect the continuous spatial change in vegetation characteristics [12]. Hence, an alternative approach to collecting trait information at the community level covering variations within and between species is of great significance [13].

Remote sensing has the potential to provide a spatially continuous representation of plant functional traits and intraspecific variations [14,15]. Previous studies have shown that a variety of plant functional traits at the species or community level in forest ecosystems can be successfully retrieved by airborne or spaceborne imaging spectrometer [16,17]. As to grassland ecosystems, although a few studies have attempted to quantify trait variations at the community level, they are either limited to a relatively homogeneous condition with few co-existing species [18], dependent on a well-managed experimental platform and observed datasets [19] or focus only on limited plant traits [20]. The conclusions drawn from these studies may not apply to heterogeneous and species-rich grasslands in natural conditions [21].

The unmanned aerial vehicle (UAV) platform can be operated flexibly according to weather and field conditions, so it has become increasingly used in ecological research [22,23]. UAV-based imaging spectroscopy is a relatively new remote sensing technology with significant benefits for high-resolution remote sensing applications, making it possible to study trait variations at the community level at a fine scale [24]. Moreover, UAV-based imaging spectroscopy could offer a link between field investigation and satellite observation, which may support the estimation of plant functional traits at the community level on a broader spatial scale [25]. However, despite the potential of UAV-based imaging spectroscopy, little research has been performed on the estimation of plant community traits directly from UAV hyperspectral imagery, particularly in heterogeneous natural grasslands. Thus, in this paper, we aim to determine whether UAV-based hyperspectral imaging could adequately estimate plant community traits in a species-rich alpine meadow. Specifically, we set out to assess the performance of four non-parametric regression models for retrieving 10 plant community traits from visible and near-infrared (450–950 nm) UAV hyperspectral imagery.

2. Materials and Methods

2.1. Study Area

Our study is within a river basin located in the northeastern Qinghai– Tibet Plateau in Qinghai Province, China. The basin covers an area of approximately 244.8 km^2 with an average elevation of 3385 m.a.s.l. (Figure 1). This area is characterised by a continental monsoon climate with a mean annual temperature of –1.1 °C and a mean annual precipitation of 485 mm. Around 80% of the precipitation falls in the growing season from mid-April to mid-October. The dominant vegetation communities are alpine meadows and alpine shrubs. These communities are very rich in plant species, ranging from 30 to 50 per square meter [26].

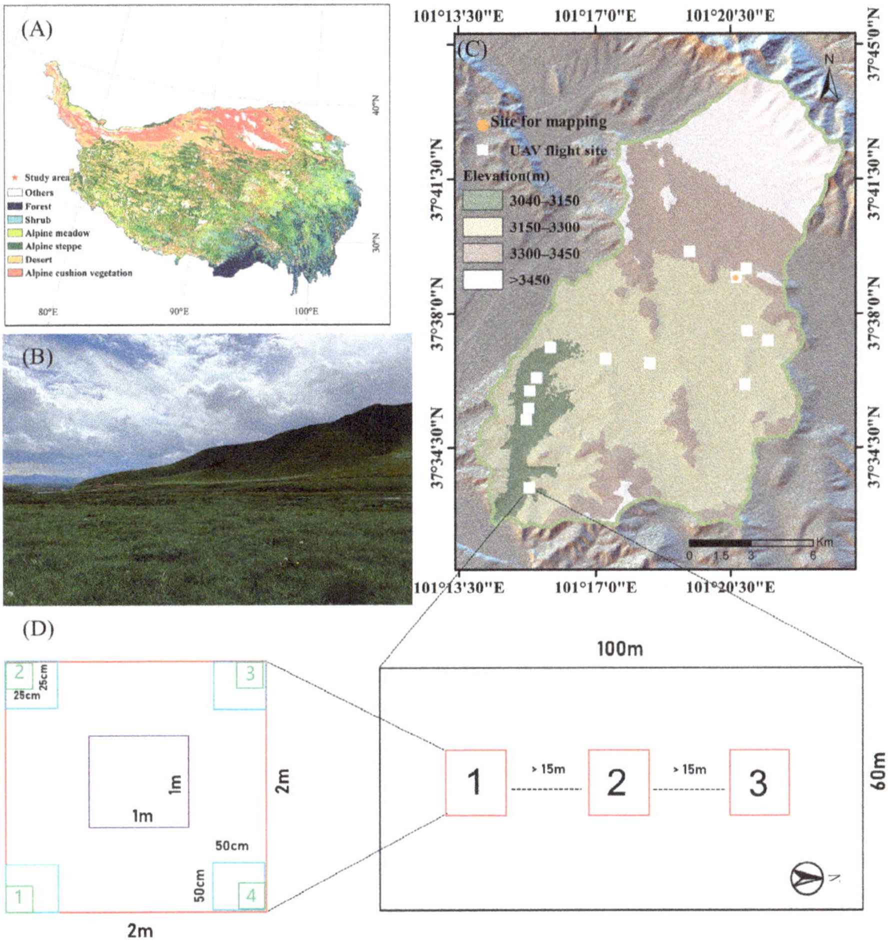

Figure 1. (**A**) Location of the study area on the Qinghai–Tibet Plateau; (**B**) photo of the study area; (**C**) study area with the location of used UAV flight sites; and (**D**) sampling design for each site in which the red boxes denote the three 2 m × 2 m sampling plots, and the green (25 cm × 25 cm), blue (50 cm × 50 cm) and purple (1 m × 1 m) boxes denote the quadrats for biomass harvesting, UAV spectra extraction and species survey, respectively, in each plot. Aboveground biomass in the subplots 1, 2 and 3; 2, 3 and 4; and 1, 3 and 4 for plots 1, 2 and 3, respectively, were harvested after the spectral measurements.

2.2. Hyperspectral Data Collection and Pre-Processing

In this study, we determined 20 survey sites according to the altitude gradients, with the altitude ranging from 3040 to 3450 m. Each survey site was 100 m × 60 m in size. UAV flight campaigns were conducted between 10 and 26 August 2021 during the growing season of the meadow. At each survey site, we collected the hyperspectral data by a Cubert UHD185 Firefly spectrometer (UHD185) equipped on a hexacopter UAV (DJI M600 PRO). DJI M600 PRO was equipped with the A3 Pro flight controller including three Inertial Measurement Units (IMU) and three Global Navigation Satellite System (GNSS) units. The UHD185 comprises 125 spectral channels and spans the spectral range from 450 to 950 nm at a 4-nm sampling interval, and the spectral resolution is 8 nm. One panchromatic band and 125 hyperspectral bands were simultaneously recorded into the UHD185 during the flight. Before each flight, we calibrated the UHD185 spectrometer using a white reference panel and a black plastic lens cap. Three 1.2 m × 1.2 m standard reference panels (with approximately 10%, 50% and 80% reflectance, respectively) were set up in the flight area for the follow-up relative normalisation [27]. To minimize the atmospheric perturbations and BRDF effects, we conducted all these flight campaigns between 11:00 and 15:00 local time on clear sunny days. The flight speed was 4.8 m/s at a flight altitude of 40 m above ground level. The UAV survey was designed to acquire 70% forward overlap and 60% side overlap. The average size of the UAV stripes was 17 m × 100 m. The spatial resolution was about 0.02 m for the panchromatic image and about 0.3 m for the hyperspectral image. The collected hyperspectral images were first fused with the corresponding panchromatic images using Cube-Pilot software (Cubert GmbH, Ulm, Germany). The entire hyperspectral image of each site was then mosaicked from the fusing images using Agisoft PhotoScan (Agisoft, St. Petersburg, Russia). As a result, 14 out of 20 mosaiced UAV hyperspectral images were retained after eliminating blurring images caused by sudden strong turbulence over the plateau.

We extracted the field spectra of three standard reference panels from each panel's centred pixels from the hyperspectral image. In addition, the reference reflectance of each panel was measured from a laboratory-integrating sphere using the full range of Analytical Spectral Devices (ASD-FR). Based on the field spectra and reference spectra of these panels, images of different study sites were calibrated using an empirical line method [24,27,28]. All images were smoothed by the Savitzky–Golay filter with a factor of 5 to remove high-frequency noise (Figure S1).

We used a 25 × 25 pixel (around 50 cm × 50 cm in size) window at four corners of the 2 m × 2 m plot to extract spectra from each plot (Figure 1). Image processing was performed with ENVI 5.3 (Exelis Visual Information Solutions, Boulder, CS, USA).

2.3. Field Data Collection

The field samples were collected on the same day as the UAV flight campaign, i.e., between 10 and 26 August 2021. At each survey site, we randomly set three 2 m × 2 m plots in a 100 m × 60 m range (Figure 1). The distance between any two plots was at least 15 m. We investigated the 1 m × 1 m area in the centre of each plot for species composition and species-wise coverage. To do so, we divided the 1 m^2 quadrat into a grid of 100 squares each representing 1% cover and then estimate the percentage cover occupied by each species in the quadrat. We marked each plot in its centre for identification in images. In this research, we sampled species that accounted for the accumulative coverage of over 80% of the entire plot in each plot.

We collected 20 fully mature leaves of each sampled species at three vertical canopy positions along the plant stem: lower (n = 6), middle (n = 6) and upper (n = 8). We mixed the sampled leaves of each species and divided them into two equal subsamples. One subsample was quickly stored in liquid nitrogen for physiological trait measurement, and the other subsample was wrapped in wet tissue and stored in an icebox for structural trait measurement. For each sampled species, we randomly selected 5–10 mature and healthy individuals for the plant height measurement and calculated the average. In each plot, four 25 cm × 25 cm subplots at corners were clockwise numbered with the southern corner ranked 1 (Figure 1). Aboveground biomass in the subplots 1, 2 and 3; 2, 3 and 4; and 1, 3 and 4 of plots 1, 2 and 3, respectively, were harvested after the spectral measurements. In total, 40 out of 60 investigated plots were considered in this study (Figure 2). The relative coverage of each species in each plot was calculated based on these sampled species according to the following formula:

$$rc_i = (C_i) / \sum_{j=1}^{n} C_j \quad (1)$$

where rc_i represents the relative coverage of the *i*th species in a given plot, C_i is the coverage of the *i*th species, C_j is that of the *j*th species in the plot, and n is the total number of all species in the plot.

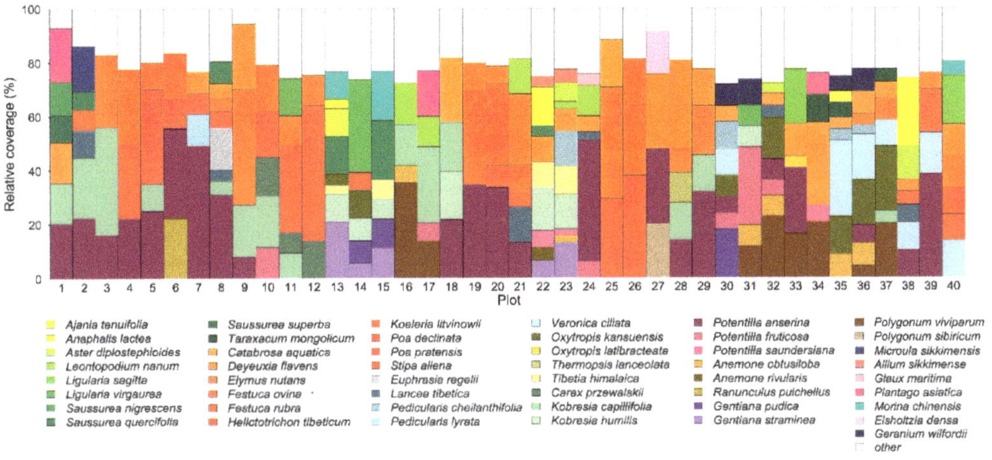

Figure 2. The sampled species composition and relative coverage in the 1 m × 1 m area (corresponding to the purple boxes in Figure 1D) in each plot used in this study.

2.4. Foliar Trait Measurements and Plant Community Trait Calculation

In this study, we measured six biochemical traits including chlorophyll *a* content, chlorophyll *b* content, carotenoid content, nitrogen content, phosphorus content and starch content as well as four structural traits including plant height, leaf thickness, leaf dry matter content and specific leaf area (Table S1).

We measured chlorophyll *a* and *b* and carotenoid contents, as well as all structural traits, with the community weighted means approach. We determined chlorophyll *a* and *b* and carotenoid contents with a UV/VIS Spectrophotometer (UV-1800PC, Shanghai Mapada Instruments Co., Ltd., Shanghai, China). Except for the plant height, which was measured during the investigation, the other structural traits, such as leaf thickness, leaf dry matter content and specific leaf area, were measured on the same day of sampling. Leaf thickness was measured by a micrometre, and fresh weight was measured by an analytical balance. After leaf thickness measurement, we scanned leaves for the fresh leaf area with a flatbed scanner and then oven-dried those leaves at 65 °C for 72 h to a constant weight to determine the specific leaf area (fresh area/dried weight) and leaf dry matter content (dried weight/fresh weight).

We calculated the CWM values of chlorophyll *a* and *b* and carotenoid contents, plant height, leaf thickness, leaf dry matter content and specific leaf area according to the following formula:

$$t = \left(\sum_{i=1}^{d} t_i \times C_i \right) / \sum_{i=1}^{n} C_i \tag{2}$$

where t represents a community-level functional trait, t_i denotes the functional trait of *i*th species, C_i is the coverage of *i*th species, d represents the total number of sampled species in a given plot and n is the total number of all species in a given plot.

In addition, we measured plant community nitrogen, phosphorus and starch contents using mixed samples. We shredded the harvested biomass, which was oven-dried at 65 °C for at least 72 h and homogenised it to mixed samples. The nitrogen content was analysed using an elemental analyser (Vario MACRO Cube, Frankfurt, Germany). The phosphorus content was measured by the molybdate–ascorbic acid method after H_2SO_4–H_2O_2 digestion [29]. Moreover, the starch content was measured by the anthrone colorimetric method using a UV/VIS Spectrophotometer (UV-1800PC, Shanghai Mapada Instruments Co., Ltd., Shanghai, China). Chlorophyll *a*, chlorophyll *b* and nitrogen contents were quantified in an area-based approach [30,31]. Area-based traits (mg/cm^2) were calculated according to the following formula:

$$trait_{area} = trait_{mass}/SLA \tag{3}$$

2.5. Mapping Plant Community Traits

Here we tested four non-parametric models' capability for the retrieval of various community-level traits (Figure 3). The two linear models are partial least square regression (PLSR) and the generic algorithm integrated with the PLSR (GA-PLSR). PLSR is a widely used algorithm in hyperspectral vegetation parameters retrieval [32,33]. As a model designed to incorporate multicollinearity problems, PLSR derives a smaller number of latent variables from the original data [34]. In this way, PLSR can eliminate the less informative variables but concentrate most explanatory variables on a few latent variables. However, the "large *p*-small *n*" problem (a large number of variables but a few samples) can still spoil the PLSR result [35]. In this condition, a variable selection pre-processing is known to improve PLSR performance [36]. Here, we adopted GA-PLSR which allows a band selection procedure in PLSR [37]. There are numerous studies showing GA-PLSR to be useful in promoting PLSR model performance [38,39]. It obeys the rule of biological evolution and natural selection to select informative features. Important features are able to survive after multiple iterations of model fitting and feature selection procedures. We selected random forest (RF) and extreme gradient boost (XGBoost) to evaluate nonlinear model performance in traits estimation. The RF model is one of the popular techniques of foliar trait prediction [40] and has been applied to map vegetation parameters at various scales [41,42]. XGBoost is an emerging machine learning algorithm showing satisfactory model performance in recent research [43].

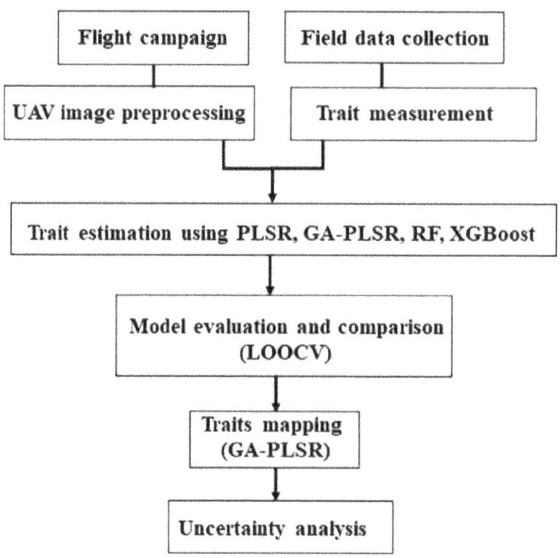

Figure 3. Overview of the study workflow.

In total, 40 plots were used as input for the four tested models. We used the leave-one-out-cross-validation (LOOCV) approach for model training and validation. Based on the *pls* package in R [44], we determined the number of latent factors used in PLSR for each community-level trait dataset by the predicted residual sum of squares (PRESS) statistic [45]. The feature selection of GA-PLSR was performed by the *plsVarSel* package in R [39]. After that, the standard PLSR routine was performed to determine the latent factors. As for the nonlinear models, we conducted a hyper-parameter optimisation process for each trait dataset (Table S2). RF models were performed with a *randomforest* package in R [46]. The number of trees (*ntree*, 100–1000 with the interval of 100) and the number of variables randomly sampled as candidates at each split (*mtry*, 1–125) were tuned, and each combination was replicated for 10 times to obtain the optimal parameters with the highest correlation coefficient. The XGBoost models were performed with the *xgboost* package in R [47], and the learning rate (*eta*, 0.1–1), maximum depth of a tree (*max_depth*, 0.1–1) and iteration rounds (*nrounds*, 1–100) were tuned to search for the best parameter in combination with the highest correlation coefficient.

We selected the model with the best performance for plant community trait mapping to visually represent plant community traits in space. Here we assessed the best model as a model with the highest R^2 among all tested traits. To test whether the selected model could be adequately applied in a species-rich meadow, we further analysed the relationship between the predicted residuals of all plant community traits and the number of dominant species in each plot. To display spatial patterns of traits, we chose one image covering an area composed of a fenced meadow and a highly disturbed meadow. The location of this image was indicated in Figure 1. Thanks to the precise spatial resolution, we excluded non-vegetation pixels by supervised classification. The image was resampled into 50-cm spatial resolution by the nearest-neighbour algorithm (Figure S2). The relative uncertainty (standard deviation/mean) was calculated based on the 40 models generated from LOOCV. Data analysis was performed with *R* 4.1.0 [48].

3. Results

Among the biochemical traits, chlorophyll *a* and *b*, carotenoid and starch contents, showed good predictive accuracy, with the highest R^2 value of four models ranging from 0.64~0.83. Phosphorus and nitrogen showed R^2 values lower than 0.60 (Figure 4). As for the structural traits, specific leaf area (highest R^2 = 0.70) and leaf thickness (highest R^2 = 0.68) were both estimated well (Figure 4), while the estimates of plant height (R^2 = 0.44) and leaf dry matter content (R^2 = 0.30) were relatively poor.

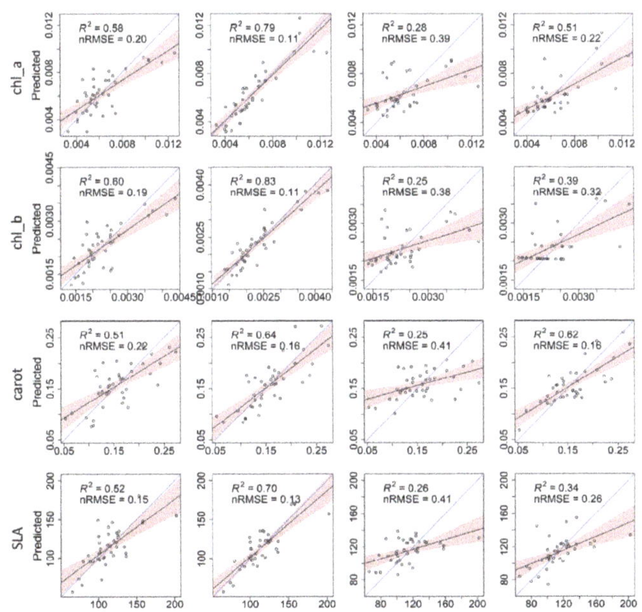

Figure 4. *Cont.*

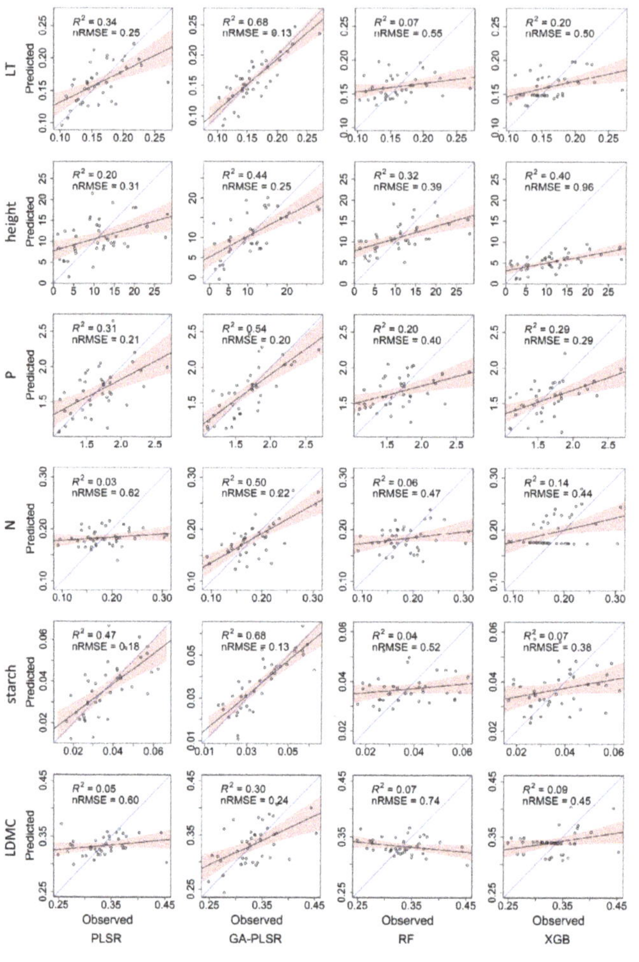

Figure 4. Predicted versus observed chlorophyll *a* (chl_a), chlorophyll *b* (chl_b), carotenoid (carot), specific leaf area (SLA), leaf thickness (LT), plant height (height), phosphorus content (P), starch content (starch), leaf dry matter content (LDMC) and nitrogen content (N) from PLSR, GA-PLSR, RF and XGBoost; the blue line denotes the 1:1 line, and the red areas denote the 95% confidence interval.

Among four estimation models, GA-PLSR proved more accurate, as it produced the highest R^2 value and the lowest nRMSE in most of the 10 plant community traits (Table 1). However, the other models displayed various performances in different traits. PLSR showed good performance in most traits, such as the biochemical traits related to photosynthesis (chlorophyll *a* and *b* and carotenoid contents) and specific leaf area. XGBoost owned model performance comparable with GA-PLSR for carotenoid content and plant height. However, it showed relatively low R^2 values for nitrogen and starch contents. RF presented the worst predictive accuracy in most traits. Moreover, GA-PLSR had nRMSE values all below 0.4, while the other three models produced higher nRMSE values for most structural traits. Among the 10 traits, the model performance for leaf thickness, nitrogen content and starch content illustrated significant differences. The R^2 values of different models varied from 0.68 to 0.04.

Table 1. Model evaluation and comparison of the coefficient of determination (R-squared) and normalised root mean-square error (nRMSE) on 10 plant community traits.

Trait	PLSR		GA-PLSR		RF		XGBoost	
	R^2	nRMSE	R^2	nRMSE	R^2	nRMSE	R^2	nRMSE
chlorophyll *a*	0.58	20.3%	0.79	10.7%	0.28	38.7%	0.51	22.0%
chlorophyll *b*	0.60	18.5%	0.83	10.7%	0.25	37.9%	0.39	32.0%
carotenoid	0.51	22.2%	0.64	16.0%	0.25	40.7%	0.62	15.8%
specific leaf area	0.52	15.1%	0.70	12.8%	0.26	41.3%	0.34	25.6%
leaf thickness	0.34	24.5%	0.68	13.5%	0.07	55.3%	0.20	50.0%
plant height	0.20	31.4%	0.44	25.3%	0.32	39.3%	0.40	95.7%
phosphorus content	0.31	20.7%	0.54	19.5%	0.20	40.3%	0.29	28.7%
nitrogen content	0.03	62.0%	0.50	22.3%	0.06	47.3%	0.14	44.1%
starch content	0.47	18.4%	0.68	13.5%	0.04	51.9%	0.07	37.6%
leaf dry matter content	0.05	59.8%	0.30	24.2%	0.07	74.4%	0.09	44.9%

As for linear models, the model performance of PLSR increased significantly with band selection (GA-PLSR). In addition, GA-PLSR conquered the weakness of PLSR in some traits' retrieval, such as nitrogen content, and yielded more satisfactory results than the two nonlinear models. For the two nonlinear models, XGBoost showed an obvious advantage over RF in all traits. However, the four tested models' prediction of leaf dry matter content all below 0.4 made it the worst prediction among all 10 traits.

In addition, we analyzed the relationships between the predicted residuals of plant community traits and the number of dominant species in each plot. It showed that there was no statistically significant relationship ($p > 0.05$) between the predicted residuals of traits and the number of species (Figure 5).

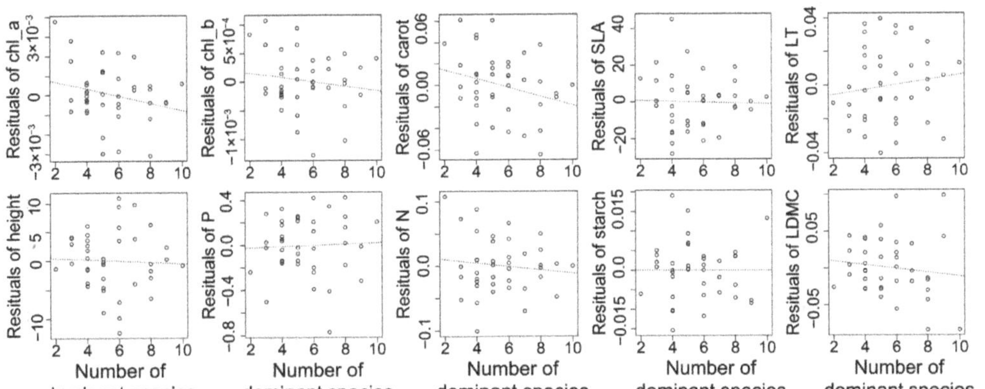

Figure 5. Relationships between the predicted residuals of plant community traits and the number of dominant species in each plot. Dotted lines denote non-significant relationships ($p > 0.05$): chl_a, chlorophyll *a*; chl_b, chlorophyll *b*; carot, carotenoid; SLA, specific leaf area; LT, leaf thickness; height, plant height; P, phosphorus content; N, nitrogen content; starch, starch content; LDMC, leaf dry matter content.

Based on these results, we mapped spatial patterns of all tested traits, applying the GA-PLSR, as it exhibited stability and outperformed all models, to map traits (Figure 6). Our results showed that biochemical traits displayed more homogeneous distributions while structural traits, especially for plant height, had obvious spatial patterns. We calculated uncertainties from the predicted maps (Figure S3). For all traits, most uncertainty values were near zero.

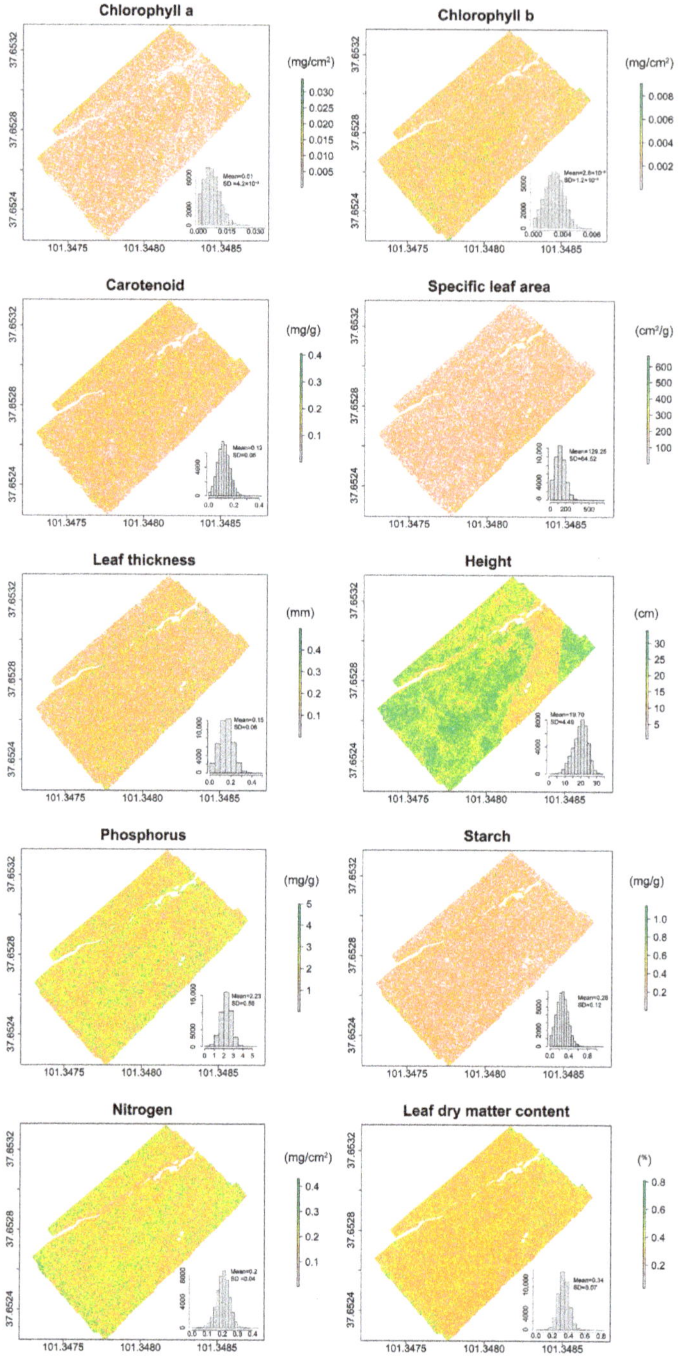

Figure 6. Spatial patterns of plant community traits mapped using GA-PLSR and the frequency distribution of pixel values. The location of this image was indicated in Figure 1. This image covers plot 32, plot 33 and plot 34 in Figure 2.

4. Discussion

In this study, we predicted 10 significant plant community traits from the near-ground UAV-based hyperspectral image in a highly heterogeneous grassland with high species richness, with 9 traits producing moderate to good accuracies. Among all tested traits, chlorophyll *a* and *b* contents, carotenoid content, specific leaf area, leaf thickness and starch content could generate a predictive accuracy comparable to that of previous studies with R^2 values greater than 0.60 [19].

The retrieval of plant height, phosphorus content and nitrogen content produced moderate predictive accuracy with R^2 values varying from 0.44 to 0.54. This may be because the nonpigmented compounds of foliage and structure correlation intervals were mostly at longer wavelengths, while the detectable weak correlation could even be influenced by strong absorption of water content in fresh leaves [49,50]. Nevertheless, the R^2 values of nitrogen and phosphorus content are comparable to other research in grasslands [18,19,51]. Similarly, all our tested models of leaf dry matter content produced poor predictive accuracies in this study, probably because leaf dry matter content reflected leaf water content. The distinct effect of foliar liquid water was at 1450 and 1950 nm, which was out of the available spectra range in this study [50]. This result suggested the relatively limited capacity of visible and near-infrared spectra in the retrieval of certain traits, which should be considered in future studies.

In this study, we tested four non-parametric models, including two linear models and two nonlinear models, in the direct retrieval of multiple traits. The results suggested that the predictive accuracy of some traits might be influenced to some extent by the choice of models. For example, an obvious increase of R^2 was observed in starch content and leaf thickness when the linear model was applied. In contrast, a few traits yielded relatively consistent predictive accuracies across four models, such as plant height and carotenoid content.

As to the linear models, PLSR produced comparative average performance among all models but failed in some traits, such as nitrogen and leaf dry matter content. The better performance of GA-PLSR was consistent with some previous studies which confirmed the effects of feature selection in PLSR [52]. On the one hand, these significant improvements in model performance may because of the "large p–small n" problem. Since PLSR will make better use of all given features than nonlinear models, redundant variables may obscure truly usable bands [39]. As a result, PLSR with relatively limited training samples cannot handle hundreds of correlated bands well [53]. On the other hand, environmental and instrumental noise was inevitably mixed with spectral data, which may weaken the predictive accuracy.

We found that the performance of GA-PLSR was not sensitive to species richness, indicating its robustness in extracting information of functional traits based on the UAV hyperspectral imaging in a fine resolution. Therefore, GA-PLSR model could adequately estimate various plant community traits in species-rich alpine meadows.

Between two nonlinear models, XGBoost reported considerable advantages in predictive accuracy over RF. This result was consistent with recently published studies [43,54]. As XGBoost is effective in high-dimension data analysis, it is becoming a reliable method in vegetation parameter modelling using UAV-based hyperspectral data [52]. This may also provide a new option when conducting similar research facing nonlinear model selection.

Overall, GA-PLSR, the PLSR model using the GA feature selection approach, outperformed the other candidates for all tested traits in this research. The trade-off between complexity reduction and information preservation is a great challenge in hyperspectral data analysis. GA-PLSR can deal with estimations of various traits from canopy spectra with relatively satisfactory predictive accuracies. Maps generated from GA-PLSR depicted various patterns at the local level that could not be achieved only by site investigation. Meanwhile, the PLSR model combined with feature selection could serve as an option for researchers facing the "large p–small n" problem.

5. Conclusions

In this study, we investigated whether UAV-based hyperspectral imaging could be used to estimate 10 different plant functional traits at the community level in a species-rich alpine meadow ecosystem on the Qinghai–Tibet Plateau. In addition, we compared the performance of four non-parametric regression models, i.e., PLSR, GA-PLSR, RF and XGBoost. Based on the result, we conclude that the UAV-based hyperspectral image can be used to adequately estimate plant community traits in a species-rich alpine meadow with moderate to high accuracy. Specifically, we show that chlorophyll *a*, chlorophyll *b*, carotenoid content, starch content, specific leaf area and leaf thickness were estimated with good accuracies, with the highest R^2 values between 0.64 (nRMSE = 0.16) and 0.83 (nRMSE = 0.11). While the estimation accuracies for nitrogen content, phosphorus content, plant height and leaf dry matter content were relatively low, with the highest R^2 varying from 0.3 (nRMSE = 0.24) to 0.54 (nRMSE = 0.20). Among the four tested algorithms, the GA-PLSR produced the highest accuracy, followed by PLSR and XGBoost, and RF showed the poorest performance. Our study demonstrates the potential of UAV-based visible and near-infrared hyperspectral imagery to directly estimate various plant community traits in a natural grassland ecosystem at a fine scale.

Supplementary Materials: The following supporting information can be downloaded at: https://www.mdpi.com/article/10.3390/rs14143399/s1, Figure S1: Examples of corrected spectra of alpine meadow from five different flight sites. Figure S2: The UAV hyperspectral image used for mapping plant community traits. The upper one is the raw image and the lower one is the corrected image shown in true colour composites. This image covers plot 32, plot 33 and plot 34 in Figure 2. The white arrows indicate the location of the three reference panels. Figure S3: Frequency distribution of relative uncertainty of the 10 plant community trait maps produced by the GA_PLSR model. Table S1: Summary statistics of the 10 plant functional traits for sampled plots (n = 40). Table S2: An overview of optimal hyper-parameters of two machine learning models for 10 plant community traits.

Author Contributions: Conceptualization, Z.T.; methodology, Y.-W.Z., T.W. and Z.T.; formal analysis, Y.-W.Z. and Y.G.; investigation, Y.-W.Z., Z.Z., R.T. and S.S.; writing—original draft preparation, Y.-W.Z., T.W. and Y.G.; writing—review and editing, T.W., A.S. and Z.T.; supervision, T.W., A.S. and Z.T. All authors have read and agreed to the published version of the manuscript.

Funding: The study was funded by the Strategic Priority Research Programme of the Chinese Academy of Sciences (grant no. XDA26010102), the National Key Research and Development Program of China (grant no. 2019YFA0606602) and the National Natural Science Foundation of China (grant no. 32025025 & 31988102) to Z.T. The work of the first author (Y.-W.Z.) was carried out at the University of Twente while she was visiting the Netherlands as a joint doctoral student, funded by the China Scholarship Council (grant no. 202106010134). The work of the fourth author (A.S.) is partially supported by the Horizon 2020 research and innovation programme—European Commission 'BIOSPACE Monitoring Biodiversity from Space' project (grant agreement ID 834709, H2020-EU.1.1).

Data Availability Statement: Not applicable.

Acknowledgments: We thank Yun-Hao Bai, Hong-Tu Zhang and Yiwei Lin for their valuable assistance during fieldwork.

Conflicts of Interest: The authors declare no conflict of interest.

References

1. Violle, C.; Navas, M.-L.; Vile, D.; Kazakou, E.; Fortunel, C.; Hummel, I.; Garnier, E. Let the Concept of Trait Be Functional! *Oikos* **2007**, *116*, 882–892. [CrossRef]
2. Laughlin, D.C.; Messier, J. Fitness of Multidimensional Phenotypes in Dynamic Adaptive Landscapes. *Trends Ecol. Evol.* **2015**, *30*, 487–496. [CrossRef] [PubMed]
3. Wright, I.J.; Reich, P.B.; Westoby, M.; Ackerly, D.D.; Baruch, Z.; Bongers, F.; Cavender-Bares, J.; Chapin, T.; Cornelissen, J.H.C.; Diemer, M.; et al. The Worldwide Leaf Economics Spectrum. *Nature* **2004**, *428*, 821–827. [CrossRef] [PubMed]

4. Díaz, S.; Lavorel, S.; de Bello, F.; Quétier, F.; Grigulis, K.; Robson, T.M. Incorporating Plant Functional Diversity Effects in Ecosystem Service Assessments. *Proc. Natl. Acad. Sci. USA* **2007**, *104*, 20684–20689. [CrossRef] [PubMed]
5. Bruelheide, H.; Dengler, J.; Purschke, O.; Lenoir, J.; Jiménez-Alfaro, B.; Hennekens, S.M.; Botta-Dukát, Z.; Chytrý, M.; Field, R.; Jansen, F.; et al. Global Trait–Environment Relationships of Plant Communities. *Nat. Ecol. Evol.* **2018**, *2*, 1906–1917. [CrossRef]
6. Gillison, A.N. Plant Functional Types and Traits at the Community, Ecosystem and World Level. In *Vegetation Ecology*; van der Maarel, E., Franklin, J., Eds.; John Wiley & Sons, Ltd.: Oxford, UK, 2013; pp. 347–386. ISBN 978-1-118-45259-2.
7. Muscarella, R.; Uriarte, M. Do Community-Weighted Mean Functional Traits Reflect Optimal Strategies? *Proc. R. Soc. B Biol. Sci.* **2016**, *283*, 20152434. [CrossRef]
8. Bjorkman, A.D.; Myers-Smith, I.H.; Elmendorf, S.C.; Normand, S.; Rüger, N.; Beck, P.S.A.; Blach-Overgaard, A.; Blok, D.; Cornelissen, J.H.C.; Forbes, B.C.; et al. Plant Functional Trait Change across a Warming Tundra Biome. *Nature* **2018**, *562*, 57–62. [CrossRef]
9. Tang, Z.; Xu, W.; Zhou, G.; Bai, Y.; Li, J.; Tang, X.; Chen, D.; Liu, Q.; Ma, W.; Xiong, G.; et al. Patterns of Plant Carbon, Nitrogen, and Phosphorus Concentration in Relation to Productivity in China's Terrestrial Ecosystems. *Proc. Natl. Acad. Sci. USA* **2018**, *115*, 4033–4038. [CrossRef]
10. Ackerly, D.D.; Cornwell, W.K. A Trait-Based Approach to Community Assembly: Partitioning of Species Trait Values into within- and among-Community Components. *Ecol. Lett.* **2007**, *10*, 135–145. [CrossRef]
11. Khalil, M.I.; Gibson, D.J.; Baer, S.G. Functional Response of Subordinate Species to Intraspecific Trait Variability within Dominant Species. *J. Ecol.* **2019**, *107*, 2040–2053. [CrossRef]
12. Thenkabail, P.S.; Lyon, J.G.; Huete, A. (Eds.) *Fundamentals, Sensor Systems, Spectral Libraries, and Data Mining for Vegetation: Hyperspectral Remote Sensing of Vegetation*, 2nd ed.; CRC Press: Boca Raton, FL, USA, 2018; ISBN 978-1-315-16415-1.
13. Pettorelli, N.; Wegmann, M.; Skidmore, A.; Mücher, S.; Dawson, T.P.; Fernandez, M.; Lucas, R.; Schaepman, M.E.; Wang, T.; O'Connor, B.; et al. Framing the Concept of Satellite Remote Sensing Essential Biodiversity Variables: Challenges and Future Directions. *Remote Sens. Ecol. Conserv.* **2016**, *2*, 122–131. [CrossRef]
14. Homolová, L.; Malenovský, Z.; Clevers, J.G.P.W.; García-Santos, G.; Schaepman, M.E. Review of Optical-Based Remote Sensing for Plant Trait Mapping. *Ecol. Complex.* **2013**, *15*, 1–16. [CrossRef]
15. Singh, A.; Serbin, S.P.; McNeil, B.E.; Kingdon, C.C.; Townsend, P.A. Imaging Spectroscopy Algorithms for Mapping Canopy Foliar Chemical and Morphological Traits and Their Uncertainties. *Ecol. Appl.* **2015**, *25*, 2180–2197. [CrossRef] [PubMed]
16. Wang, R.; Gamon, J.A.; Emmerton, C.A.; Li, H.; Nestola, E.; Pastorello, G.Z.; Menzer, O. Integrated Analysis of Productivity and Biodiversity in a Southern Alberta Prairie. *Remote Sens.* **2016**, *8*, 214. [CrossRef]
17. Asner, G.P.; Martin, R.E.; Anderson, C.B.; Knapp, D.E. Quantifying Forest Canopy Traits: Imaging Spectroscopy versus Field Survey. *Remote Sens. Environ.* **2015**, *158*, 15–27. [CrossRef]
18. Zhao, Y.; Sun, Y.; Lu, X.; Zhao, X.; Yang, L.; Sun, Z.; Bai, Y. Hyperspectral Retrieval of Leaf Physiological Traits and Their Links to Ecosystem Productivity in Grassland Monocultures. *Ecol. Indic.* **2021**, *122*, 107267. [CrossRef]
19. Wang, Z.; Townsend, P.A.; Schweiger, A.K.; Couture, J.J.; Singh, A.; Hobbie, S.E.; Cavender-Bares, J. Mapping Foliar Functional Traits and Their Uncertainties across Three Years in a Grassland Experiment. *Remote Sens. Environ.* **2019**, *221*, 405–416. [CrossRef]
20. Li, C.; Wulf, H.; Schmid, B.; He, J.-S.; Schaepman, M.E. Estimating Plant Traits of Alpine Grasslands on the Qinghai-Tibetan Plateau Using Remote Sensing. *IEEE J. Sel. Top. Appl. Earth Obs. Remote Sens.* **2018**, *11*, 2263–2275. [CrossRef]
21. Pajares, G. Overview and Current Status of Remote Sensing Applications Based on Unmanned Aerial Vehicles (UAVs). *Photogramm. Eng. Remote Sens.* **2015**, *81*, 281–330. [CrossRef]
22. Rossi, C.; Kneubuehler, M.; Schuetz, M.; Schaepman, M.; Haller, R.; Risch, A. Spatial Resolution, Spectral Metrics and Biomass Are Key Aspects in Estimating Plant Species Richness from Spectral Diversity in Species-Rich Grasslands. *Remote Sens. Ecol. Conserv.* **2021**, *8*, 297–314. [CrossRef]
23. de Castro, A.I.; Shi, Y.; Maja, J.M.; Peña, J.M. UAVs for Vegetation Monitoring: Overview and Recent Scientific Contributions. *Remote Sens.* **2021**, *13*, 2139. [CrossRef]
24. Aasen, H.; Honkavaara, E.; Lucieer, A.; Zarco-Tejada, P.J. Quantitative Remote Sensing at Ultra-High Resolution with UAV Spectroscopy: A Review of Sensor Technology, Measurement Procedures, and Data Correction Workflows. *Remote Sens.* **2018**, *10*, 1091. [CrossRef]
25. Jetz, W.; Cavender-Bares, J.; Pavlick, R.; Schimel, D.; Davis, F.W.; Asner, G.P.; Guralnick, R.; Kattge, J.; Latimer, A.M.; Moorcroft, P.; et al. Monitoring Plant Functional Diversity from Space. *Nat. Plants* **2016**, *2*, 16024. [CrossRef]
26. Zhang, C.; Willis, C.G.; Klein, J.A.; Ma, Z.; Li, J.; Zhou, H.; Zhao, X. Recovery of Plant Species Diversity during Long-Term Experimental Warming of a Species-Rich Alpine Meadow Community on the Qinghai-Tibet Plateau. *Biol. Conserv.* **2017**, *213*, 218–224. [CrossRef]
27. Hadley, B.C.; Garcia-Quijano, M.; Jensen, J.R.; Tullis, J.A. Empirical versus Model-based Atmospheric Correction of Digital Airborne Imaging Spectrometer Hyperspectral Data. *Geocarto Int.* **2005**, *20*, 21–28. [CrossRef]
28. Arroyo-Mora, J.P.; Kalacska, M.; Løke, T.; Schläpfer, D.; Coops, N.C.; Lucanus, O.; Leblanc, G. Assessing the Impact of Illumination on UAV Pushbroom Hyperspectral Imagery Collected under Various Cloud Cover Conditions. *Remote Sens. Environ.* **2021**, *258*, 112396. [CrossRef]
29. Jones, J.B., Jr. *Laboratory Guide for Conducting Soil Tests and Plant Analysis*; CRC press: New York, NY, USA, 2001; ISBN 1-4200-2529-5.

30. Roelofsen, H.D.; Van Bodegom, P.M.; Kooistra, L.; Witte, J.-P.M. Trait Estimation in Herbaceous Plant Assemblages from in Situ Canopy Spectra. *Remote Sens.* **2013**, *5*, 6323–6345. [CrossRef]
31. Wang, Z.; Skidmore, A.K.; Darvishzadeh, R.; Heiden, U.; Heurich, M.; Wang, T. Leaf Nitrogen Content Indirectly Estimated by Leaf Traits Derived From the PROSPECT Model. *IEEE J. Sel. Top. Appl. Earth Obs. Remote Sens.* **2015**, *8*, 3172–3182. [CrossRef]
32. Schweiger, A.K.; Cavender-Bares, J.; Townsend, P.A.; Hobbie, S.E.; Madritch, M.D.; Wang, R.; Tilman, D.; Gamon, J.A. Plant Spectral Diversity Integrates Functional and Phylogenetic Components of Biodiversity and Predicts Ecosystem Function. *Nat. Ecol. Evol.* **2018**, *2*, 976–982. [CrossRef]
33. Durán, S.M.; Martin, R.E.; Díaz, S.; Maitner, B.S.; Malhi, Y.; Salinas, N.; Shenkin, A.; Silman, M.R.; Wieczynski, D.J.; Asner, G.P.; et al. Informing Trait-Based Ecology by Assessing Remotely Sensed Functional Diversity across a Broad Tropical Temperature Gradient. *Sci. Adv.* **2019**, *5*, eaaw8114. [CrossRef]
34. Rosipal, R.; Krämer, N. Overview and Recent Advances in Partial Least Squares. In *Subspace, Latent Structure and Feature Selection*; Saunders, C., Grobelnik, M., Gunn, S., Shawe-Taylor, J., Eds.; Lecture Notes in Computer Science; Springer: Berlin, Heidelberg, Germany, 2006; Volume 3940, pp. 34–51, ISBN 978-3-540-34137-6.
35. Mehmood, T.; Liland, K.H.; Snipen, L.; Sæbø, S. A Review of Variable Selection Methods in Partial Least Squares Regression. *Chemom. Intell. Lab. Syst.* **2012**, *118*, 62–69. [CrossRef]
36. Dormann, C.F.; Elith, J.; Bacher, S.; Buchmann, C.; Carl, G.; Carré, G.; Marquéz, J.R.G.; Gruber, B.; Lafourcade, B.; Leitão, P.J.; et al. Collinearity: A Review of Methods to Deal with It and a Simulation Study Evaluating Their Performance. *Ecography* **2013**, *36*, 27–46. [CrossRef]
37. Hasegawa, K.; Miyashita, Y.; Funatsu, K. GA Strategy for Variable Selection in QSAR Studies: GA-Based PLS Analysis of Calcium Channel Antagonists. *J. Chem. Inf. Comput. Sci.* **1997**, *37*, 306–310. [CrossRef]
38. Santos-Rufo, A.; Mesas-Carrascosa, F.-J.; García-Ferrer, A.; Meroño-Larriva, J.E. Wavelength Selection Method Based on Partial Least Square from Hyperspectral Unmanned Aerial Vehicle Orthomosaic of Irrigated Olive Orchards. *Remote Sens.* **2020**, *12*, 3426. [CrossRef]
39. Mehmood, T.; Sæbø, S.; Liland, K.H. Comparison of Variable Selection Methods in Partial Least Squares Regression. *J. Chemom.* **2020**, *34*, e3226. [CrossRef]
40. Moreno-Martínez, Á.; Camps-Valls, G.; Kattge, J.; Robinson, N.; Reichstein, M.; van Bodegom, P.; Kramer, K.; Cornelissen, J.H.C.; Reich, P.; Bahn, M.; et al. A Methodology to Derive Global Maps of Leaf Traits Using Remote Sensing and Climate Data. *Remote Sens. Environ.* **2018**, *218*, 69–88. [CrossRef]
41. Loozen, Y.; Rebel, K.T.; de Jong, S.M.; Lu, M.; Ollinger, S.V.; Wassen, M.J.; Karssenberg, D. Mapping Canopy Nitrogen in European Forests Using Remote Sensing and Environmental Variables with the Random Forests Method. *Remote Sens. Environ.* **2020**, *247*, 111933. [CrossRef]
42. Chemura, A.; Mutanga, O.; Odindi, J.; Kutywayo, D. Mapping Spatial Variability of Foliar Nitrogen in Coffee (*Coffea arabica* L.) Plantations with Multispectral Sentinel-2 MSI Data. *ISPRS J. Photogramm. Remote Sens.* **2018**, *138*, 1–11. [CrossRef]
43. Yang, X.; Yang, R.; Ye, Y.; Yuan, Z.; Wang, D.; Hua, K. Winter Wheat SPAD Estimation from UAV Hyperspectral Data Using Cluster-Regression Methods. *Int. J. Appl. Earth Obs. Geoinf.* **2021**, *105*, 102618. [CrossRef]
44. Mevik, B.-H.; Wehrens, R. The Pls Package: Principal Component and Partial Least Squares Regression in R. *J. Stat. Softw.* **2007**, *18*, 1–23. [CrossRef]
45. Chen, S.; Hong, X.; Harris, C.J.; Sharkey, P.M. Sparse Modeling Using Orthogonal Forward Regression with PRESS Statistic and Regularization. *IEEE Trans. Syst. Man Cybern. Part B* **2004**, *34*, 898–911. [CrossRef] [PubMed]
46. Breiman, L. Random Forests. *Mach. Learn.* **2001**, *45*, 5–32. [CrossRef]
47. Chen, T.; He, T.; Benesty, M.; Khotilovich, V.; Tang, Y.; Cho, H.; Chen, K.; Mitchell, R.; Cano, I.; Zhou, T.; et al. *xgboost: Extreme Gradient Boosting*; R Package Version 1.6.0.1; 2022; Available online: https://CRAN.R-project.org/package=xgboost (accessed on 24 May 2022).
48. R Core Team. *R: A Language and Environment for Statistical Computing*; R Core Team: Vienna, Austria, 2019.
49. Ferwerda, J.G.; Skidmore, A.K.; Mutanga, O. Nitrogen Detection with Hyperspectral Normalized Ratio Indices across Multiple Plant Species. *Int. J. Remote Sens.* **2005**, *26*, 4083–4095. [CrossRef]
50. Ollinger, S.V. Sources of Variability in Canopy Reflectance and the Convergent Properties of Plants. *New Phytol.* **2011**, *189*, 375–394. [CrossRef] [PubMed]
51. Ramoelo, A.; Skidmore, A.K.; Cho, M.A.; Schlerf, M.; Mathieu, R.; Heitkönig, I.M.A. Regional Estimation of Savanna Grass Nitrogen Using the Red-Edge Band of the Spaceborne RapidEye Sensor. *Int. J. Appl. Earth Obs. Geoinf.* **2012**, *19*, 151–162. [CrossRef]
52. Zhang, J.; Cheng, T.; Guo, W.; Xu, X.; Qiao, H.; Xie, Y.; Ma, X. Leaf Area Index Estimation Model for UAV Image Hyperspectral Data Based on Wavelength Variable Selection and Machine Learning Methods. *Plant Methods* **2021**, *17*, 49. [CrossRef]
53. Rivera-Caicedo, J.P.; Verrelst, J.; Muñoz-Marí, J.; Camps-Valls, G.; Moreno, J. Hyperspectral Dimensionality Reduction for Biophysical Variable Statistical Retrieval. *ISPRS J. Photogramm. Remote Sens.* **2017**, *132*, 88–101. [CrossRef]
54. Zhao, X.; Pan, X.; Ma, Y.; Yan, W. Research on Forage Hyperspectral Image Recognition Based on F-SVD and XGBoost. *MATEC Web Conf.* **2021**, *336*, 06027. [CrossRef]

MDPI AG
Grosspeteranlage 5
4052 Basel
Switzerland
Tel.: +41 61 683 77 34

Remote Sensing Editorial Office
E-mail: remotesensing@mdpi.com
www.mdpi.com/journal/remotesensing

www.ingramcontent.com/pod-product-compliance
Lightning Source LLC
Chambersburg PA
CBHW041950260926
51850CB00019B/2282
* 9 7 8 3 7 2 5 8 6 4 6 4 5 *